GeoJournal Library

Volume 114

More information about this series at http://www.springer.com/series/6007

Qisheng Pan • Jason Cao
Editors

Recent Developments in Chinese Urban Planning

Selected Papers from the 8th International Association for China Planning Conference, Guangzhou, China, June 21 - 22, 2014

Springer

Editors
Qisheng Pan
Texas Southern University
Houston, TX, USA

Jason Cao
University of Minnesota, Twin Cities
Minneapolis, MN, USA

ISSN 0924-5499 ISSN 2215-0072 (electronic)
GeoJournal Library
ISBN 978-3-319-18469-2 ISBN 978-3-319-18470-8 (eBook)
DOI 10.1007/978-3-319-18470-8

Library of Congress Control Number: 2015949153

Springer Cham Heidelberg New York Dordrecht London

Printed on acid-free paper

Springer International Publishing AG Switzerland is part of Springer Science+Business Media (www.springer.com)

Preface

China, the most populous country in the world, has experienced unprecedented urbanization in a relatively short period. During the past decades, urbanization in China has centered on land development through industrialization and investment, but it has largely ignored the prosperity and well-being of the people. Livable cities are not just those with magnificent buildings and infrastructure; they are great places where people want to live. China's recently inaugurated leaders have proposed a new model to actively and prudently enhance the quality of urbanization through compact, intelligent, and low-carbon development. It symbolizes the departure from land-centered urban development to a form of people-oriented urbanization, as China's Premier, Li Keqiang, has advocated. Quality of life should be one of the key guides for the new model of urbanization.

This new model offers a platform for planning researchers and practitioners to tackle urbanization challenges, such as social equity, environment, energy, ecological and historic preservation, affordable housing, and externalities of megacities. Further, the people-oriented urbanization calls for public participation and stakeholder engagement in the planning process.

This book discusses various planning issues that Chinese planners and citizens are facing in the era of the rapid urbanization. This book includes 20 peer-reviewed and rigorously edited papers, which were presented at the 8th International Association for China Planning (IACP) Conference held June 21–22, 2014, at South China University of Technology (SCUT) in Guangzhou, China, on the theme Urban Regeneration: New Model and New Practice.

In chapter "Research of Beijing Automobile Exhaust Treatment Strategies Based on I/M System Program," Zhehao Jin and Qun Huang investigate the problems of Beijing automobile exhaust treatment. Using fixed-point monitoring data about automobile exhaust in Beijing and drawing on foreign experience, they explore problems and trends of existing measures in the process of urban development. They propose an I/M-centered treatment strategy system cooperating with multi-perspective policy by means of economy, administration, technology, and law.

Pin Wang and Qinglin Meng examine the heat island problem in Guangzhou in chapter "Analysis About Key Indicators of High-Density Blocks in the New Central District from the Perspective of Mitigating Urban Heat Island: A Study at Guangzhou." By examining the factors influencing thermal environment, they propose three key indicators affecting the heat island intensity, including floor area ratio (FAR), building density (BD), and greenery ratio(GR). Based on an empirical study in Zhujiang New Town in Guangzhou, they confirm that the summertime heat island problem in the new central district is severe. Their study also finds that FAR and GR are critical in improving thermal environment in urban areas.

Chapter "Study on Relationship Between Beijing Residential Building Insolation Standards, Improvement of Residential Environment, and Urban Ecological Benefits" by Linfei Han, Jiahua Mu, and Haofei Zhang reviews the Chinese residential building insolation standards and examines their impacts on residential environment and urban ecology. After a comparison of residential building insolation standards between China and other countries, it concludes that current natural insolation standards in China are not in line with international standards and hygiene and health, and the original intentions of the standard formulation are not fully realized either. From the view of technology, it proposes a solution by modifying the current Beijing residential building insolation standards, removing the "lost spaces" in residential areas, and increasing the "active spaces" for the purpose of improving the ecological benefits of the whole city.

In chapter "The Methods and Approaches of Green Infrastructure in Promoting China' Urban Transformation—A Case Study of Dongguan National Wetland Park," Shaoping Guan, Xi Zhang, and Jingjun Wu discuss the potential role for green infrastructure to promote China's urban transition. First, this study reviews the important role of Dongguan National Wetland Park in Dongguan City, which has an urgent demand of ecology and society transition. Then, it discusses how Dongguan National Wetland Park can help Dongguan City to achieve its transition by improving urban ecological environment, promoting industrial transformation and upgrading, and easing social conflicts. Next, it analyzes the mechanism and development pattern of Dongguan National Wetland Park for promoting Dongguan City transition. Finally it discusses the extension of this mechanism and pattern to other cities in China.

Chapter "The Symbiotic Strategies Study of Low-Carbon Eco-City Based on Multi-symbiosis Theory" by Yangyue Zhou and Xia Zhu argues that it is imperative to transform Chinese urban development model to promote multi-symbiotic low-carbon eco-city. From the perspective of "multi-symbiosis," the chapter explains the symbiosis theory, illustrates the concepts of a "low-carbon eco-city," and describes the characteristics of multi-symbiotic low-carbon eco-city. It also proposes the key symbiotic interfaces and a variety of symbiotic strategies to realize sustainable development and coexistence under different symbiotic dimensions.

China's urbanization is transforming from the land-oriented development to the people-oriented development while the optimal allocation of land resources is a key issue for China's future urbanization development. In chapter "The Paradoxes of

Land Resource Allocation's Game Playing: From Zhoukou Grave Events," Ze Zhang and Ying Gu examine the problems of land resource allocation in Chinese cities. First, it reviews the Henan Dig Graves Movement, which is considered as a conflict between urban development and cultivated land resource protection. Then, it explores the difficulties of local government in the allocation of land resources. Finally, it proposes some suggestions for regional development based on the case study of Common Agriculture Policy (CAP), which is the agricultural policy of the European Union (EU).

Chapter "Conflicts in the Urban Renewal of the Historic Preservation Area—Based on the Investigation of Nanbuting Community in Nanjing" by Tian Ruan and Meicheng Wang examines a community renewal case of a deprived historic block named Nanbuting, located about 1.4 km away from the CBD of Nanjing. By investigating a typical urban renewal case in the Nanbuting community, this chapter illustrates how the multi-stakeholder conflict is generated and exacerbated. It drew on a framework from the planning and policy-making process, linking multi-stakeholder conflict assessment to potential suggestions for participatory planning.

As an emerging urban form, megacity region is a quite new concept for planning theorists and practitioners in China. In chapter "Planning for Emerging Megacity Regions in China: A Preliminary Research Within a Socioeconomic Framework," Hao Li and Qian Zhu attempt to track the development of planning concepts from megacity to megacity regions by focusing on Asian and Chinese cases. The rise of megacity regions within China is discussed in a comprehensive socioeconomic framework. The current planning and governance approaches for this kind of regions are critiqued. The chapter concludes with a discussion on the policy implications for the development of China's megacity regions.

In chapter "Study and Prospects on the Self-Organizing Evolution Stage of County Urbanization in Hubei," Lingyun Liu introduces the self-organization theory for a system that can be organized and created and evolve by itself without outside instructions. He employs the theory to investigate the historic urbanization process of counties in Hubei Province. The study analyzes four temporal phases from 1949 to present. It explains the reasons why the present county urbanization in Hubei falls behind others. In addition, it also predicts that Hubei Province may meet an unprecedented turnaround in terms of county development with the possible changes in the national strategy and global situation.

In China's new-type urbanization, it is important to improve the efficiency of urban land use development. Chapter "From Three-Old Reconstruction to Expansion and Promotion: The Strategic Transformation for County-Level City in Rapid Urbanization Area: A Case Study of Guangdong Province in Mainland China," by Yikeng Luo, employs Guangdong Province as an example to describe the basic characteristics and the main challenges of urban land use development in China. It explicitly points out the necessary and inevitable changes of strategies from "three-old reconstruction" to "expansion and promotion" ("kuo rong ti zhi") to improve development efficiency. Furthermore, it suggests to explore the interacting and restricting relationships of "expansion and promotion" from the perspectives of priority, subjectivity, and hierarchy in strategic implementation in order to achieve a balance of the development of incremental land ("zeng liang tu di") and the redevelopment of stock land ("cun liang tu di") in the rapid urbanization areas of county-level cites.

The uneven economic growth challenges China's regional administration system and sustainable development efforts in the long term. Chapter "Spatial Structure of Regional Economic Development in Henan Province, China" by Li Hao and Qian Zhu explores the dynamic spatial structure of regional economic development in Henan Province during the period of 1999–2009. By utilizing advanced geomatic methods, it attempts to gain a better understanding of the spatial dynamics of economic growth in the province. The empirical study shows the fast economic growth of the whole region and increasingly uneven geographic development between the southeast and the northwest areas. The study highlights the importance of reducing the growing spatial polarization within a region, which calls for a more balanced and inclusive regional planning strategy in the near future.

Beijing, the capital city of China, has experienced a rapid urban expansion in the decade of 2004–2013. In chapter "Understanding Beijing's Urban Land Use Development 2004–2013 Through Online Administrative Data Sources," Xiao Rong, Ying Jin, and Ying Long report their recent research that examines the effectiveness and potential of using online administrative data sources published by the municipal government bureaus/commissions in understanding Beijing's urban expansion. They use the novel online land plot provision data collected from the website of Beijing Municipal Bureau of Land and Resources to examine the patterns of land supply for housing and employment. They first report the assembly of administrative data sources through web and GIS techniques and then analyze the growth pattern of Beijing's land supply through statistics analysis and mapping. Through a comparison of the new data with published statistics, they assess the strength and weaknesses of the online data. The study suggests that housing land is decentralizing at a faster pace than employment land, and in particular the developments for offices and institutions tend to focus on the central city area. It points out that the decentralization trend may worsen the job-housing balance across the municipality and exacerbate the pressure on radial transport corridors from the city center.

Shanghai, the largest city of China, has expanded its high-capacity metro network very rapidly in the past two decades. The substantial change in accessibility induced by the metro system in such a short period is parallel with massive increases in population, land use intensification, and urban diversification. In chapter "A Configurational Accessibility Study of Road and Metro Network in Shanghai, China," Lingzhu Zhang, Alain Chiaradia, and Yu Zhuang aim to identify the spatially disaggregated micro-macro relative accessibility relationships between road and metro-line network design, metro stations and bus stop locations, commercial land use location distribution, and station usage in Shanghai. Using GIS and spatial Design Network Analysis (sDNA) software to perform multilevel accessibility on each network link, they find that most of metro stations, bus stops, and commercial land use are located at the road network with the highest level of micro-macro accessibility, which indicates the multiplier effect between metro stations, land use, and multilevel spatial accessibility. These findings suggest an appropriate way of evaluating the effectiveness of land use planning in response to the micro-macro accessibility change due to the change in transportation system.

In chapter "Research on Interaction Between Traffic Improvement Around the Old Railway Station and Urban Land Utilization: A Case Study in Hohhot Railway Traffic Regulation," Linfei Han, Jianmin Guo, and Junyan Han employ the Hohhot railway station built in 1921 as an example to analyze the problems that old railway stations have been suffering in recent urban development. They propose the mixed urban spatial pattern combined with traffic improvement and land utilization as a solution to solve those problems in revitalizing old railway station areas. They also illustrate the relationship between traffic improvement and land utilization and discuss the improvement of urban transportation infrastructure and urban quality.

"Study on the Redevelopment Model of Multistory Residence in the Area of City Center" investigates the important role that multistory residence plays on improving the living conditions of urban residents in China. In this chapter, Hu Rui and Zhenyu Che select three communities constructed at different times with different backgrounds as empirical cases. Based on their empirical studies, the authors propose three different types of redevelopment models, including upgraded redevelopment, evolutionary redevelopment, and revitalization redevelopment, which offer guidance for renovating multistory residence in city centers.

Chapter "Elementary Analysis of the Impact of Large-Scale Sports Events on Space Regeneration of the Hosts: From Cases of 2008 Beijing Olympic Games and 2012 London Olympic Games" by Ying Gu and Ze Zhang examines the impact of large-scale sports events on the space regeneration of the host. It explores the effective methods of space regeneration in the case of large-scale sports events such as the Beijing Olympic Games and the London Olympic Games. First, it analyzes the impact of large-scale sports events on urban spatial pattern reconstruction, land value increment, and the change of land use of the host cities. Second, it studies the impact of infrastructure construction for large-scale sports events on space reconstruction of the hosts. Third, it discusses the impact of subsequent utilization of relative facilities of large-scale sports events on the evolution of urban spatial pattern. Finally, it makes recommendations for the hosts to take advantage of large-scale sports events to promote their space regeneration.

Recognizing the importance of healthcare facilities to the fitness of residents, in chapter "Study on Planning of Healthcare Facilities in High-Density Urban Residential Areas: Taking a New Community in Beijing as an Example," Jing Li and Tian Chen select three new communities in Beijing to analyze the distribution, design, and development of healthcare facilities in large cities. They recognize that the health issues caused by high-density population and limited outdoor activity space are unique issues encountered by Chinese people that cannot be solved through foreign experience. They explore how to improve the effectiveness of healthcare facilities in Chinese cities by taking into account of residents' age structure, living pattern, transportation space, residential area development mode, healthcare facilities distribution, as well as actual demands.

In chapter "Job-Housing Balance: The Right Ratio for the Right Place," Qian Wu, Ming Zhang, and Daniel Yang aim at providing theoretical and empirical evidence of job-housing balance and examining the applicability of job-housing ratio for different planning purposes in local context. Based on a rich literature review, the study removes the "deceptive simple concept" of job-housing balance on the surface and gathers insights on job-housing balance from the existing case studies. In the absence of a consensus of a good job-housing ratio, the study intends to present the possible ways of measuring and defining job-housing balance in complex urban development. It analyzes existing job-housing balance in Austin, Texas, and shows the limitation of job-housing ratio in guiding the distribution of employments and housing. It suggests that local municipalities might consider more factors in terms of the application of job-housing ratio in local context.

Chapter "The Implication of City Form Transformation in Rongcheng Area of Jieyang City" by Jianwen Huang and Ying Xu examines the change of urban form in Rongcheng area, Jieyang City, in three typical development periods, including the Qing Dynasty, the Republic era, and today. Qualitative and quantitative analyses are made on the morphological characteristics of the street and block of the urban area. It also discusses the patterns and forces of the urban transformation as well as a series of problems local planners are facing today.

Recognizing urban design as an important means of promoting urban-built environment in China, Liangping Hong and Wenzhu Tao introduce the urban design practice in Wuhan, Hubei Province, and analyze the urban design guidelines of the city in chapter "Urban Design Practice Towards Planning Management in China: Urban Design Guidelines in Wuhan City." First, they introduce the status and formulation levels of urban design in the planning system of Wuhan City. Then, they summarize the contents of urban design guidelines and design control elements with some specific cases. Finally, they explore the major approaches for incorporating urban design into urban planning management process.

As the proceedings of the 2014 IACP Annual Conference, the audiences of the book include, but are not limited to, the faculty members, students, practitioners, and the general public interested in the subjects of urban and regional planning, land use and urban design, environment planning, transportation planning, geographic information system (GIS) applications, urban studies, and geography. The key contribution of the book is to present recent developments in Chinese urban planning with a summary of new models and new practices for urban regeneration. The discussion will contribute to the advancement of urban planning in China as well as the world.

Houston, TX, USA — Qisheng Pan, Ph.D
Minneapolis, MN, USA — Xinyu (Jason) Cao, Ph.D

Contents

Chapter 1
Research of Beijing Automobile Exhaust Treatment Strategies Based on I/M System Program

Zhehao Jin and Qun Huang

1.1 Introduction

Air pollution is a serious urban environmental problem because of its negative effects on human health, plant growth, cultural relics and visibility. According to the research on the factors affecting air quality, one of the main sources of air pollution in cities is automobile exhaust.

With the recent surge in car ownership, pollutants in vehicle exhaust are an increasingly serious environmental concern. As of July 2013, the number of cars in Beijing has risen to 5.354 million and the air quality was rated a rare level 6, meaning hazardous with pollution reading over 301. Many developed countries have also struggled with these problems. Most of them have taken a strategy of "Pollution first, Treatment later" approach, allowing air quality to deteriorate before attempting to manage pollution (Qiang Ning 2003). This strategy destroys ecological environment and threatens the survival of human beings. Los Angeles, California has significantly improved "Photochemical Pollution" by implementing a systematic and improved I/M (Inspection and Maintenance) detection system aimed at vehicle inspection. I/M detection system, which has been analyzed in a series of thorough and systematic papers by foreign scholars such as Kahn (1997), Englert (2004), is gradually introduced to China and responded with a greater expectation by experts and scholars. I/M systems are currently the subject of a number of studies and discussions, especially focusing on legislation, system design, cost effectiveness, brittleness, and so on.

Z. Jin (✉)
School of Government, Central University of Finance and Economics, Beijing, China
e-mail: vjin2011@gmail.com

Q. Huang
School of Taxation, Central University of Finance and Economics, Beijing, China
e-mail: sophie.cufe@gmail.com

Q. Pan, J. Cao (eds.), *Recent Developments in Chinese Urban Planning*,
GeoJournal Library 114, DOI 10.1007/978-3-319-18470-8_1

However, air quality in Beijing is not only far worse than in European and North American cities, but also relatively poor compared with major cities in China (Huang Cheng et al. 2003). Instead of copying foreign practice blindly and mechanically, we should conduct a specific and in-depth policy program that is based on the beneficial practices from abroad but in line with our national conditions. Using fixed-point monitoring data about automobile exhaust in Beijing and drawing on foreign experience, this paper continues to explore problems and trends of existing measures in the process of urban development. We propose an I/M-centered treatment strategy system cooperating with multi-perspective policy by means of economy, administration, technology and law.

1.2 The State of Quote of Beijing's Automobile Exhaust Gas Treatment

1.2.1 The Current Treatment Measures of Automobile Exhaust in Beijing

Beijing has implemented 16-stage countermeasures for air pollution control centering on automobile exhaust prevention since 1998. Those measures, mainly focusing on formulating new vehicle emissions standards, in-use vehicle management, and vehicle replacement, have achieved initial results.

Beijing has effectively controlled new vehicle pollution since 2008 with the full implementation of national IV vehicle-emission standards (Luo Yiping 1999). In accordance with relevant provisions of Traffic Safety Code, Beijing has strictly implemented a car-driving registration system and mandatory write-off standard (Ma Ning et al. 2010). Meanwhile, Beijing has regularly checked the access of new automobiles from car manufacturers and organized the relevant units to carry out annual consistency checks on new vehicles and in-use vehicle conformity spot tests in order to ensure standard vehicle emissions as source control.

In-use vehicle management in Beijing, mainly focusing on daily management and cooperating with annual inspection measures, has strengthened in-use vehicle pollution control. Beijing, insisting the principle of "Prevention first, regular inspection, Mandatory maintenance, Condition-based Repair", carries on the inspection of small and micro non-operational passenger cars (Sun Guolian 2005; Xiguo 2002). Meanwhile, clean fuel and alternative raw materials have been promoted especially in vehicles used with high frequency. Additionally, peak restrictions and odd and even number rules on working days have been fully implemented for high-polluting vehicles.

At the same time, Beijing is accelerating the process of phasing out "yellow label cars" (heavy-polluting vehicles) through China's newly adopted economic incentives. By accelerating the retirement process of old taxis, a larger proportion of vehicles have achieved the stricter standard of motor safety technical inspection. The municipal government has invested about ¥ 500 million to offer a 1–2 years loan discounts to consumers who buy eco-cars. With the promotion of "Green Team" policy, there will come at a time when "yellow label cars" have been cleaned up off the road.

1.2.2 Beijing Automobile Exhaust Governance Problems and Causes Analysis

Admittedly, the full implementation of a series of policies and measures governing automobile exhaust has promoted the environmentally friendly mode of the auto industry. However, it would likely do little to aid the overall exhaust gas pollution.

There is still a technical gap between the actual vehicle detection ability and project requirements. According to the data from Beijing Traffic Management Bureau, Beijing has already established 43 vehicle detection spots. Though the number of A-level comprehensive performance testing station in the overall population is relatively high, the utilization rate of detection equipment is low due to the lack of standardization of detection. Furthermore, new vehicle emission standards in Beijing are in line with national standards which are formulated mainly referring to European rules and regulations, while in-use vehicle emission standards mainly refer to American laws and regulations. Obviously, those measures are not in line with local conditions of Beijing. The uneven professional quality of detectors, directly resulting in the gap between vehicle inspection level and requirements, make the situation worse.

In terms of policy making, there is a negative bias towards the policy of "emphasis on new vehicle emission standards over in-use vehicle standards". Since 1998, almost all the measures have focused on new vehicle emission standards. However, emissions from in-use vehicle make up 50–60 % of vehicle emissions, far exceeding pollution from new vehicle emissions. Therefore, the maintenance and repair of in-use vehicles is exactly to be the most efficient method to reach the requirements and then reduce air pollution.

Ineffective management negatively influences the system implementation effort. On the one hand, since the EPA and Beijing Traffic and Detection Spots are located far away from each other, timely communication between agencies is hard to achieve. On the other hand, every department plays its own role and independent functions are executed among transport agencies, public security departments, technical supervision departments, revenue departments as well as the Environmental Protection Agency. Though "The regulations for the control of air pollution in Beijing", expressly states that unified management provided by both environmental protection administrations and the supervision departments was implemented implementation in March 2014, there still remains the issue of the lack of effective coordination in practice resulting from several potential problems such as unclear responsibilities between departments, maladministration, and disunity in identification procedures. According to the problems and issues addressed above, the current management system in Beijing cannot adapt to the special requirements of vehicle exhaust prevention and control.

1.3 The Basic Meaning and Significance of I/M System

1.3.1 I/M System Connotation

Inspection and Maintenance (I/M) is a way to check whether the emission control system on a vehicle is working correctly. I/M helps improve air quality by identifying high-emitting vehicles in need of repair (through visual inspection, emissions testing, and/or the downloading of fault codes from a vehicle's onboard computer) and requiring them to be fixed as a prerequisite to vehicle registration within a given non-attainment area. The 1990 Amendments to the Clean Air Act made I/M mandatory for many regions across the USA, based upon various criteria, such as air quality classification, population, and/or geographic location. This program was then promoted and advanced in Europe and Japan (Liu Junmin and Yang Yin 2003). A well-designed I/M program remains the single most effective and cost-effective way to achieve major reductions in vehicle pollution.

1.3.2 I/M System Benefits

The I/M system in the USA is recognized as the most effective way of in-use vehicle emission control with an 87 % overall citizen satisfaction, according to the result in the 1997–1999 California roadside tests. California is the most strictly-managed state in the USA. The chief goal of I/M systems in California is to reduce exhaust emissions by 100 t every day (Zhao Futang et al. 2005). In practice, California I/M system can be characterized as follows: (1) the California region is divided into three categories (Improved area, Basic area, main area), with different policies; (2) high-emitting vehicles, whose emissions are two times higher than the standard of vehicles stated by California Bureau of Automotive Repair (BAR), are intensively monitored with high threshold detection; (3) a certain inspection exemption system is included; (4) BAR routinely releases test results and health effects associated with air pollution exposure.

Compared to the highly-focused new vehicle emission standards in Beijing, I/M program emphasize the control of in-use vehicle emissions. All new passenger cars and trucks sold in the United States today must meet stringent pollution standards, but they remain low-polluters if the emission controls and the engine are both functioning properly. I/M system is a state-operated program designed to ensure that vehicles stay clean after they leave the factory and are in use by the driving public. Through periodic emissions performance checks and repairs for those vehicles that fail emissions tests, I/M encourages proper vehicle maintenance and discourages tampering with emission control devices. This program will be much more efficient than the Annual Test in China.

The I/M program is a cost-effective method of controlling harmful automobile-related pollutants. Congress acknowledged the program's utility by providing for it in the Clean Air Act. The Environmental Protection Agency has in theory accepted the notion that I/M programs can effectively achieve attainment by at least threatening to impose sanctions against communities who refuse to implement such programs. According to a research on cost-effectiveness of I/M program held by Fred Harris Jr (1989), the costs of cutting CO emissions are as little as $110 per ton since drivers do not need to change the original technical parameters and equipment of the car, or install new purification devices which can be very expensive. Of course, such programs can never be the sole control mechanism; but there seems no better way to control the vehicle emissions with such advantages of high efficiency, low cost and high benefits.

The US Environmental Protection Agency and state air agencies' estimates imply that I/M program has been highly effective. They derive these estimates from simulations of fleet wide emissions with and without I/M program. (California I/M Review Committee, 1993; Environmental Protection Agency, 1992) Using this approach, the state of California's I/M Review Committee estimated that its 7 I&M programs reduced vehicle emissions of HC, CO, and NOx during 1992 by 18.2 %, 9.8 %, and 3.9 %, respectively (California I/M Review Committee, 1993). Per vehicle emissions in California were far lower than the average in the USA (Tom Wenzel, 2001). According to AIRNow, in 2013, the AQI value of 38 in California represented good air quality with little potential to affect public health. Thus, it is likely that the I/M program played a substantial role in the improvement of the ambient air quality and can be well taken as a reference for Beijing, considering the resembling density of population and climatic environment. Additionally, the annual testing in China has many similarities with I/M program. Therefore, the introduction and advancement of I/M program to Beijing, where conditions mature introducing this program, will likely have significant environmental consequences.

1.4 Beijing Automobile Exhaust System Program Improvement Strategy

1.4.1 I/M Program Core Detection System

The current vehicle inspection and maintenance program in Beijing and I/M system have commonalities in technology and management with I/M program, but it pays much attention to the vehicle performance test but lacks basic control of pollutant emission and vehicle maintenance. Drawing lessons from foreign I/M system standardization process will allow Beijing to make real progress of in-use vehicle energy conservation and emissions reduction. To carry out comprehensive security system plan smoothly and ensure the actual results, the I/M system-centered program of Beijing should actively cooperate with supporting policies concerning technological innovation, market regulation, department coordination and social participation.

1.4.1.1 I/M Inspection and Maintenance of Core Processes

Learned from the American I/M system as it shows on the left side of Table 1.1, Beijing automobile exhaust treatment system requires new vehicle information and in-use vehicle raw data import at first place. As the recorded automobile information matches it current situation, the automobile should go through an emission testing process. If the testing results meet all requirements, the car will get the road permit or else it will go for the fault diagnosing process and will be forced to perform maintenance until it could pass the test. On the basis of American I/M system, Beijing automobile exhaust treatment system adds a key process of real-time monitoring which can be seen on the right side of Table 1.1, with the help of detecting techniques of vehicle-mounted QR code. The vehicle will go through punishing, problem diagnosing, repairing as well as second testing processes if it doesn't pass the real time detection. All information of vehicle inspection and maintenance, including the information of passing vehicles from other provinces, will be recorded to the data center at Beijing Traffic Administration Bureau.

Regular testing and random testing should proceed simultaneously: (1) Regular testing by the transportation and the environmental protection department should detect vehicle fuel consumption and emissions with distinction between operating vehicles and non-operating vehicles; (2) Random testing of fuel consumption and emissions such as spot checks of vehicles in parking areas or on the sides of street, and doing sampling inspections of corporate business vehicles.

Testing process includes six steps: (1) Vehicle registration: Confirm the discharge standards of the test vehicles; (2) Visual inspection: Once unqualified vehicle is found according to the test items during this step, the vehicle user will be required to repair his/her vehicle within a time limit. (3) Function test; (4) Exhaust pollutant detection; (5) Fuel consumption measurement; (6) Result judgment and re-inspection: When a vehicle passes all of the above tests, it shall be judged to be qualified and evaluated of the corresponding level with emission environmental symbol. If it fails in the steps of (3), (4), or (5), the vehicle will need to make "fault diagnosis and forced repair" to get qualified on the road.

Non-local vehicle and transit vehicles exhaust pollutants and fuel consumption detection should be conducted according to the standards of their respective districts. Non-local vehicles detained in Beijing within 3 months shall meet the transit vehicle requirements. Accordingly, non-local vehicles detained more than 3 months in Beijing should meet the local vehicles requirements.

Implementation of green and yellow flag system for motor vehicle fuel consumption and exhaust pollutant detection is divided into three levels, depending on

Table 1.1 Beijing Automobile Exhaust Treatment Strategies

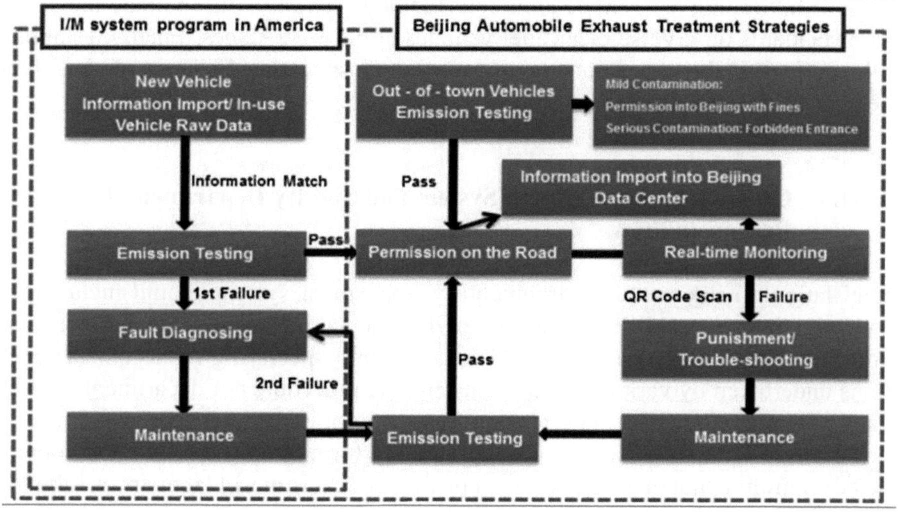

the emission: Issue a green flag to standard low-emission vehicles; issue a yellow flag to standard high-emission vehicles; issue no flags to the substandard vehicle. The flag issuing process comes with annual application. If excessive fuel consumption and emissions are found at road sampling, police have the obligation to revoke the flag and require compliance. After the car gets repaired at a designated maintenance location, it should go through the test at a testing organization, and it will receive the environmentally friendly flag after regaining eligibility. Green flag should be carried on the car and prepared for the inspection. Vehicles on the road with no flag will be subject to penalties by traffic police.

Real-time dynamic monitoring station is based on existing P.M 2.5 monitoring sites. Additional stations should be added in road congestion and traffic-intensive areas in order to form a digital monitoring network. Meanwhile, setting up peripheral monitoring sites at the toll booths around Beijing as the safety valves can help to improve the detection efficiency. Dynamic network management can be ensured by uploading test results.

1.4.1.2 Develop I/M System Testing Standards, Improve the Car Pollution Prevention Policy

Currently, Beijing regulates only the sale and registration of light-duty gasoline vehicles. Since there are no specific regulations and emission standards for clean-fuel vehicles, this regulation has not met "Beijing V" vehicle emission standards. So actions should be taken to improve the current exhaust treatment. Firstly, it is important to refine the clean fuel and clean fuel vehicle management system.

Secondly, the blue environment-friendly flag will be issued for the vehicle compliance with "Beijing V" vehicle emission standards. Thirdly, differentiate the emission standards on diverse clean fuel vehicles at different stages. Finally, establish pilot projects in the ozone nonattainment area to inspect the efficiency of clean fuel vehicle emission pollutants control.

1.4.1.3 Create I/M Management System and Specify Departmental Responsibilities

Legislation on integrated environmental management system should include the following items: establishing status of environmental management department and constitution of mechanism (Zeng Xiangang 2009), specifying management functions undertaken by various departments and the procedures of department coordination, cooperation and supervision. Responsibilities of different districts and departments have to be specified, progressively forming a systematic and coordinated environmental management system. Specific divisions of labor are as follows:

(1) Environmental Protection Agency: develop vehicle emission pollution control plans and objectives; Evaluate system implementation effects; establish and manage the inspection and maintenance program network; aggregate data and publish the status of vehicle emissions to the public; advocate the idea of energy-saving and environmental protection. (2) Transportation Department: set up exhaust gas monitoring stations along the road; identify and manage the maintenance equipment and maintenance organizations; supervise and manage vehicle maintenance quality. (3) Public Security Department: road sampling vehicles; levy fines from substandard vehicles and manage the files of inspection results. (4) Technical Supervision Department: identify and manage the test organizations and test equipment; control and supervise the quality of vehicle detection. (5) Tax Department: regulate and use the funds of levied congestion charges; pay for green subsidies; register and audit the financial condition of testing organizations and maintenance organizations.

1.4.1.4 Establish I/M System Information Management System, Conduct Network Management

Establish a vehicle database, including basic statistics, emissions testing results, the fuel consumption testing, maintenance control record, etc. A high efficiency of information management can be achieved through standardized management of electronic, automated information. Through the sharing and transmission of data updates, information about vehicle fuel consumption and exhaust emissions testing and maintenance governance can be reached by owners, testing stations and maintenance stations and environmental protection, public security, transportation department, technical supervision, industry and commerce, taxation department and other departments in a timely fashion. The transparency and accessibility of information will help process system implementation issues and make sure effective implementation of the system.

1.4.1.5 Establish I/M System Evaluation Model, Hold the Hearing to Attract Public Opinions

Tendering for academic research in the early stage will trigger widespread concern and discussion about the impending policy programs. The most reasonable and effective solutions must be a complex system containing the front opinion of professional scholars, and recommendation from the official representatives, stakeholders and other social circles representatives. Based on the methodical assessment of past policy programs, it will ensure all demands of the different stakeholders reflected and conduce to explore and promote a more perfect new system model.

1.4.2 Automobile Exhaust Inspection and Maintenance Program Auxiliary Policy Programs

1.4.2.1 Strengthen Exhaust Pollution Prevention Policy and Legal Responsibility

Beijing V standards in accordance with EU standards should be strictly implemented, while strengthening the guidance of overall urban traffic planning construction. The law should clarify the required legal liability of automobile exhaust pollution prevention policies so that the significance of vehicle exhaust pollution governance could be effectively developed. Administrative punishment permission should be given to exhaust treatment of the administrative department to strengthen the independence and authority of law enforcement. Besides, with increasing fines and broadening the means of punishment, the object of punishment should be extended from the individual to auto makers, maintenance companies and supplier with violations, in order to ensure the absolute authority of the law.

1.4.2.2 Adopt Preferential Taxation and Other Economic Policy Instruments

Combined with the vehicle inspection and maintenance testing results and related ratings, the governance department should adjust the downtown parking fees and parking violation fines and charge downtown additional congestion surcharges by level (Wang Binhui 2004). A charge levied on related projects such as congestion should be proportionately reserved as special financial funds for green tax subsidies, or be poured into advancing the traffic network and facility construction as well as promoting clean industries. Lower excise tax and subsidies on energy-saving cars could relieve the burden on consumers who buy the clean vehicles product in line with the related standard. The deduction and exemption of value added tax (VAT) and import duties levied on car cleaning technology and import equipment will encourage private capital for clean technology and clean fuel infrastructure investment and participation.

1.4.2.3 Strengthen the Urban Transport Master Plan and Develop Public Transport

Rationalized urban planning and balanced allocation of urban resources will lead to further improvement of the road network structure. Public transport priority should be incorporated into urban transport development strategy. Three-dimensional rapid rail transit districts are going to be established in accordance with plans and functional division. Connection with the transportation system and parking problems must be considered during the development of real estate and commercial construction along traffic arteries. Bicycles can play a role in a fast, convenient and comfortable public transport system and should be encouraged through the establishment of bicycle parking lots near the trunk bus station and public bike rental systems in the city center (Zhang Yu et al. 2013).

1.4.2.4 Implement Technological Innovation and Improve the Operational System Efficiency

Beijing will conduct fuel cell bus demonstrations, promote the transition to alternative fuel vehicles, and ultimately attain the popularity of fuel cell vehicles. Paste the two-dimensional exhaust testing code label on all vehicles in Beijing. These QR codes store all the data from all detection systems used to be repeatedly copied, such as automobile engine numbers, license plate numbers and the last detection parameters. The promotion of two-dimensional code scanner device configuration could improve inspection efficiency and reduce detection errors. Regulate various non-lead anti-knock additives, while strictly controlling detection index. Try to develop a special "city oil", optimize the structure of the components of gasoline, in order to meet the policy objectives and strengthen the supervision of gas detection.

1.4.2.5 Strengthen Environmental Education and Public Participation

Educate citizens in order to assure public knowledge, public participation and public scrutiny on automobile exhaust governance. Recruit environmental volunteers from the community to promptly report black smoke and failing scrap cars to monitoring sites. Include environmental education in public schools, traditional media and new media. The public also has a duty to use more public transport, not to buy car with excessive exhaust emissions.

1.4.3 Feasibility Analysis of Improvement Plan

1. Macro policy guidance: "China's 12th Five-Year Plan" about air pollution control has established the goal of an average annual PM2.5 concentration descent speed of 6 % in Beijing-Tianjin-Hebei region by 2015. The State Council "atmospheric pollution prevention plan of action", referred to as atmospheric "country of ten", further clarifies the strict control of PM2.5 reduction. Beijing plans to reduce PM2.5 concentrations to about 60 micrograms per cubic meter by 2017, decreasing more than 25 % compared to 2012. Relying on financing matchmaking, key projects fairs and other ways, the government will invest 200–300 billion Yuan and soak up social investment of nearly one trillion Yuan within 5 years to control air pollution. Above all, the guiding role of macro policy with government support will benefit exhaust governance, while adequate funds will facilitate more effective operation of programs and accelerate the implementation of the relevant incentive policies.
2. International experience: Western countries generally conduct I/M system as pollution control measures, and its advancement, reasonableness, economy, operability and effectiveness of pollution control and other aspects are wildly accepted. This successful management experience will provide the guidance for Beijing exhaust governance practices. Current development of vehicle pollution control technology will help in achieving Beijing's goal of building a sustainable developing and world-class city.
3. Mature conditions: After years of effort, Beijing vehicle emission management, emission testing, and motor vehicle repairing industry have developed to a certain level where it is ready for the implementation of I/M system. To begin with, the current car maintenance program provides a basis for the implementation of I/M system. On the hardware side, automobile repair business and comprehensive performance monitoring stations provide the technical support for the operation of I/M system. On the software side, Beijing has established a series of car maintenance regulations, standards and improved institutional system of car maintenance.
4. Advanced technology: Beijing will effectively control and reduce car pollution through the introduction of advanced scientific detection methods and accurate judgments about vehicle emissions status. The gradual establishment of advanced scientific, economic and efficient exhaust pollution control system meets the needs of air pollution control and economic and social development in Beijing. It will also contribute to further rationalizing the motor vehicle pollution control management system and establishing a scientific and systematic management mechanism.

1.5 Conclusion and Prospect

Vehicles are the main source of ambulatory pollution and have led to progressively heavier pollution problems. Pollution is severely harmful to the ecological environment and people's health and also hinders the sustainable development of economy and society. Due to this grim reality, in-depth study of automotive emission control and governance strategy has become an important global issue.

In recent years, Beijing has taken a series of automobile exhaust pollution control measures which have achieved substantive results. Nevertheless, some problems have been exposed, such as imperfect technical conditions, inefficient scientific management system, and unscientific policies. International experience shows that the establishment and implementation of I/M system is an effective way to achieve the greatest energy-saving emission reduction.

On the basis of Beijing's existing automobile inspection and maintenance program, we should improve the governance program while regulating the core inspection and maintenance procedures. On the one hand, we need to advance inspection and maintenance testing standards and perfect the management system to fully develop a feasible I/M system well adapted to Beijing's present management method. It is equally necessary to establish a network of information management systems and an approach to evaluation and feedback. We should add a detection procedure for out-of-town vehicles. Since the reduction in automobile exhaust is a complex technical and social issue, automobile exhaust gas treatment inspection and maintenance program should be supported by a wide and diverse mix of policy instruments.

The auxiliary policies take legal and administrative means as their mainstay, based on planning and technical progress and supplemented by the market economy and social advocacy. Enhancing implementation performance by stricter standards and administrative penalties, along with the use of green taxes, subsidies and other economic policy tools, will protect and advance enforcement of automobile exhaust pollution control. In addition, through innovating fuel technologies and strengthening urban transport planning, we can reduce exhaust pollution caused by road congestion, enhance citizen environmental education and promote citizen participation in exhaust pollution control consciously. Those processes will accordingly prevent acts of violation of exhaust emissions to some extent.

The appropriate vehicle exhaust pollution prevention strategies will help achieve the goal of reducing vehicle emissions effectively, thus improving people's quality of life, protecting social and ecological sustainability and providing the necessary foundation for building a world-class city and a livable Beijing.

References

Englert, N. (2004). Fine particles and human health—A review of up ideological studies. *Toxicology Letters Journal, 14*(9), 235–242.

Huang Cheng, Bing Yan, Chen Changhong, et al. (2003). A comparative study of air quality in cities at home and abroad, Shanghai. *Energy Research and Information Journal, 19*(3), 165–171.

Kahn, M. E. (1997). Particulate pollution trends in the United States. *Regional Science and Urban Economics Journal, 27*, 87–107.

Liu Junmin, & Yang Yin. (2003). Foreign I/M system in the environmental aspects of the application. *Automobile and Society Journal, B01*, 54–56.

Luo Yiping. (1999). Beijing air pollution control measures to optimize research. *Environmental Protection Journal, 9*, 26–28.

Ma Ning, Liu Min, & Liang Wannian. (2010). Effect of Beijing Olympic Games on Beijing air quality. *Capital Public Health Journal, 4*(3), 103–110.

Qiang Ning. (2003). *Urban air quality management and control*. Beijing: Science Press.

Sun Guolian. (2005). The direction and management of automobile maintenance industry. *Auto Repair Journal, 7*, 6–7.

Wang Binhui. (2004). OECD and China's green tax system reference. *Urban issues Journal, 5*, 59–62.

Xiguo, L. (2002). Study on short tests used in enhancing I/M program in Shanghai. *Shanghai Environmental Sciences Journal, 21*(7), 397–400.

Zeng Xiangang. (2009). Local government environmental management system analysis. *Teaching and Research Journal, 1*, 34–39.

Zhang Yu, Liu Xuemin, & Zhang Hong. (2013). Chinese urban public bicycle system: Situation, problems and countermeasures. *Chinese Development Journal, 13*(5), 74.

Zhao Futang, Qu Hua, & Xie Jianguang. (2005). *Modern automobile detection and diagnosis and repair* (p. 19). Beijing: Beijing Institute of Technology Press.

Chapter 2
Analysis About Key Indicators of High Density Blocks in the New Central District from the Perspective of Mitigating Urban Heat Island: A Study at Guangzhou

Pin Wang and Qinglin Meng

2.1 Introduction

Urban Heat Island (UHI) is an important issue, as a consequence of urbanization (Oke 1982), UHI was first documented by British scholar Howard L in 1818. Over the last 100 years, UHI has been a prominent field of research and many achievements have occurred (Rizwan et al. 2008). In China, Xie Kekuan translated The Climate of The Cities (Das Stadtklima) (Kratzer 1956), written by German scholar Kratzer, into a Chinese version named "Urban Climate"; this was the first book introducing urban climate. Zhou Shuzhen published "An Introduction to Urban Climatology" which analyzes the UHI features in Shanghai city. Besides these prominent authors, many Chinese scholars have done research (Chen et al. 2009; Lin et al. 2010; Ng et al. 2012) relative to thermal environment.

According to previous research, UHI is always influenced by planning and design. Unger (2004, 2006) and Bottyán et al. (2005, 2003) both took Szeged as the studied city, and found the UHI impact factors of urban areas within a 500×500 m region; factors included built ratio, water area ratio, sky view factor and building height. Hart et al. (2009) found canopy cover, total floor area and road length were the most important factors influencing the UHI of urban areas of a 300×300 m area in Portland. In Beijing, Zhao et al. (2011) found that the green coverage ratio, floor area ratio and building density were key factors influencing the UHI of urban areas of within a 500 m radius.

Primarily researchers targeted a certain area of a city, while other researches focused on a regional scale. Urban residential is one of the most common regions, and it's thermal environment is always a hotspot. Giridharan et al. (2004, 2007) found that sky view factor and terrain were the main cause of air temperature

P. Wang (✉) • Q. Meng
School of Architecture, South China University of Technology, Guangzhou, People's Republic of China
e-mail: theresa_1015@126.com

Q. Pan, J. Cao (eds.), *Recent Developments in Chinese Urban Planning*, GeoJournal Library 114, DOI 10.1007/978-3-319-18470-8_2

change inside residential areas. High building density and anthropogenic heat hindered the optimization of the outdoor thermal environment, on the contrary, high arbor coverage could effectively reduce heat island intensity. Yang et al. (2010) studied the relationship between UHI and variables in Shanghai's residential areas, e.g. sky view factor, green coverage ratio and surface albedo. Chen et al. (2004) and Kubota et al. (2008) conducted a similar study.

In addition to residential areas, other functional regions, such as industrial areas and business districts need attention.

2.2 New Central District and Thermal Environment

2.2.1 New Central District

The urban center is the core of a city. New central district is the area of centralized building and population. At the neighborhood level, the blocks in new central district are characterized as high density..The new central district has large numbers of tourists and a high residential population. Therefore, the outdoor spacial quality is important.

The urban center is developed with detached close-set high-rise buildings, and with a large aspect ratio and high imperviousness. Such underlying surface influences local microclimate environment. Unique microclimate environment of urban center was shown by a series of urban climatic map studies at home and abroad (Ren et al. 2011). Figure 2.1 is the thermal environment map of Tokyo, and Fig. 2.2 shows the climatic map of Hong Kong. We found that the temperature of high density commercial and residential areas was higher in Tokyo, and high density blocks underwent bad geothermal environment. This leads to a heat island problem.

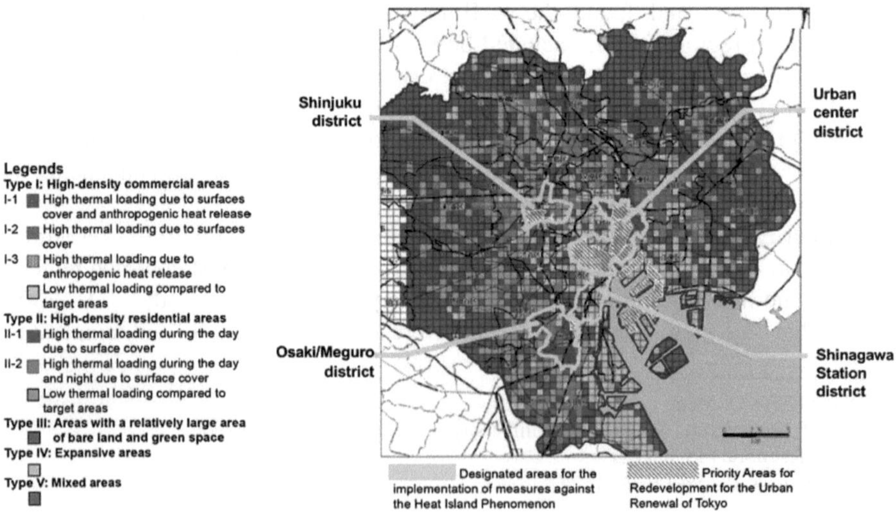

Fig. 2.1 Thermal environment map of Tokyo

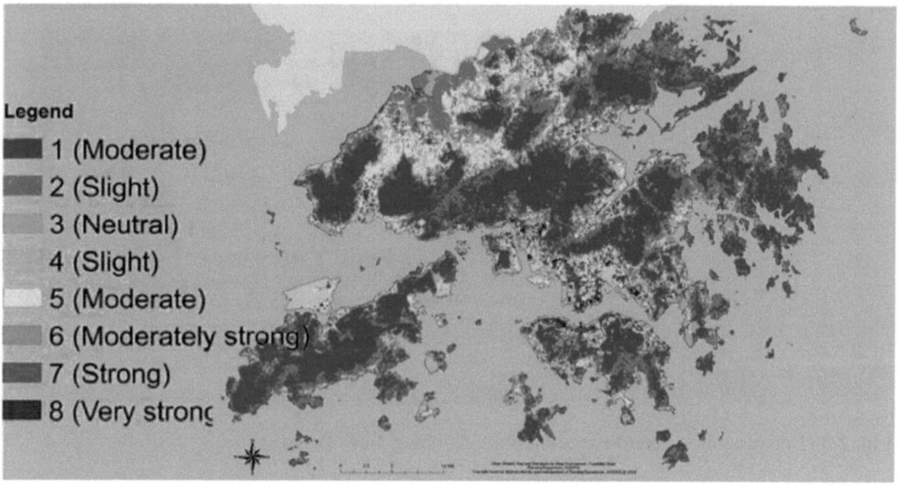

Fig. 2.2 Climatic map of Hong Kong

Hence, maintaining the land use efficiency as well as optimizing its thermal environment, has important research value. However, like other functional regions with the exception of urban residential, little research has been done regarding heat island problem in new central district. One of the main reasons may be that the new central district usually has no clear boundaries. So in this paper, city blocks were selected as the research scale.

2.2.2 Empirical Study

In order to empirically show the local UHI problem, a field experiment was conducted in Zhujiang New Town, a representation of the new central district in Guangzhou city, in July 11–13, 2013 (Fig. 2.3). It turned out that summertime heat island problem was severe. Take the data of July 13 as an example (Fig. 2.4), heat island occurred all day, and the average UHII of ten measure points reached to a maximum of 5.5 °C, which happened at 2:00. During the daytime (6:00–20:00), the average UHII was 1.77 °C.

2.3 Planning Indicators and UHI

2.3.1 Regulatory Detailed Planning and UHI

Regulatory detailed planning is an important stratum of urban planning system, which takes indices, stipulation and statutory plan as a control, ensure both land use and intensity of construction are under control.

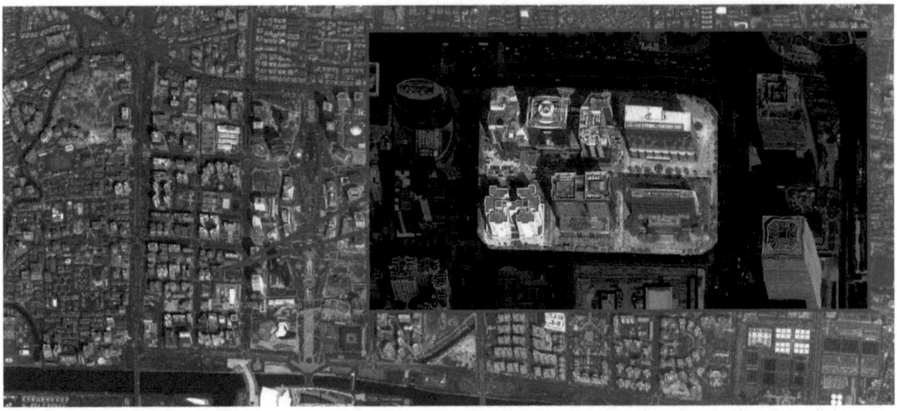

Fig. 2.3 Location of the field experiment in Zhujiang New Town

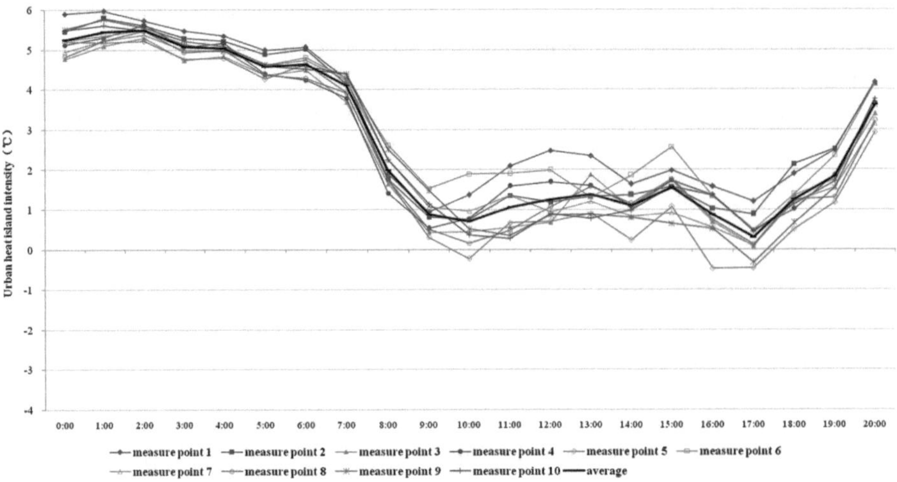

Fig. 2.4 UHII of ten measure points in July 13, 2013

Since urban construction is the cause of urban heat island effect, and the statutory regulatory detailed planning is the basis of both urban construction and government management, there may be a certain relationship between regulatory detailed planning and heat island intensity. If we can find the key indicators of regulatory detailed planning that have the greatest influence on the thermal environment, the heat island problem may be effectively controlled by determining a reasonable range for each indicator.

2.3.2 Key Indicators

Artificial control of factors influencing the urban thermal environment is mainly divided into three categories: spatial geometry, underlying surface physical property and anthropogenic heat (Davies et al. 2008; Mirzaei et al. 2010). By reviewing the related literature at home and abroad, Floor Area Ratio (FAR), Building Density (BD), and Greenery Ratio (GR) are found to be important factors influencing the heat island effect, and are the key indicators of regulatory detailed planning at the same time.

2.3.3 Control Requirements of the Key Indicators

Some technical regulations of urban planning administration(i.e. standards and guidelines) of different provinces or cities were studied to summarize the control requirements of the key indicators from the perspective of government management. We find that the upper limit of FAR regarding public projects is mainly between 4 and 8, BD upper limit is mainly between 30 and 60 %, and GR lower limit is mainly between 20 and 25 %. As for Guangzhou city, FAR upper limit of high-rise public projects is 12, BD upper limit is 55 %, and GR lower limit is 20 % (Fig. 2.5).

In addition, we surveyed Zhujiang New Town, and the statistic result showed the average FAR of high density blocks was about 8.4, BD was mainly around 30–40 %, and the average BD was 35 %. It is worth mentioning that GR was only between 10 and 20 %, and the average GR about 15 %. In reality, GR of many blocks can not meet the required limit. The existence of central axis with more public greenery space leading to less interior greenery space in blocks may be the reason.

Next, a typical block of Zhujiang New Town was chosen as the prototype, and simulation cases are established.

2.4 Parametric Studies

2.4.1 A Brief Introduction of ENVI-Met

ENVI-met is a three-dimensional microclimate model designed to simulate thermal environment with a typical resolution of 0.5–10 m in space and 10 sec in time. The model is a tool for studying the surface-plant-air interactions in the urban environment at the microclimate scale (The Hitchhiker's Guide to ENVI-met 2015). Some recent studies have used ENVI-met to simulate urban microclimate (Ng et al. 2012;

Ali-Toudert et al. 2007), demonstrating that ENVI-met could describe the urban micro-environment relatively accurately.

The results of ENVI-met simulation (Fig. 2.6) and observed data in above-mentioned field experiment were strongly correlated($R^2 = 0.917$, $P < 0.001$), and this numerical simulation software was shown capable of predicting thermal environment of high density blocks (Fig. 2.7).

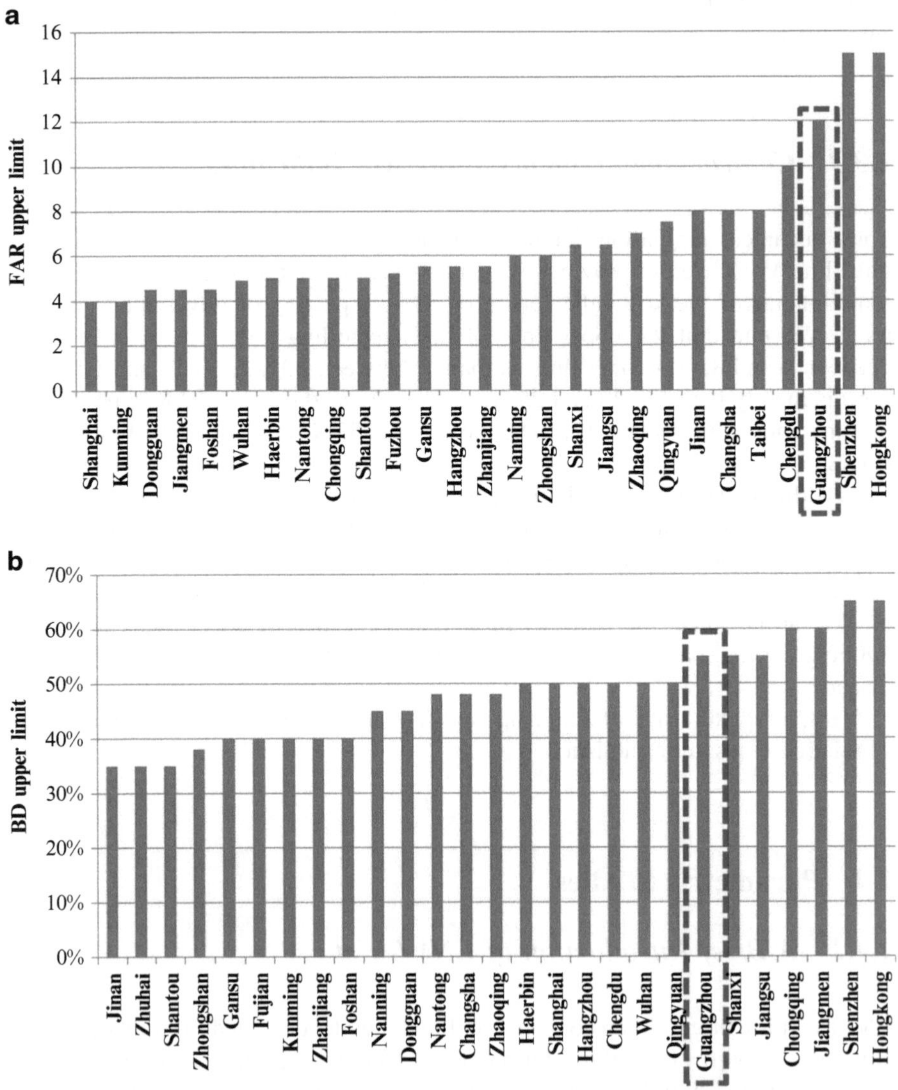

Fig. 2.5 Control requirements of the key indicators in technical regulations of urban planning administration, standards and guidelines of different provinces or cities (**a**) FAR, (**b**) BD, (**c**) GR

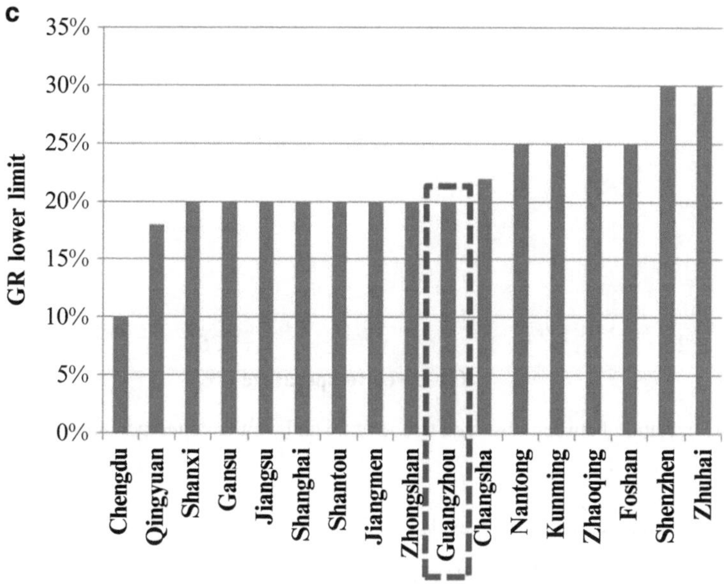

Fig. 2.5 (continued)

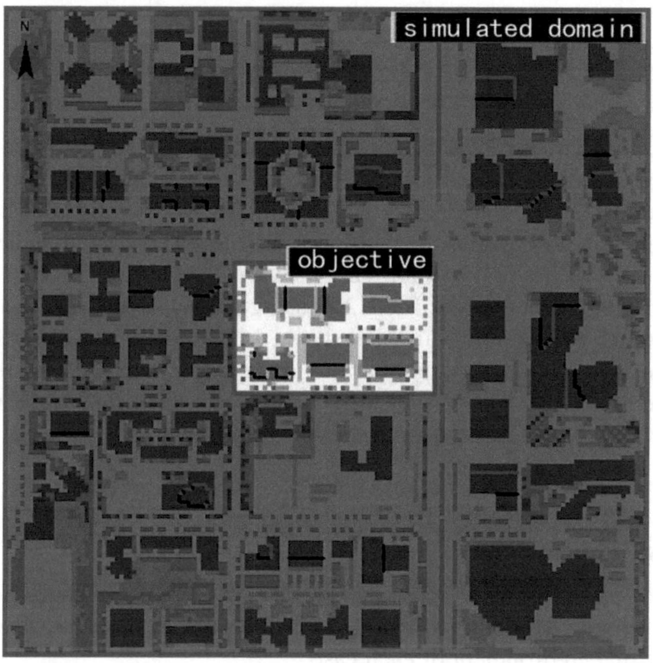

Fig. 2.6 Simulation and target domain in ENVI-met

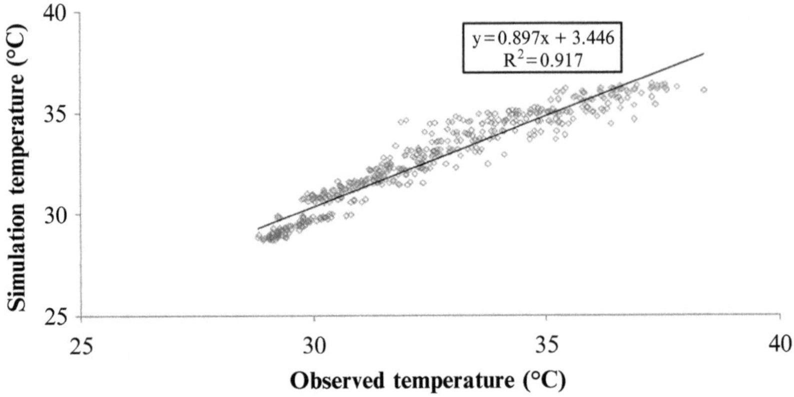

Fig. 2.7 Correlation between observed temperature and simulated temperature

Table 2.1 Variables and changing levels

Changing level	TBD (%)	BD (%)	FAR	GR (%)
1	20	35	4	30
2	24	40	8	20
3	28	50	12	10
4	32	55	16	0

2.4.2 Setting of Cases by Experimental Design

The I2 block of Zhujiang New Town, which was also the location of the above field measurement, was chosen as the prototype. An experimental design was adopted. Variables are BD, FAR and GR. What's more, the tower building density (TBD) is also changed with four levels, distinguished with the podium building density (BD). Changing levels of the four variables are listed in Table 2.1.

A total of 16 cases were set (Table 2.2) using a table of orthogonal array of L16 (4^2) with four levels of four variables above. In each case, trees inside the grass were placed near the streets, and the ratio of the trees were 50 % of the greenery areas. Meteorological data of typical summer day of Guangzhou is used as the boundary condition when we simulated with ENVI-met.

2.5 Results and Discussions

Average daytime UHII during 8:00–18:00 was taken as the evaluation index for outdoor thermal environment of each case.

Table 2.2 Cases set by the experimental design

Case number	TBD (%)	BD (%)	FAR	GR (%)	Schematic diagram	Case number	TBD (%)	BD (%)	FAR	GR (%)	Schematic diagram
case 1	20	35	4	30		case 9	28	35	12	0	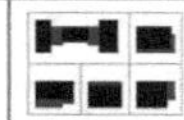
case 2	20	40	8	20		case 10	28	40	16	10	
case 3	20	50	12	10	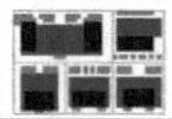	case 11	28	50	4	20	
case 4	20	55	16	0		case 12	28	55	8	30	
case 5	24	35	8	10		case 13	32	35	16	20	
case 6	24	40	4	0		case 14	32	40	12	30	
case 7	24	50	16	30		case 15	32	50	8	0	
case 8	24	55	12	20	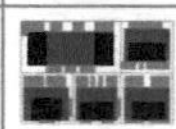	case 16	32	55	4	10	

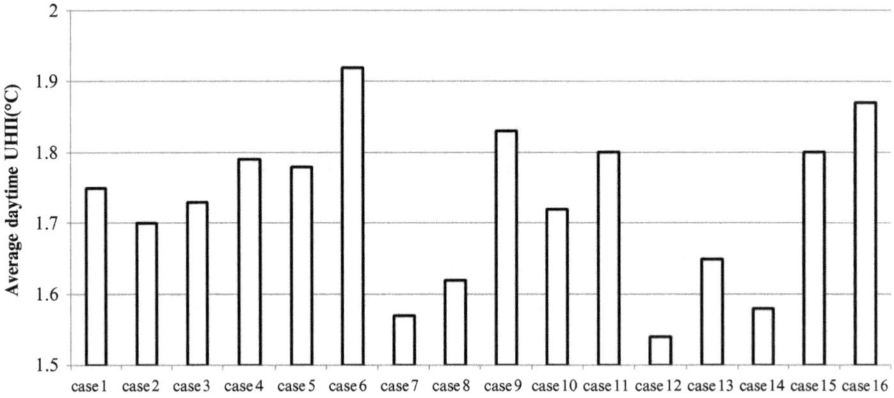

Fig. 2.8 Simulation results of each case

2.5.1 Visual Results

Simulation results are presented in Fig. 2.8. The highest UHII is from case 6, whose FAR is smallest and GR is 0. While the lowest UHII is from case 12, whose BD and GR are the largest. We can figure out that more building shade and greening can effectively mitigate the heat island problem.

When the variable TBD was 20 %, there were four cases (case 1, 2, 3, 4). Average their evaluation index and the result meant the thermal state of TBD-20 %. All levels of each variable were computed in the same way. Figure 2.9 reports the computed results. The results show the larger BD, FAR and GR, the better the outdoor thermal environment.

TBD is the variable that has smallest impact on the thermal environment, while GR has the most significant impact. When GR rises from 0 to 30 %, the total UHII reduction is nearly 0.25 °C. When BD rises from 35 to 55 %, the UHII reduction is only 0.05 °C. Building density does not have a large influence on thermal environment. As for FAR, average daytime UHII reduces more than 0.1 °C when FAR rises from 4 to 8, while the UHII reduction lowers to 0.02 °C when FAR rises from 8 to 16.

FAR is the most important indicator adopted in urban planning management. As we know, in many provinces and cities, the upper limit of FAR in public projects documented in urban planning administration, standards and guidelines is between 4 and 8, while in reality the average FAR of high density blocks was about 8–9. At the same time, high FAR means more tall buildings and more shade to reduce solar radiation influence, and may be effective in mitigating UHI. But when FAR increases to a certain extent(i.e. 8) its effect on improving thermal environment becomes smaller. That is to say, there exists a reasonable FAR range, because smaller FAR may lead to under-utilization and UHI problem, and larger FAR (10) may lead to other issues, such as the strong wind in the pedestrian area, far

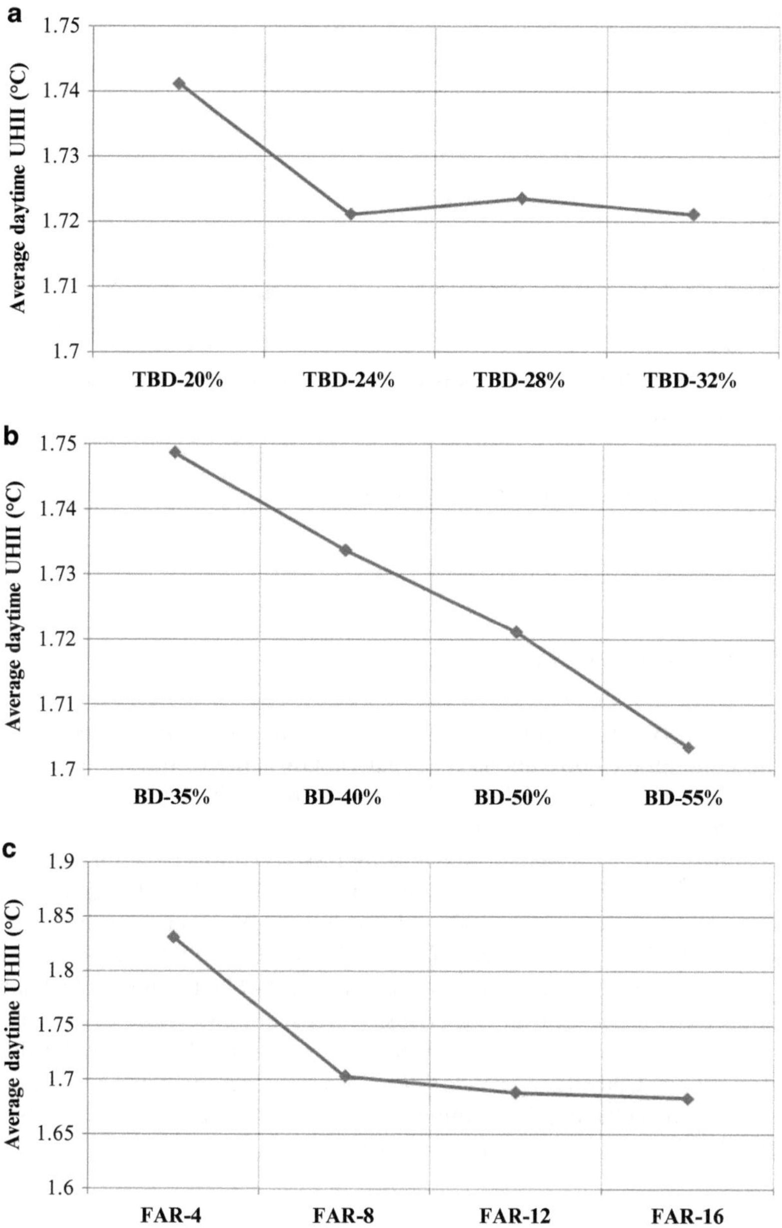

Fig. 2.9 Average daytime UHII of each level of each variable (**a**) TBD, (**b**) BD, (**c**) FAR, (**d**) GR

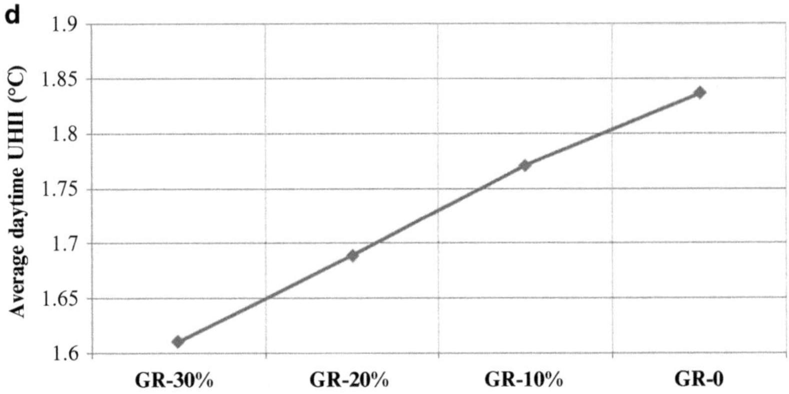

Fig. 2.9 (continued)

beyond the benefit of controlling UHI. Thus, the present FAR value is more conducive on this front.

2.5.2 *Statistical Results*

Correlations between variables and the evaluation index are presented in Fig. 2.10. We could find all the correlations are negative. GR is more correlated ($R^2 = 0.618$, $P < 0.05$) than the other variables. When GR increases by 10 %, average daytime UHII decreases by close to 0.1 °C.

The above four variables are stepwise regressed with UHII, with BD and TBD being ruled out. Fitting results show that a R^2 value of 0.859. Significance test results show that F value is 39.502, and P value <0.05. Therefore the established regression equation, $UHII = 1.96 - 0.758 \times GR\text{-}0.012 \times FAR$, can be considered valid. According to this equation obtained above, GR is the most outstanding in mitigating UHI. In view of the high density block in new central district, creating more green space, especially more lush trees, within the block may be the most suitable method.. Considering actual GR values are generally between 10 and 20 %, which can not meet the required limit of governmental documents, there exists great potential for GR to improve thermal environment.

2.6 Conclusion

Based on the empirical study in Zhujiang New Town, we confirm that the summer-time heat island problem in the new central district is severe. In this paper, three key indicators, FAR, BD and GR, were found important in regulatory detailed planning

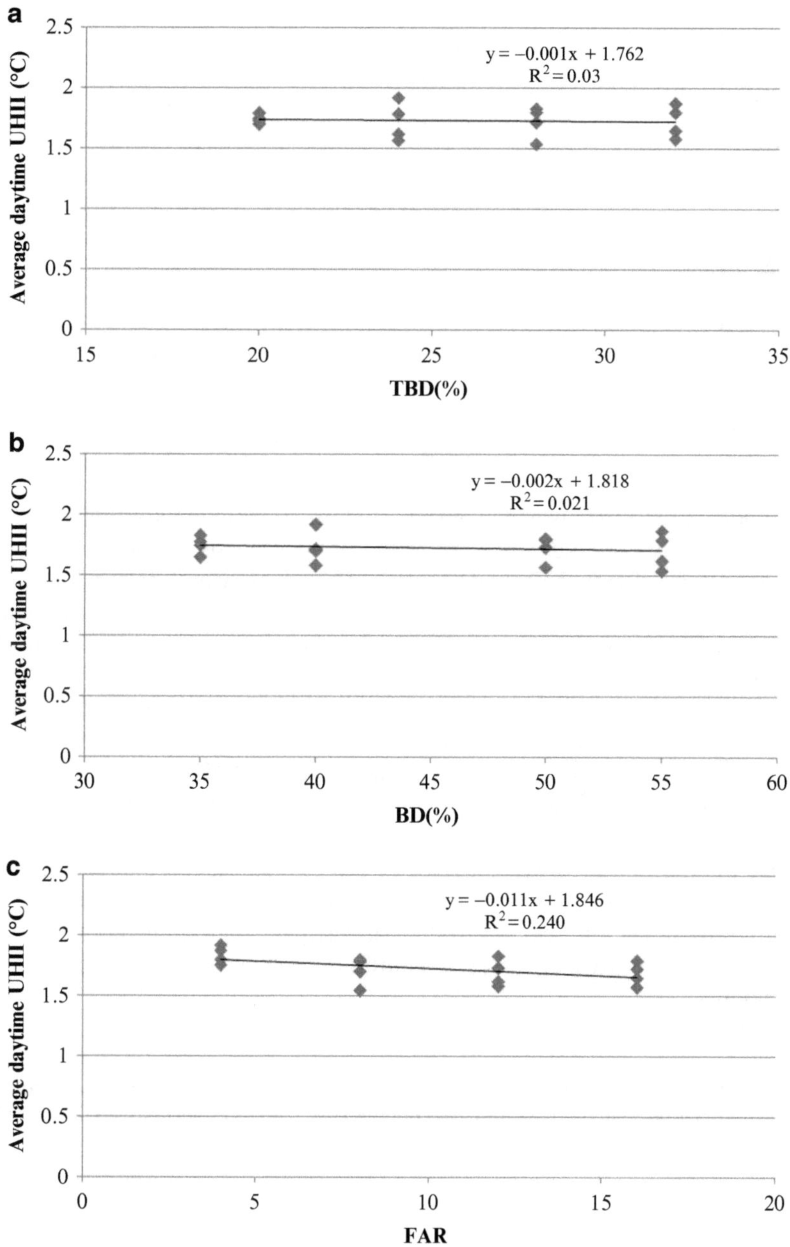

Fig. 2.10 Correlation between variables and average daytime UHII (**a**) TBD, (**b**) BD, (**c**) FAR, (**d**) GR

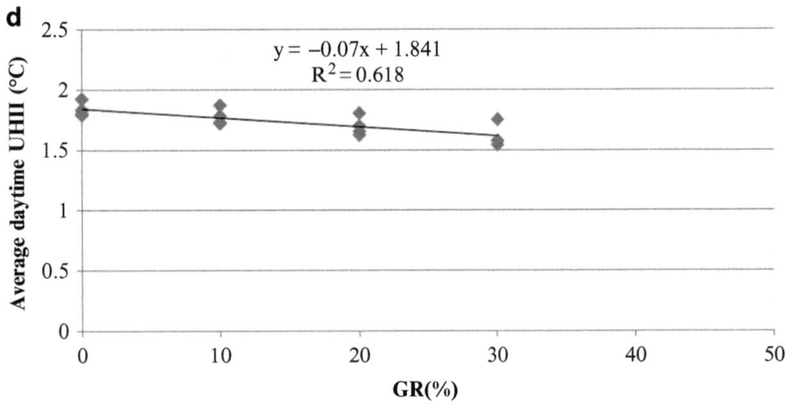

Fig. 2.10 (continued)

and influencing the heat island. Some related urban planning specification files of different provinces or cities were combed to summarize their control requirements, and then compared with the actual situation. Then the key indicators were variables with changing levels. Cases were set by experimental design and simulated by ENVI-met. The respective contribution of the above key indicators to control heat island intensity was evaluated.

On the premise of urban high density development, the present FAR value is more conducive, and GR is the most outstanding in improving thermal environment. When GR increases by 10 %, average daytime UHII decreases by close to 0.1 °C. Considering GR value in actual blocks in new central district, GR has great potential to mitigate UHI problem.

Since the ratio of trees was set 50 % of the greenery areas, and greening forms were unitary, in the future work, we will consider different kinds of greening forms and different ratios of trees, to further investigate their influences on thermal environment, and provide more scientific reference for urban planners and governments.

Acknowledgements The author wishes to give special thanks to the students in Building Environment and Energy Laboratory who took part in the field experiment. The project was supported by National Key Technology R&D Program of China (Project No. 2012BAC21B03).

References

Ali-Toudert, F., et al. (2007). Effects of asymmetry, galleries, overhanging facades and vegetation on thermal comfort in urban street canyons. *Solar Energy, 81*(6), 742–754.

Bottyán, Z., et al. (2003). A multiple linear statistical model for estimating the mean maximum urban heat island. *Theoretical and applied climatology, 75*(3), 233–243.

Bottyán, Z., et al. (2005). The relationship between built-up areas and the spatial development of the mean maximum urban heat island in Debrecen, Hungary. *International Journal of Climatology, 25*(3), 405–418.

Chen, H., et al. (2004). Study on outdoor thermal environment of apartment block in Shenzhen, China with coupled simulation of convection, radiation and conduction. *Energy and Buildings, 36*(12), 1247–1258.

Chen, H., et al. (2009). Study on mitigation measures for outdoor thermal environment on present urban blocks in Tokyo using coupled simulation. *Building and Environment, 44*(11), 2290–2299.

Davies, M., et al. (2008). Strategies for the modification of the urban climate and the consequent impact on building energy use. *Energy Policy, 36*(12), 4548–4551.

Giridharan, R., et al. (2004). Daytime urban heat island effect in high-rise and high-density residential developments in Hong Kong. *Energy and buildings, 36*(6), 525–534.

Giridharan, R., et al. (2007). Urban design factors influencing heat island intensity in high-rise high-density environments of Hong Kong. *Building and Environment, 42*(10), 3669–3684.

Hart, M. A., et al. (2009). Quantifying the influence of land-use and surface characteristics on spatial variability in the urban heat island. *Theoretical and applied climatology, 95*(3), 397–406.

Kratzer, A. (1956). *The climate of cities (Das Stadtklima)*. London: Air Force Cambridge Research Laboratories.

Kubota, T., et al. (2008). Wind tunnel tests on the relationship between building density and pedestrian-level wind velocity: Development of guidelines for realizing acceptable wind environment in residential neighborhoods. *Building and Environment, 43*(10), 1699–1708.

Lin, T. P., et al. (2010). Shading effect on long-term outdoor thermal comfort. *Building and Environment, 45*(1), 213–221.

Mirzaei, P. A., et al. (2010). Approaches to study urban heat island–abilities and limitations. *Building and Environment, 45*(10), 2192–2201.

Ng, E., et al. (2012). A study on the cooling effects of greening in a high-density city: An experience from Hong Kong. *Building and Environment, 47*, 256–271.

Oke, T. R. (1982). The energetic basis of the urban heat island. *Quarterly Journal of the Royal Meteorological Society, 108*(455), 1–24.

Ren, C., et al. (2011). Urban climatic map studies: A review. *International journal of climatology, 31*(15), 2213–2233.

Rizwan, A. M. (2008). A review on the generation, determination and mitigation of urban heat island. *Journal of Environmental Sciences, 20*(1), 120–128.

The Hitchhiker's Guide to ENVI-met. (2015). http://www.envi-met.com. Accessed 23 January 2015.

Unger, J. (2004). Intra-urban relationship between surface geometry and urban heat island: Review and new approach. *Climate Research, 27*(3), 253–264.

Unger, J. (2006). Modelling of the annual mean maximum urban heat island using 2D and 3D surface parameters. *Climate Research, 30*(3), 215–226.

Yang, F., et al. (2010). Summertime heat island intensities in three high-rise housing quarters in inner-city Shanghai China: Building layout, density and greenery. *Building and Environment, 45*(1), 115–134.

Zhao, C., et al. (2011). Urban planning indicators, morphology and climate indicators: A case study for a north-south transect of Beijing, China. *Building and Environment, 46*(5), 1174–1183.

Chapter 3
Study on Relationship Between Beijing Residential Building Insolation Standards, Improvement of Residential Environment and Urban Ecological Benefits

Linfei Han, Jiahua Mu, and Haofei Zhang

3.1 Chinese Residential Building Insolation Standards

Residential insolation standards are one of the important factors of the living environment. They are also a significant part of the mandatory provisions of Chinese technical standards for construction. But China has not yet established a separate set of documents for insolation standards, with the standards only reflected in those of related fields (to varying degrees). For example: ***Code for design of civil buildings (GB 50352–2005)*** stipulates: Insolation standards should be in keeping with related rules of the current natural standard ***Code of urban residential areas planning & design*** that at least one living space of each residence should be bright. ***Design code for residential buildings (GB 50096–1999)*** stipulates: At least one living space of each residence should be bright, and when there are more than four living rooms in an apartment, two of them should be bright. There are detailed quantifications and calculations for insolation standards in the ***Code of urban residential areas planning & design (GB50180–93, the 2002 edition)*** that describe the rules for particular situations (Table 3.1):

① Residential buildings for elderly people should have no less than 2 h of sunshine on the Winter Solstice.

L. Han (✉)
School of Architecture and Design, Beijing Jiaotong University, Beijing, China
e-mail: usi2000@126.com

J. Mu
School of Architecture and Design, Beijing Jiaotong University, Beijing, China

Real Estate Management Department, Power China, Beijing, China
e-mail: moujh@powerchina.cn

H. Zhang
Real Estate Management Department, Power China, Beijing, China

Q. Pan, J. Cao (eds.), *Recent Developments in Chinese Urban Planning*,
GeoJournal Library 114, DOI 10.1007/978-3-319-18470-8_3

Table 3.1 Chinese residential building insolation standards

The climate zone of building	I, II, III, VII Climatic region		IV Climatic region		V, VI Climatic region
	Large city	Small-medium city	Large city	Small-medium city	
The standard day for sunlight	The Great Cold Day			The Winter Solstice Day	
Sunshine duration (h)	≥2	≥3		≥1	
Effective sunshine time zone (h)	8–16			9–15	
Starting point of sunlight time calculation	Bottom window sill surface				

Data source: Code of urban Residential areas planning & Design (GB50180-93, 2002 edition)
Note: Bottom window sill surface refers to a point 0.9 m high from the indoor floor

② Any additional infrastructure of original design should not reduce the insolation standards of adjacent houses.
③ The insolation standards of new residences in old district reconstruction projects can be reduced according to practice, but the new insolation standards should not be less than 1 h of daylight on the Great Cold Day (The State Bureau of Technology Supervision and Ministry of Construction syndication 2002).

3.2 The Current Situation of the Living Environment in Beijing

As an important reference index of the city environment, city green rate is a direct reflection of the city living environment and the degree of residential comfort.

The coverage rate of the built-up area green zone in Beijing is 46.2 %. The area of green space in the entire city is about 65,540 ha, and green space in parks is about 21,178 ha (China Statistical 2013). Typical daily use of green areas for recreation is extremely short. Also, these parks are mostly located in the suburbs, so the green space per capita of the city's central area, where population is most concentrated, is only 7.24 m^2.

3.2.1 The Lack of Green Space Per Capita in the Center of Beijing

With the accelerating process of urbanization and rapid population growth, newly constructed residential areas in Beijing have been high density, high plot ratio, high-rise, etc. A deficiency of natural and green space is the major environmental defect in Beijing, especially in the center of city.

3.2.1.1 Where Has the Green Space in Beijing Gone?

The residential population of Beijing in 2012 is 20,693,000 (Wudi 2013). Per capita green space is 31.7 m^2. However, statistical results from 2012 show a park green space per capita of 11.87 m^2 in Beijing. According to these data, residents of Beijing already seem to have much green space. Actually, these green areas are in the suburbs, and the green space that residents can enjoy is still insufficient. The majority of space for walking is occupied by cars and roads, meaning that the residents who need green space the most must spend a lot of time traveling to it. These areas are therefore not available green space for residents to exercise and enjoy their free time. Residents all want to ask, "Where has the green space gone?"

3.2.1.2 Green Space in the Central Areas of Beijing

The total area of central Beijing (East District and West District) is 92.39 km^2. Statistical data from 2011 shows a rate of woody plant cover in East District of 19.1 %; the rate in West District is 14.6 % (Beijing Municipal Bureau 2012). Residential population density in Beijing is 23,271 person/km^2 (The working population is much larger than this.). 2,150,000 people live in the limited area of East District and West District. But the green space is only 15.57 km^2, or 7.24 m^2 per capita. These green areas contain some places that common people cannot use, such as important heritage gardens or state compounds (the Forbidden City, Sea Palaces). If we remove these latter areas, how much green space remains available for common people?

3.2.1.3 Accessibility Analysis for City Parks in Beijing

The study by Qiukuanbiao of the Chinese Academy of Forestry (Qiukuanbiao 2011) shows that in Beijing, according to the measure of walking speed, high accessibility areas are distributed in the northeast and northwest more often, and are less often distributed in the southeast near the places between Ring 4 and Ring 5 (Fig. 3.1).

The picture below shows that people living within Ring 2 generally spend 30–120 min to reach the gardens. Such a walking distance creates barriers between people and the green space in parks, particularly for the elderly and children.

Based on the analysis above, there is little park space that residents of central districts can reach and use in daily life. At the same time, there is another important factor.

The green spaces inside communities have become private spaces within the communities. They cannot be used by members of the public. This phenomenon has exacerbated the lack of green spaces for central city residents.

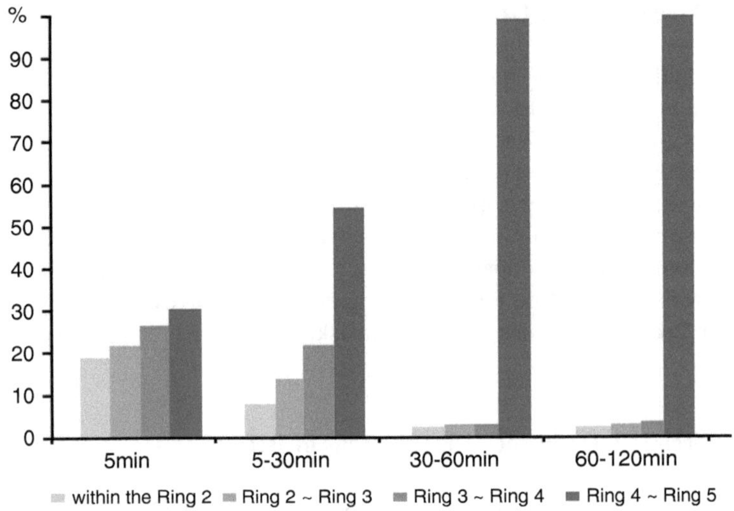

Fig. 3.1 Data sources of major parks green areas' accessibility in Beijing: a preliminary study on the city vegetation pattern and the value of parks green space ecosystem service within Beijing City Ring 5

3.2.2 The Situation of Green Space in the Communities of Beijing

By observing and researching community public space, the authors discovered the existence of a large number of so-called "lost spaces" (Portland Thicke et al. 2008) which few people enter. The major reasons are as follows: These green spaces have no sunshine perennially and they are semi private because of their location within the communities. These spaces finally become "lost spaces" when the residents inside the community do not use them and residents outside the community cannot approach the spaces. It leads to a waste of community green space and limits the improvement of the whole residential environment.

3.2.3 The Building Insolation Standards of Beijing Lead to a Large Number of "Lost Spaces"

According to the statistics, from 2007 to 2013, there were a total of 8,540 ha of residential land sold in Beijing which resulted in about 146,160,000 m^2 of residential construction. The quantities of residential land sold annually are about 220 ha, and the average capacity rate is 1.7. See the figure below (Fig. 3.2) (Table 3.2):

According to the current residential land remise scale of Beijing and the 30 % green rate of the city planning requirements, this should create 2,562 ha of green

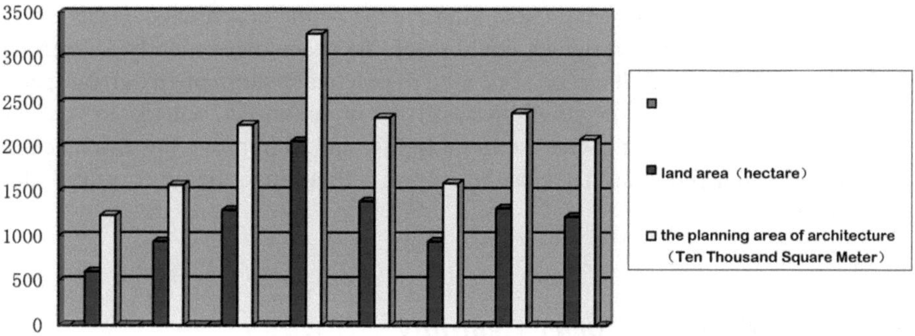

Fig. 3.2 The residential land statistics of Beijing from 2007 to 2013 (Drawn by author)

Table 3.2 The residential land statistics of Beijing from 2007 to 2013

The residential land statistics of Beijing from 2007 to 2013.									
Year	2007	2008	2009	2010	2011	2012	2013	Total	Average
Land area (hectare)	600	940	1291	2064	1390	938	1314	8540	1220
Planning construction area (thousand square meters)	1233	1571	2244	3267	2328	1591	2379	14,616	2088

Data sources: Beijing City Land Resources Bureau website http://www.bjgtj.gov.cn/publish/portal0/

space. On the basis of the green space setting requirements in the ***Code of urban residential areas planning & design*** it should be not less than 1/3 green space area outside the standard architecture sun shadow line range. We can calculate that there have been 1,708 ha of green space under the shadow of the insolation standards in these nearly 7 years in Beijing, which account for 66.7 % of residential green space. These areas receive little light, and the growth of plants is limited, and so do not create good activity space. Gradually, these valuable land resources become the "lost spaces" which are closed and cannot be fully used in the residential communities. It is an objective waste under the current insolation standards.

3.3 The Investigation for Eliminating the "Lost Space"

As described in the paragraphs above, green space within communities has become lost space generally because of lack of sunlight, and the number of these spaces is increasing year after year along with the city's development. City planners must consider that the problem of how to reduce the growth of these spaces.

The authors assume that moving the space cannot be used adequately within the communities outside. It can make these spaces get abundant sunlight, and form many pocket parks everywhere near the streets. This would not only solve the problem of lost space, but would also improve the whole appearance of the city at the same time. Plus, city residents could reach the space conveniently to exercise and enjoy entertainment. It realizes one kind of practical pattern of city green space.

In order to move the negative space in the communities outside to become positive space that residents can use in daily life, and to improve the efficiency of land use, the first problem that must be solved is to arrange the green spaces that have been moved outside.

3.3.1 Set the Green Space Centrally

Investigating several typical communities in Beijing, through the analysis of a questionnaire and interviews, 63.9 % of residents think that it is better to set the community green space centrally, because this can concentrate to set service facilities and better communicate with neighbors. 26.7 % of people would rather have green space dispersed, and another 9.4 % of residents do not care about the way to arrange green space (Fig. 3.3). This shows in a way that the small areas of green spaces which are encircled by residential buildings have trouble fulfilling the requirements of daily life.

Research conducted by Songpeihao, master of Henan Agricultural University, shows that centralized green space is superior to distributed green space in improving city microclimate (Songpeihao 2013). The insulation and cooling effect of centralized green space is better than that of decentralized green space. Having green space dispersed could improve the accessibility of green space, but is not conducive to the formation of complex plant communities. Secondly, the ventilation effect of centralized green space is better than that of decentralized green space. It is good for the exchange of air, blowing air pollution and relieving the effects of pollution.

Based on the above research, the centralized use of the city green space is more conducive to reduce the urban heat island effect and air pollution. This is why the temperature of places near the parks is lower. From this we can see that setting community public green space centrally will effectively improve the city environment.

3.3.2 The Lack of "Pocket Parks" in Beijing

Pocket parks appeared in the 1960s, and were also known as Bag parks. They mean the small open spaces in the city and they are usually massively distributed or hidden in the city structure, serving the local residents.

Because pocket parks have features like flexible location, small area, and discrete distribution, they can seize every opportunity to appear in the city. They look like an oasis in the desert for the crowded city, improving the city environment

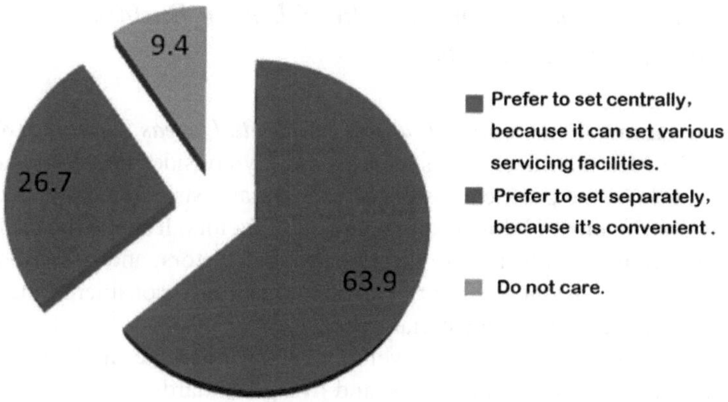

Fig. 3.3 The sensitivity analysis of the public green space arrangement (Drawn by author)

to a great degree and fulfilling the demand for parks of people who live in the high-density districts of the city center.

Beijing has high population density in the central district, so the parks have a relatively long distance between one another, and green space is distant from the residents. Therefore pocket parks can become a good addition and form entertainment spaces that can be used daily.

3.4 Discussions on Rationality of Insolation Standards

Based on the contemporary situation of society and city development, China's insolation standards were rational when they were originally formulated. However, with nearly 20 years of development, a lot of conditions which seemed reasonable at that time have changed, especially with big cities such as Beijing and Shanghai emerging with a unique side to their development–that contradiction between residential living environment and the scarcity of land resources is increasingly prominent. As an important key regulation to guide residential district plans, insolation standards should be applicable and reasonable, and should be more forward-looking in the aspects of future city development. However, according to the current implementation, insolation standards have restricted residential environment improvement in Beijing area to a certain extent.

3.4.1 The Original Intention of the Chinese Building Insolation Standards

The interpretation of the ***Code of urban residential areas planning & design*** introduces the idea that insolation standards mainly consider the city location and climate conditions, the practical results of residential district planning, and protection of hygiene, health, building heating, and other factors. It should be said that it is scientific and rational to formulate these referential factors, and is consistent with the situation of city development at that time, which also considered the expectations of social development in a certain period.

However, with rapid economic development in China, the mode of city development has experienced great changes, and living standards and the way of life of residents changed drastically in the final decades of the last century. Especially in large cities such as Beijing, Shanghai, and Guangzhou, residential living needs and city development level have changed greatly. With a view of present city development and the expectation for the future city development, we must rethink and reposition the significance of residential insolation standards for city development.

3.4.1.1 Review of National Situation

Code of urban residential areas planning & design introduces that Chinese insolation standards mainly reference insolation standards in the former Soviet Union, Europe, and America. But a large number of data show that the southern former Soviet Union districts (Latitude 38°–48°) used the Rain Day (February 19th) as the standard day for sunlight. The standard day for sunlight in Germany is nearly the same as the Rain Day, but America and London (England) regard March 1st as the standard day.

The latitudes of the places in the northern area of North China and most of the northeast area are roughly equal to the latitudes of the southern former Soviet Union districts. But the standard day is defined as The Great Cold Day (around January 20th). In contrast, in similar conditions of latitude and climate, the insolation standards in China are higher than that of the former Soviet Union. In addition, Europe, America, and the former Soviet Union have much land and small populations. The land resources are relatively abundant. But in China, especially in the large cities like Beijing, land resources are scarce, requiring a higher standard. Thus it can be seen that the current insolation standards have not fully learned from the international experience, but also did not consider natural situations of land resources.

3.4.1.2 Review of Hygiene and Health Requirements

From a historical point of view, the proposition and constitution of insolation standards are largely concerned with residential hygiene and health.

Research shows that ultraviolet rays in the sunshine, which are regarded as a "health line," (Wumei and Zhangwei 2011) have the effect of destroying bacteria and promoting wound healing. So in the establishment of insolation standards, it stipulates that city residents should enjoy sunshine indoors with the purpose of protecting the sanitation and health indoors.

But recent studies have found that ultraviolet UVC, namely the wavelength of 200–275 nm, also known as the short wave ultraviolet light, has the weakest ability to penetrate, and is unable to penetrate the great mass of transparent glass and plastic.

Wumei and Zhangwei who belong to Beijing Maternity Hospital, affiliated with Capital Medical University, measured and found that in the same season of the same area of Beijing, the UV not across the glass is several times the intensity of the UV across the glass (Table 3.3) (Lijingwei et al. 2011). This illustrates that glass has a great capacity to absorb UV and that the disinfection effect of sunlight is very limited indoors because of glass barriers.

We can see from the study and practice above that indoor sunshine cannot achieve the sterilization effect well because of the glass. Obviously, there are some limitations of considering the sterilization effect of sunlight at the beginning of the standard formulation.

3.4.1.3 Review of Original Intention to Save Energy

The initial establishment of the insolation standards considered infrared rays in sunlight having a productive heating effect, and that sunlight has an effect on building heating. For the internal environment of buildings, the heating effect of sunlight can warm the indoor space directly, reduce power consumption in winter, and achieving the purpose of saving energy.

In the 1980s and 1990s, Chinese building energy conservation technology and building materials technology are both incomplete. With nearly 20 years of development, building energy-saving technology has been able to meet the demands of heat preservation inside building to a great degree. In recent years, the newly-built residences in the north area have achieved an appropriate heating condition which can ensure a comfortable temperature for the human body. With the development of building technology in the future, especially in the case of introducing advanced construction technology and concepts from foreign countries, it makes us rethink whether building heating of residential architecture will still rely on the sunlight.

Secondly, the number of annual sunshine hours of Beijing has decreased obviously in the past 50 years, and the annual sunshine hours have an approximate reduction of 13.6 % (450 h). There are many reasons for sunshine hours to decrease,

Table 3.3 Comparison between UV intensity of direct sunlight and that of the isolated sunlight in Beijing (four seasons)

The comparison of fine day UV intensity which are 0.5 m far from the window between direct sunlight and that of the isolated sunlight in every seasons								μw/cm^2
	Spring		Summer		Autumn		Winter	
Time	0.5 m direct sunlight	0.5 m isolated sunlight	0.5 m direct sunlight	0.5 m isolated sunlight	0.5 m direct sunlight	0.5 m isolated sunlight	0.5 m direct sunlight	0.5 m isolated sunlight
8:00	0.39 ± 0.07	0.08 ± 0.04[a]	0.47 ± 0.06	0.09 ± 0.02[a]	0.28 ± 0.03	0.10 ± 0.00[a]	0.00 ± 0.00	0.00 ± 0.00
10:00	0.90 ± 0.15	0.19 ± 0.02[a]	2.27 ± 0.41	0.52 ± 0.19[a]	1.03 ± 0.21	0.39 ± 0.13[a]	0.14 ± 0.05	0.08 ± 0.03[a]
12:00	1.47 ± 0.39	0.80 ± 0.26[a]	3.47 ± 0.39	1.53 ± 0.16[a]	1.77 ± 0.45	0.68 ± 0.10[a]	0.38 ± 0.08	0.21 ± 0.07[a]
14:00	0.95 ± 0.23	0.24 ± 0.08[a]	1.38 ± 0.25	0.60 ± 0.13[a]	0.96 ± 0.19	0.14 ± 0.05[a]	0.21 ± 0.04	0.09 ± 0.02[a]
16:00	0.44 ± 0.05	0.07 ± 0.05[a]	0.47 ± 0.12	0.11 ± 0.05[a]	0.19 ± 0.09	0.04 ± 0.02[a]	0.00 ± 0.00	0.00 ± 0.00

Data sources: Wumei and Zhangwei (2011)

Note: [a]The comparison between isolated sunlight and direct sunlight in the same season, $P < 0.05$

but research shows that the decline of the sunshine hours in large cities like Beijing has become a trend. So how much functionality remains for sunlight and the indoor heating of residences?

3.4.2 *Discussions on the Reality of Sunshine Days' Number*

In accordance with the specification requirements, China's housing should meet the requirement that there are at least 2 h of sunshine on The Great Cold Day. If this data is converted into annual amount of sunshine, each house should meet the sunshine requirements about 300 days per year. This is 50 % higher than the 200 days per year which is demanded in France, and 25 % higher than 240 days required in Germany. The quantities of sunshine days are so great. Is it that the residents in western developed country do not need sunshine? Or are their living standards lower than ours? The answer is no; it indicates that Chinese insolation standards lack the theory and scientific evidence needed at the beginning of the formulation.

3.4.2.1 The Current Insolation Standards of Beijing Is a Little High

Beijing is located in north latitude 39°56′, and the latitudes of large cities such as London, Berlin, Paris, Moscow and that in other European countries are higher than Beijing. Their sun angle is lower than Beijing in winter, and winter climate characteristics are similar to Beijing, but they use the Rain Day, or even the Spring Equinox or the Autumnal Equinox as the standard day for sunlight, the standards of which are relatively lower.

3.4.3 *Summary*

In conclusion, our current natural insolation standards are not in line with international standards at the beginning of formulation. Recent research and experiments also show that hygiene and health, the original intentions of the standard formulation, are not fully realized either.

The current insolation standards cannot be well implemented in the northern area districts, and at the same time they cannot play a positive role in environmental improvement of large cities like Beijing. Thus the current sunshine standard should be revised in order to accommodate the needs of development both in the present and in the future.

3.5 Discussions on Revising Residential Insolation Standards of Beijing

The major factors which influence the effectiveness of sunshine are the standard day for sunlight, effective time zone, sunlight hours, sunlight coverage rate and so on. The effective time zone, sunlight hours, and sunlight coverage rate of insolation standards in China has been basically closed to the international standards, so there have been no discussions about revisions in this essay at present. Choosing the standard day for sunlight is the major factor of sunlight hours, so it is a key point to discuss the rationality of the standard day for sunlight in this essay.

3.5.1 The Debate of Choosing the Insolation Standard Day in Beijing

The revision of the standard day for sunlight should be in accord with the city location and features, and in accordance with the current situation of city development and the requirements of future developments. So the authors try to discuss the choice of the Beijing standard day for sunlight. According to the solar elevation change regulation of Beijing, each solar term, when standard revisions are carried out, the first thing to consider is the difference between the adjacent solar term and the current standard, as well as the remaining adjustable space (etc.), then try to discuss defining the Rain Day, Spring Equinox, or the Autumnal Equinox to be the standard day for sunlight (Table 3.4).

Beijing is located in north latitude 39°57′, and similar to the latitude of the southern former Soviet Union, whose area was between latitude 28°–48°. The standard day for sunlight of the areas in the former Soviet Union is the Rain Day. At the same time, Paris is located in north latitude 49° and Berlin is located in North latitude 52°30′. The two cities' insolation standard days are also near the Rain Day.

Large cities like Beijing have development levels and needs that are completely different from small and medium-sized cities. Referring to the standard day division of the former Soviet Union, it would be propitious to the current development situation of Beijing to choose the Rain Day to be the standard day for sunlight. This can also help relieve the contradiction between city development and land resources.

3.5.2 The Discussion About the Feasibility to Revise the Insolation Standard Day

The adjustment of the insolation standard day will certainly lead to a reduction of building interval. The factors restricting the building interval include lighting,

Table 3.4 The relationship between solar elevation at midday and solar term in Beijing area

Date	January 20th	February 18th	March 20th	April 20th	June 21st	September 23rd	October 23rd	November 22nd	December 21st
Solar Elevation (°)	29.7	38.9	50.2	62.0	72.8	50.3	39.0	29.9	26.5
Solar Term	The Great Cold Day	The Rain Day	The Spring Equinox	The Grain Rain Day	The Summer Solstice	The Autumnal Equinox	The Frost's Descent Day	Slight Snow Day	The Winter Solstice
Note	The coldest day in a year		The average height of the sun		The maximum height of the sun	The average height of the sun			The minimum height of the sun

Table 3.5 The comparison of midday shadows in The Great Cold Day, the Rain Day and the Spring Equinox

The insolation standard day	Time	Solar elevation	The rate of midday shadows	Possible building interval
The Great Cold Day	January 20th	29°49′	1.745	1.75
The Rain Day	February 18th	39°30′	1.213	1.2
The Spring Equinox	March 20th	50°03′	0.84	0.85

Drawn by author

ventilation, fire protection, pipeline laying, visual health, and space environment. The following content will discuss the factors that may influence building interval in the condition of different sunlight standard days.

Assuming division of the revised residential insolation standard day into the two times, the Rain Day and the Spring Equinox, the restricted conditions about the building interval are discussed.

From Table 3.5 we can see that the sunlight shadow of the Great Cold Day is 1.44 times as long as that of the Rain Day, and is 2.1 times longer than the one of the Spring Equinox Day. The revised insolation standards will narrow the building interval greatly. The following will discuss the relationship between building interval and the factors which restricted building interval respectively, with the two different building intervals.

As the three factors of lighting, ventilation, and space environment usually can all be satisfied when the requirements of insolation standards have been met, it is not discussed in detail below. Taking a residential district under construction in Beijing as an example, just two factors, pipelines burial and fire protection requirements, are discussed.

(A) The Rain Day:

Taking a residential district in Beijing as an example, we analyze the revision of the insolation standards. The building height in the residential district is limited to 60 m, and the plot ratio is 2.5. It can be seen in Figs. 3.4 and 3.5 that the building interval has changed to about 40 m from 60 m of the Great Cold Day.

Through analysis of pipeline distance and the calculation of fire protection requirements, 40 m distance can already guarantee the requirements above, so it is feasible to choose the Rain Day as the standard day for sunlight in the technical specifications of Beijing area.

(B) The sunlight situation of the Spring Equinox and the Autumnal Equinox:

After adjustment, the minimum distance for sunlight of the spring equinox and the Autumnal Equinox has become 30 m (from 60 m before). Through analysis of pipeline distance and the calculation of fire protection requirements, it can be found

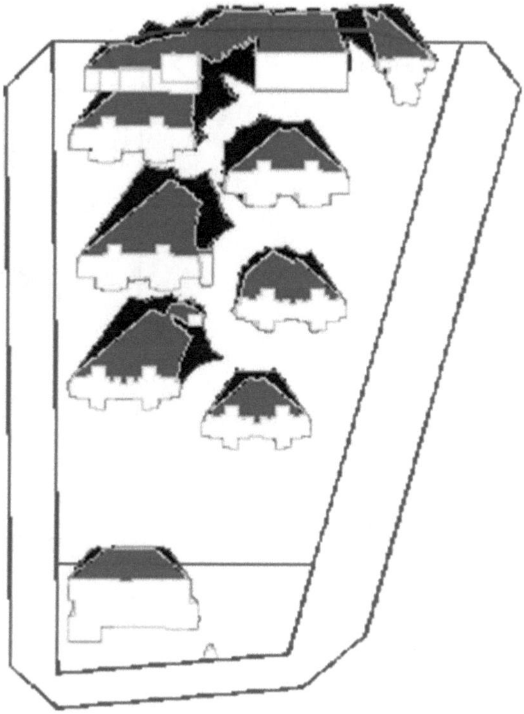

Fig. 3.4 The brief diagram of the Rain Day (Drawn by author)

that 30 m distance is not reasonable. So it is not feasible to choose the Spring Equinox or the Autumnal Equinox as the standard day for sunlight of Beijing area.

3.5.3 The Revision of Insolation Standards Is Good for Eliminating "Lost Space"

After discussion of the technical conditions of the actual cases, it can be seen that the building interval which has been reduced is feasible for the Rain Day. It can compress the building interval appropriately after the adjustment of the standard day for sunlight, and can guarantee that the community building density and the plot ratio remain unchanged. It is more conducive to centralize the public green space in communities, and can provide more community infrastructure and communication space, and also ensure the sunlight requirements of the public green space.

The adjustment of sunlight distance cannot solve all problems in the construction of the residential community, but it is an effective solution to improve the residential environment. Reasonable insolation standards can provide practicable conditions to create a good living environment.

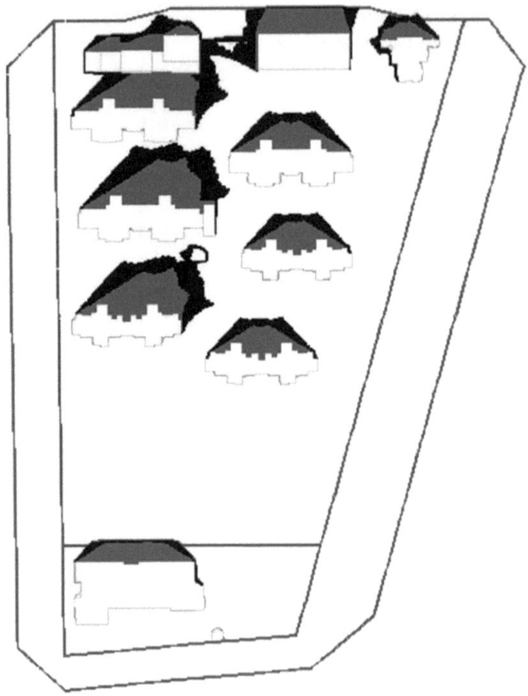

Fig. 3.5 The brief diagram of the Spring Equinox (Drawn by author)

The revision of insolation standards provides a likelihood of reducing residential building distances, makes that the construction of central public green spaces for communities possible, and offers a kind of solution to eliminate the "lost spaces" within communities. The revision of insolation standards cannot eliminate the "lost spaces" entirely, but may set positive space effectively and extinguish the technical barriers of improving the whole city environment.

3.6 Conclusion and Suggestions

3.6.1 Conclusion

1. Compared with other countries and regions in the world, considering the situation of economic development and city construction in China, the current residential building insolation standards are high. Especially for large cities like Beijing, the standard sunlight day is also high.
2. The performance of the natural residential insolation standards leads to a large number of "lost spaces" inside communities. The city governor cannot solve the contradiction between the city environment and the requirements of residents.

3. The current insolation standards have restricted the likelihood of constructing green space centrally in communities.

3.6.2 Suggestion

1. Revise the current natural insolation standards, and set insolation standards according to the type of area and city. On the basis of local climate city construction in Beijing, it is suggested to revise the insolation standards to 2 h in the Rain Day.
2. Keeping the building plot ratio and building density unchanged, city planning administration departments need to establish regulations of community green spaces in central settings in order to move from separate green spaces to central public green spaces and take advantage of the agglomeration effect from plants, which is good for reducing the urban heat island effect and decreasing pollution and fog in the city. This can solve the contradiction between the city pollution in development and the environmental improvement. The city planning guidelines designed by city planning departments should define the form and the position clearly.

Acknowledgement The Work is Supported by NCET Program (NCET-10-0217).

References

China Statistics Press. (2013). *2013 China statistical abstract*. Beijing: China Statistics Press.

Lijingwei, Zhangzhiyou, Zhounianqing, & Gekenan. (2011). The analysis of climate change of Beijing in recent 50 years [J]. *Proceedings of Conference on Environmental Pollution and Public Health (CEPPH2011)*, (pp. 315–318). Wuhan, China.

Portland Thicke, Zhu Ziyu, et al. (2008). *Looking for the lost space* [M]. Beijing: Chinese Architecture Industry Press.

Qiukuanbiao. (2011). *A preliminary study on the city vegetation pattern and the value of parks green space ecosystem service within Beijing City Ring 5* [D]. Dissertation, Chinese Academy of Forestry.

Songpeihao. (2013). *The micro climate characteristics and simulation of two kinds of green space layout*. Dissertation, Agricultural University of He'nan.

Beijing Municipal Bureau (2012) http://www.bjstats.gov.cn/sjfb/bssj/ndsjcs/201308/t20130828_255760.htm

The State Bureau of Technology Supervision and Ministry of Construction syndication. (2002). *GB Code of urban residential areas planning & design 2002*. Beijing: China Architecture & Building Press.

Wudi. (2013). http://zhengwu.beijing.gov.cn/gzdt/bmdt/t1333801.htm

Wumei, & Zhangwei. (2011). Intensity of ultraviolet radiation in different seasons of Beijing and by diverse way to expose [J]. *Hygiene Research, 2*, 233–235.

Chapter 4
The Methods and Approaches Green Infrastructure that will Promote China's Urban Transformation: A Case Study of the Dongguan National Wetland Park

Shaoping Guan, Xi Zhang, and Jingjun Wu

4.1 Introduction

On September 10, 2013, the international management consulting firm Accenture and the Chinese Academy of Science released an investigation report of 73 Chinese cities. The report assessed the economic performance and environmental sustainability, as well as the degree of dependence on natural resources; this report revealed that only 16 cities can achieve coordination between economic development and the usage and protection of resources and environment (Accenture and Chinese Academic of Sciences (CAS) 2013).

The report also revealed that there are 25 cities with high emissions and low development level that rely on heavy industry, such as some of the cities in the Pearl River Delta and Yangtze River Delta that give priority to manufacturing. These 25 cities have high economic growth, but at the same time they are facing many problems, such as overcrowding, air pollution, too much rubbish and water shortages.

Economic and social development at the expense of the environment is a weak and unstainable model. Environmental problems are generated in the process of urban development and the process of urban development can also solve these problems. Environmental restoration and the growth and development of city are

S. Guan (✉)
College of Design, South China University of Technology (SCUT), Guangzhou 510641, China

College of Architecture and Urban Planning, Tongji University (TJU), Shanghai 200092, China
e-mail: guanshaoping2000@163.com

X. Zhang
College of Design, South China University of Technology (SCUT), Guangzhou 510641, China

J. Wu
Guangdong Dongguan Eco-Industrial Park Administrative Committee, Dongguan 523668, China

Q. Pan, J. Cao (eds.), *Recent Developments in Chinese Urban Planning*,
GeoJournal Library 114, DOI 10.1007/978-3-319-18470-8_4

complementary and can be promoted mutually. Therefore, China's city transition must seek to solve environmental problems. We can achieve economic and social transformation by ecological restoration, and correspondingly, the transformation of the economy and society can fundamentally improve the environment.

4.2 Methods

First of all, this paper analyzes the urgent demand of Dongguan City transition. Then, we study the means, approaches and features of several American city transitions and find the particularities of China's urban transition by comparing the national conditions between China and US. Further, we propose green infrastructure as an important way to promote a sustainable city transition for Dongguan City. Next, we study the means and approaches of green infrastructure to promote city transition by looking at the case study of Dongguan National Wetland Park. Finally, we analyze the general characteristics between the city transition of Dongguan and other Chinese cities.

4.3 The Demand for the Economic and Social Transformation of Dongguan

Dongguan, located in the Pearl River Delta, has depended on manufacturing to drive rapid economic development. The city relies on cheap labor and natural resources to attract foreign investment. This extensive economic growth mode has established Dongguan as the "world's semi-manufactured factory" and many problems have followed, such as fast resource consumption, low energy efficiency, environmental degradation, the influx of migrant populations, and social conflicts.

Firstly, since China's reform and opening up, the cost of resource and land have increased rapidly. In the process of urbanization was supported by an increase in the number of manufacturing plants which required large numbers of migrant workers. Inefficient city development resulted in a reduction in the amount of cultivated land (Fig. 4.1). At present, the city's actual available cultivated land is just about 26,667 ha. The land resource in Dongguan will be depleted in the coming decades if we keep consuming land at the same rate.

Secondly, extensive economic growth has caused a serious waste of resources. In 2003, every kilowatt hour produced 3.34 RMB in Dongguan, far below the national average level which was 6.19 RMB per kilowatt hour. The shortage of electricity has greatly influenced productivity and life. At the same time water pollution resulted in a shortage of water resources. It is unsustainable if we keep following traditional development practices that consume large amounts of resources (Chen Guiming and Li Ganqiu 2005).

Thirdly, the extensive economic growth has led to ecological deterioration. Table 4.1 compares environmental indicators for 1978 and 2007. These indicators show the fast economic growth and the environmental cost. According to the indicators we can see that the amounts of three industrial wastes are higher in

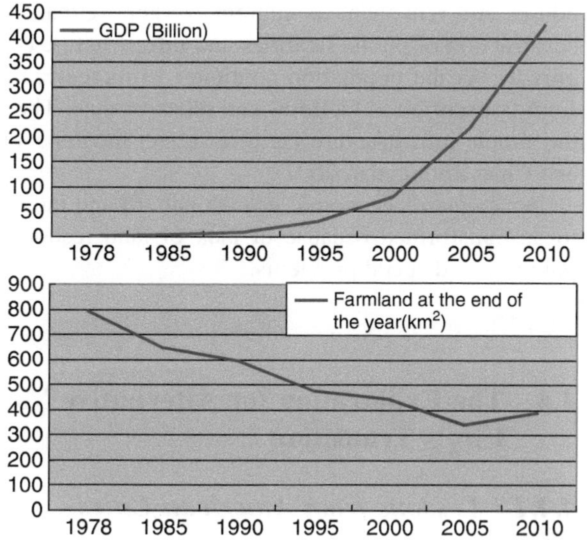

Fig. 4.1 The GDP and cultivate area variation of Dongguan

Table 4.1 The statistics of GDP and environmental indicators of Dongguan in 1978 and 2007 and Guangzhou in 2007

	GDP (billion)	The amount of waste water caused by every billion GDP (thousand tons)	The amount of waste residues caused by every billion GDP (thousand tons)	The amount of waste gas caused by every billion GDP (billion standard cubic meters)
Dongguan in 1978	56.1	550	3	0.6
Dongguan in 2007	315.1	3600	7	0.7
Guangzhou in 2007	821.5	256	0. 002	0.0002

Source: Dongguan city statistical yearbook (1978–2012)

Dongguan than in Guangzhou. Therefore, the economic growth mode in Dongguan is unsustainable.

At last, in addition to the resource, energy and environment problems, the process of rapid urbanization in Dongguan is also accompanied by many social problems, such as an overcrowded immigrant population, social contradictions, and poor city appearance. The statistics show that Dongguan's immigrant population in 2009 is as high as ten million and the proportion of immigrant population to local population is 6:1, what's more, most of the migrants are young. The average age of population in Dongguan is only 28 years old (Mo Anda et al. 2009). These immigrant population are in a state of "economy absorb and society rejection" because of their marginalization; so they become the origin of social instability and crime due to their feeling of not belonging. According to statistics, the number of crimes done by immigrants is about 97 % of total crime (Dongguan City Bureau of Statistics 2009). In the industrialization process of Dongguan, villagers build houses and rent them to migrant workers. Urban space layout is messy and the construction of public facilities and infrastructure cannot meet the requirements of growth. As the population continues to increase, the city is also facing pressure from transportation facilities and other aspects. The chaos in residential space and the simple infrastructure facilities affect the appearance of city (Liang Liangfeng and Chen Zhongnuan 2007).

Donggguan's economic growth pattern and industrial development model need to be transformed because of lack of land resources and energy, environmental pollution and social problems.

4.4 The Exploration for Alternative Modes for Dongguan City's Transition

4.4.1 Lessons from American Cities

As shown in Table 4.2, Toffler divided American society transition into three stages by the characteristics of American cities in the process of the preindustrial age to the industrial age and the post-industrial age (Alvin Toffler and Heidi Toffler 1994). Ruchelman viewed the fourth stage of American society transition as the age of human capital, with globalization as the driving force (Leonard I Ruchelman 2007).

The path of American urban transition is mainly the realization of the city's economic, social and cultural transformation, through industrial upgrading and transformation. From the industrial age, industry upgrading and transformation became a green-oriented development process where industry developed from manufacturing to services and then high-end knowledge-based services and cultural services. Specifically, the process from the second transition can be summed up the following three aspects.

First, industries transformed from manufacturing to service. The American Manufacturing Industry developed rapidly because it took advantage of good locations and was boosted by the industrial revolution in nineteenth century. Thriving manufacturing sped up the urbanization of American cities with many cities quickly becoming important world cities. However, its world economic standing was weakened by Europe and Japan's economic recovery in 1960s, and followed by the oil crisis of the 1970s. These forces led to a sharp recession in American manufacturing (Qian Wei 2011). To cope with the plight of the recession, America upgraded technology and focused on advanced manufacturing. It also used the policies and the dual role of market to improve industry upgrading and structural adjustments. These measures reduced the negative effects caused by the manufacturing recession to a certain extent. At the same time, America sought ways to transform secondary industries to tertiary industries and vigorously developed the Service Industry to promote economic development and prosperity (Shuai Ping and Zhou Lijun 2013). Driven by the post-industry period, the development of

Table 4.2 Four transformations of American cities

Transition stage	Driving force of transition	Urban transition	The basic content of transition
First stage	Agricultural technology revolution	Pre-industrial city (1760–1820)	Low level of urbanization and small urban population scale.
Second stage	Industrial revolution and manufacturing	Era of industrial city (1820–1970)	Increased urbanization, population and resources gathered into city. Manufacturing develop rapidly and form a concentrated and centralized industrial metropolis. Later, the suburbanization trend is obvious because the development of vehicles, communications technology and high-voltage transmission technology and the expansion of factory.
Third stage	Knowledge economy, advanced technology and innovation is the core	Era of metropolis at post-industrial age (1970s to now)	The manufacturing cities become service oriented cities based on the development of information and communication technology. Traditional single center metropolitan replaced by dispersed multi-center urban areas. With the rapid development of economy, information and transportation and urban problems intensified in city center, centralized city become disperse and spread.
Forth stage	Globalization	Global city reconstruction	Characterized by multi-location production process, high value-added activities, new international divisions of labor and the increasing human capital resources. Labor and capital is less important than company and industry itself, the production can be spread to distant places and we can coordinate commodity production and sales through information technology. The city became economic activity center; city and region became basic unit of global reconstruction. City development and transition characterized by self-rebuild and has different features.

Source: Gu Chaolin (2011)

professional services, support services and computer services achieved the American city transformation from manufacturing to service-based economies. The American urban transition led the global-integrative knowledge-based economy further upgrading the industrial structure. Depending on the applied information technology and modern management concept, American service industry transformed to knowledge intensive and technology intensive high-end services, this transformation fully embodied the modern service cultural. On the one hand, the knowledge and technology in traditional service industry has been upgraded, its service function has been strengthened and new function and form has been derived from it. On the other hand, the development of information technology forced the advent of high-end services, such as computer and software service, mobile communication service, information consulting service, health industry, ecology industry, international business and modern logistics. What's more, the American system has advantages, the first-class culture of science; technology and education propel the transformation and development of knowledge intensive modern services. Therefore, culture industry, innovation industry and high-tech development becomes the main features of urban transition in America. In the process of urban transition, America paid close attention to industry structural adjustments, new technology industries and tertiary industry development, thus the transformation and development of services provided jobs and then enhanced international competitiveness and attractiveness. At last, the industry transformation is green, focused on environment-protection.

The healthy development of the environment is fundamental to human welfare. We should reduce the intensity of urban energy consumption and carbon emissions to cope with global climate change. It is the job of cities to construct low carbon, green, ecological and livable places. In the process of pursuing value realization, leisure, and entertainment, the city transition must be focused on the improvement of the city environment and the protection of the environment. Green development has become the important direction and basic trend of American urban transition, the main measures include the development of energy conservation and emission reduction technology and green industry, the construction of green infrastructure, and advocating of low-carbon lifestyle and so on.

The above three aspects can be reflected in the transition of Chicago and Los Angeles.

Before the 1980s, Chicago and Los Angeles were the two biggest manufacturing cities in America. In the 1960s and 1970s, manufacturing was impacted and their economies declined. In the 1980s Chicago abandoned heavy industries like steel and metallurgy and set the development goal of a "service-led diversified economic system". Chicago supported the competitive industries like food, printing, metal processing and other light industry by upgrading them. In addition, it introduced investment and the research, development and management department of high-end industries like software and biopharmaceutical (Qian Wei 2011). The famous universities in Chicago like Chicago University, Northwest University, University of Illinois, and Illinois Institute of Technology provided rich human resources for these high-tech industries. In addition, Chicago adjusted city land use and optimize the spatial structure for economic diversification. It started a large number of reconstruction projects in downtown to optimize the development environment and promote urban vitality. In the mid-to late 1990s, it finished adjustments to economic structure. Chicago not only successfully went through the economic trough, but also formed the agglomeration effect and promoted the application of the new technologies in manufacturing. Chicago advanced the development of tertiary industries like commercial trade, finance, traveling and exhibition industry, these measures greatly enriched the diversity of industry structure and achieved economic vitality and prosperity. Chicago became an important node in economic globalization again and showed its status as an international city.

Los Angeles found a lot of oil in 1892 and it became the main oil production base in America in the 1920s. The city thrived with the development of the oil industry, the film industry and manufacturing industries for iron, rubber, automobiles, aviation, shipbuilding and machine. In the late of 1930s, Los Angeles formed multi-manufacturing structure and developed an economy that was not dependent on one industry. During the period from World War II to the Cold War, the national defense industry strongly stimulated and promoted the development of manufacturing and high-tech industries. Unlike the industry fluctuation in many manufacturing cities, Los Angeles' industry spiraled upward to new and better industries. Because Los Angeles adjusted its industrial structure earlier, it made full use of its advantages and mainly developed high-tech and modern service industries when faced with the economic recession. Thus it became the biotechnology center of America and a world-famous design center, the locomotive of the entertainment industry and America's second largest financial center behind New York. With the push of high-tech and service industries, Los Angeles achieved the transformation from manufacturing to high-tech industry and modern services (Qian Wei 2011). In the transition period, Los Angeles paid close attention to the protection of the environment to improve its attractiveness, by doing so it became one of the first-choice cities for immigrants and successful people. It also reconstructed industrial space through measures like "the revival of central" and "the rise of peripheral city" and achieved smart growth by carrying out many measures. What's more, Los Angeles improved infrastructure and ensured the supply of strategic resources like water and electricity (Wang Lixin 2011). After persistent efforts, Los Angeles created a good environment for science, technology, culture and ecology and it successfully became a global city.

In the period of urban transition, American cities have also had many serious problems, like environmental pollution, suburbanization and urban sprawl. In the second stage of transition, cities sprang up and developed quickly, fast social change led to many problems. Cities became overcrowded and had serious environmental and social problems because of large populations, factories in the city, laissez-faire policies, and a lack of long-term planning (Li Qingyu and Zhou Guiyin 1994). Urban deterioration resulted in middle and upper classes moving out to surrounding suburbs. Suburbanization reduced the city pressure and relieved the environmental problems to some extent, but it also led to "industry hollowing", economic recession, social space differentiation and racial segregation. Although government provided support for infrastructure construction, tax and housing, the problems caused by suburbanization have not been relieved.

In the transition period from post-industrialization to globalization, new city problems appeared. Urban sprawl is one of these problems. People in cities moved to suburbs to seek for healthier environment and a lower-cost of living. Urban sprawl also developed because of information technology, transportation and other technologies (Gu Chaolin 2011). Urban sprawl caused destruction to ecology and culture and reduced the utilization of public facilities; it aggravated the social stratification and the decline of inner cities. By 1990s, these problems got great attention and the government, city planners and environmental organizations have initiated many actions to prevent urban sprawl. Among them, New Urbanism and Smart Growth have the most significant impact; they become the leading force to prevent urban sprawl. New Urbanism and Smart Growth propose that the living environment be constructed with humanistic concern, suitable for walking and with intensive growth using mixed land-uses and large capacity bus systems oriented urban development. These measures seek to prevent urban sprawl (Wang Dan and Wang Shijun 2007).

4.5 The Exploration of Dongguan City Transition Mode

American urbanization started before China's and its urbanization level is higher than China's. China's city development is faced with problems and challenges that American cities faced in their transition. So it is very important for China's urban transition to learn lessons from American urban transition. Nevertheless, the historical process and stages of China and its resources, population, technology and education is different from America and our cities faced more pressure in the process of transition. Firstly, China has a shortage of land resources and the urban sprawl intensified the situation. Secondly, the condition of too much population led to less available resource and energy per capita, this gives special requirements on the industry development model at present which is high investment, high consumption and low output. Thirdly, there is a big gap between China's education, technology, science and talents and America's. At last, the environmental pollution in China's cities is more serious than America's. Therefore, when using the lessons from America, China should explore the ways and methods of urban transition that suit its own condition.

As one of the industry cities that developed rapidly after China's reform and openness, Dongguan has the typical characteristics of a Chinese industrial city: extensive economic development with high development, high consumption, high emission and high expansion. Problems with population, resources and the environment are especially serious in Dongguan. Dongguan has its own particularities. Firstly, the simple and low-end manufacturing in Dongguan is a low added-value model and has put great pressure on resources and the environment. This economic development model is also weak under the conditions where international industries change frequently. The most important is that environmental capacity cannot support this extensive economic development model. Secondly, as labor intensive and low-skilled industry, low-end manufacturing also brought a large migrant population to work in the city. At last, unlike the single center diffusion urban development model in most Chinese cities, Dongguan's city development model has patches of development. Every town has its own style and does things in its own way. The industries in different towns are homogeneous. This urban spatial pattern and development model is bad for land use efficiency, diversity and structural adjustment.

In order to explore the ways and methods of urban transition that suit Dongguan, we should combine the experiences of the American urban transition and Dongguan's situation at present. We can conclude some points of urban transition from American experiences. Firstly, we need scientific strategy and initiative. It is the first condition for us to know the cause of urban development issues, determine the ways and methods for transition and seize opportunities to achieve a smooth transition. Secondly, we should focus on the global economy and industry division of labor. Globalization can direct a city's economic transition. Thirdly, we should coordinate the development of traditional industries and new industries and deal with the connection of them. This is the basic condition of a stable city and the foundation of new industry development. Fourthly, we should construct an ecological system that benefits from an innovative economy. Like ecological systems, the elements in an economic system influence and constrain each other and evolve constantly. The growth of any new innovative economic element needs the exchange of information and energy with relevant factors like enterprise, finance, scientific research and education. Urban transitions driven by innovation need the support of strong economic and ecological systems. So, we should construct a diverse industrial base and form the multiple and complementary industrial structure to achieve sustainable economy development. Fifth, we need talent and resources. Talent and resources are the strong impetus of city industry transformation and innovative economy. Sixth, we should coordinate economic development and environmental protection. An ecological environment is the basic life support; we should get a balance between economic development and environmental protection to achieve sustainable city development. In addition, a beautiful and comfortable living and working environment can attract a talented workforce as well as investment and jobs. Therefore, it is an important condition to promote an urban transition that creates a great environment. When we consider these six aspects in the background of Dongguan at present and the ways and methods of city transition in Chicago and Los Angeles, we will discover the effective methods to promote Dongguan City's transition. Specifically, first, we should improve the ecological environment and increase efficiency of resource consumption. Second, we should upgrade and transform the existing manufacturing industry and eliminate some high energy consumption and high pollution manufacturing. Third, we should introduce high-tech and high value-added industry and its corresponding talents. Forth, we should develop services. Fifth, we should optimize urban space structure and land use patterns. Sixth, we should introduce and develop high education and research institutions to provide high quality human resources. Seventh, we should improve the job environment and construct healthy and rich space for living and entertainment. Among these aspects, the ecological environment, resources and talents are the first things to consider in Dongguan City's transition.

Green development is the key to solve resources and environment problems in economic growth. The construction of green infrastructure is an effective method to promote Dongguan city transition and it can support the above ways and methods that promote city transition. Green infrastructure contains natural areas with internal connectivity, the open space network and the accompanied engineering facilities. This network has the function and value of natural ecological system and can provide natural spaces for human and wildlife, like habitats, clean water and migration channels. They construct the ecological structure that ensures a sustainable environment, society and economy (Benedict and McMahon 2006). Just like the water and electricity network, green infrastructure is the basis for community development, good quality of community life, the existence and development of other facilities and sustainable development (Li Kairan 2009). So, considering its function and features, the construction of green infrastructure should be considered first. As shown in Fig. 4.2, we can conclude the ways and methods green infrastructure promote Dongguan city transition as follows.

Green infrastructure can protect and restore the ecological environment. It can reconstruct Dongguan's natural ecological cycle network and support circular utilization of resources to achieve a sustainable ecological environment.

Green infrastructure can connect scattered towns and promote the connection of industries among towns to achieve the information, technology and talents sharing. In this way, it can improve the facilities' utilization efficiency and enhance the interpersonal communication among different classes and departments. Thus, it benefits the formation of diversity, multiple industrial ecosystems in a large district. It can also promote intensive land use based on the formation of a diverse industrial chain and the construction of circular economy and then achieve smart growth.

Green infrastructure attracts high-end industry and a talented workforce, by improving city's environment and then promotes industry upgrading and transformation (Richard Florida 2003). Specifically, after the high-end industry and creative workforce move to this area, we can develop industries that are high-tech and high value-added, like tourism and some creative industries. Thus, it can promote multiple and diversity industry development and form good industrial structure to achieve a resilient economy.

Green infrastructure can promote the formation of a beautiful and friendly natural environment and then change people's life behavior. The homogenization of a high-quality environment is conducive to social equity and can reduce social unrest. It is also conductive to the development of diverse economies and cultures to promote all-round development of society and it will do good to promote and develop quality of material and spiritual life and then achieve a resilient society.

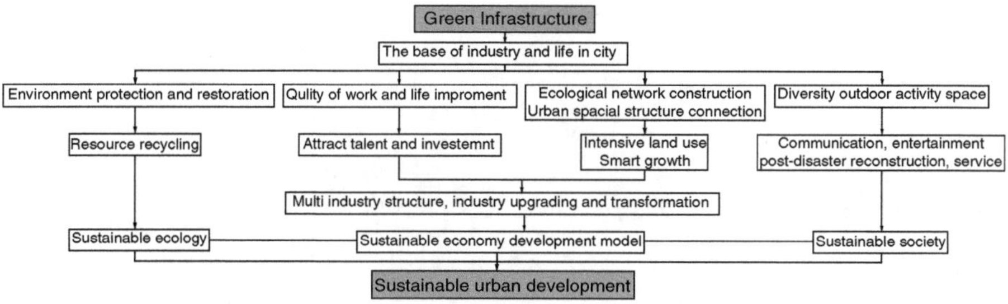

Fig. 4.2 Green infrastructure's mechanism of action

4.6 How the National Wetland Park Promote the Economic and Social Transformation of Dongguan City

Dongguan's national wetland park located in the east of Dongguan City, it is in Dongguan's Eco-industrial Park. This area is at the edge of six towns include Liaobu Town Dongkeng Town, Chashan Town, Shipai Town, Hengli Town and enclosed by them, as shown in Fig. 4.3.

Dongguan National Wetland Park system transforms the 31 km^2 low-lying areas which are enclosed by six towns and are seriously polluted by emissions from the towns. The pollution flows into several wetland parks including Yanling wetland, central water system ecological island group, Yuehu wetland, Xiasha wetland, and the Dazhenpu wetland. The total area of Dongguan National Wetland Park is 651.1 ha; the area of water surface is 342.7 ha (Lin Yudong 2014). As shown in Fig. 4.4, Dongguan National Wetland Park forms an ecological core area and becomes the green infrastructure in this area. It connects the original scattered land and reshapes the road network and ecological network in order to achieve regional integration. The project includes connections between the road network within the park and the road network in the surrounding towns to optimize the regional traffic and promote regional transportation integration. And then, as a green infrastructure network, the wetland park system connects six towns and forms a large organic area. In this way, it slowly promotes industrial connections between the six towns. It also extends green infrastructure to form a large ecological framework that can support the sustainable ecological, social and economic development of the six towns or even the larger region.

4.7 Dongguan National Wetland Park System Repairs and Manages the Ecological Environment

Dongguan National Wetland Park system forms a network of wetlands with a multifunctional green water system. This system can restore natural circulation and participate in the Urban Recycling Economy System through many ecological functions of wetland eco-systems and other engineering measures. The ecological functions

Fig. 4.3 The location of Dongguan's Eco-industrial Park

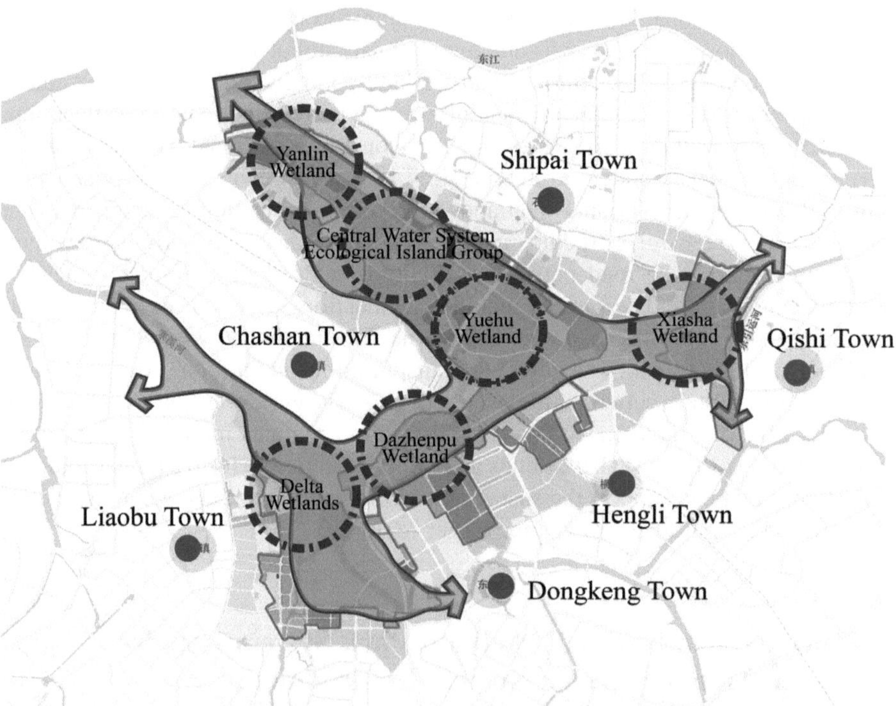

Fig. 4.4 Dongguan National Wetland Park system

include restoring biological communities, wastewater treatment, natural water recovery and flood control; the engineering measures include intercepting external pollutants, dredging and canal expansion, circulating and supplementing water, a sewage treatment plant, a tail water recycling system and other green projects.

Dongguan National Wetland Park system could help to repair and manage the natural water environment. Firstly, this project achieves flood control by increasing regional flood detention capacity. Engineering projects include dredging and the expansion of ponds into lakes to increase the capacity of flood storage by three times while reducing the area of water surface from 60 to 30 %. This project also includes the construction of two new floodwater drainage stations. These measures increase the area's standard of flood control to deal with 20-year storms. Secondly, by building habitat and other measures, the National Wetland Park system becomes a water ecosystem in this area to achieve water purification. At last, the water purification system of National Wetland Park can achieve regional water recycling. As shown in Fig. 4.5, there are many water treatment facilities in the National Wetland Park and they form a water treatment system for water recycling. First, the waste water from six surrounding towns should be cut off and then disposed by Nanshelang Sewage Treatment Plant. Then the treated water should be further treated by vertical flow system and the natural flow system in Yanlin wetland. With this system the water quality could be up to the fourth class water quality standard and can be up to third class in the long term. It could be recycled as the water source in the Eco-industrial Park, such as it will be discharged into the central water system. It also can supply water for surrounding enterprises to achieve water recycling.

The repairing and management effect of National Wetland Park to ecological environment also include purifying air and protecting animals. Firstly, National Wetland Park System improves air condition in the region by using 51 % green rate and the infiltration of greenways into the surrounding towns. Secondly, National Wetland Park System creates a suitable natural environment for animals with the construction of water environments, and ecological management measures. The amount of land vascular plant species has increased from 133 to 356, the amount of water vascular plant species has increased from 25 to 70, the green area has increased about three times and the amount of animal species has increased from 194 to 326 and the total amount of animals has increased by about 2 times.

4.8 Dongguan National Wetland Park System Promotes Industry Upgrading and Transformation and Leads to Dongguan City Economy Transformation

1. The mechanism of industry upgrading and transformation

Dongguan National Wetland Park system integrates 31 km^2 area and implements intensive development strategy. It achieves regional spatial intensive development by its attractive environment and achieves regional industry intensive development by through industrial driving effects.

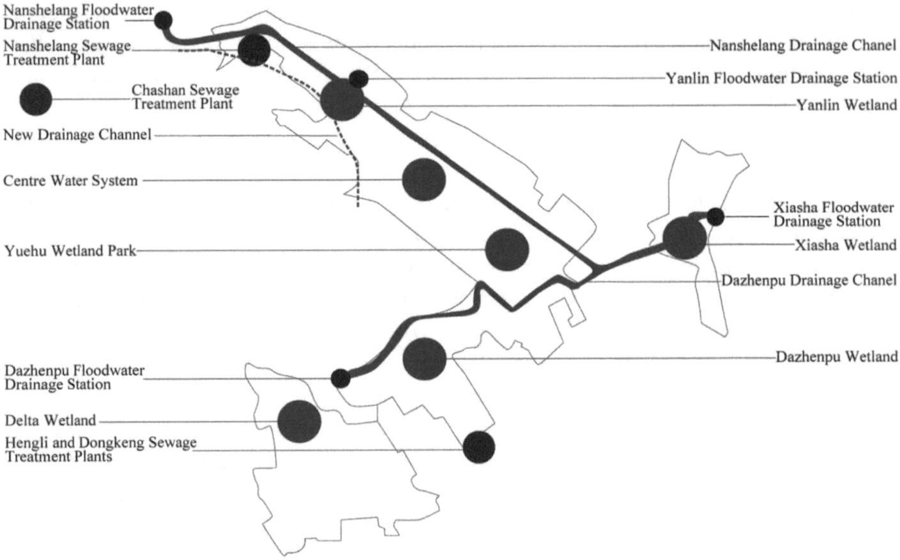

Fig. 4.5 Water treatment projects of Wetland Park system

Firstly, Dongguan National Wetland Park system improves the regional natural environment by the above engineering and non-engineering measures. Then it attracts high-end industry and a creative workforce to the Eco-industrial Park because there is a good quality of life and a good working environment. In this way, it has changed the area from an abandoned and centrifugal area to a beautiful and centralized area with a good quality of life, as shown in Fig. 4.6.

Thus, beginning the construction of Eco-industrial Park on the basis of green infrastructure can promote industry upgrading and transformation and a natural transformation. The high-end industry and creative workforce will come to Eco-industrial Park because of the environmental attractiveness of the National Wetland Park. The Eco-industrial Park should pick the industries which don't compete with but drive, serve and improve the industry in surrounding towns. And then, we should avoid similar industries and construct complimentary industries in surrounding towns. In this way, we could achieve intense regional development. Specifically, we put the core industries in the Eco-industrial Park and the supporting industries in the surrounding towns and the connecting industries in the edge of each town. Therefore, the Eco-industrial Park can connect the surrounding towns to form a large area with a diverse and multi-level industry to promote upgrading and the transformation of the area.

2. The implemention of industry upgrading and transformation

Currently, the towns in Dongguan depend on a few industries and with several industries repeating among these towns. The industrial development of each town does not consider each town's characteristics and resources, therefore the main

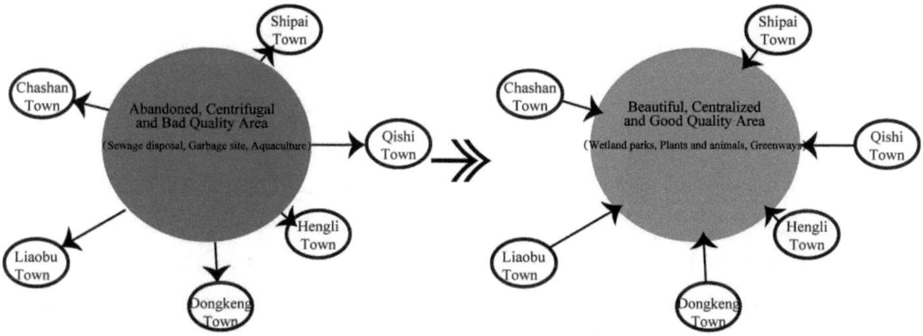

Fig. 4.6 The area has been changed from centrifugal development to centralized development

industry types of each town are electronics, apparel industry and automobile assembly. The National Wetland Park System introduces energy conservation and environmental protection industries, such as high-end information electronics, high-end equipment manufacturing, modern services and leisure and tourism. The direction of industrial development is established by considering the advantages and characteristics of each town. Specially, we preserve the original distinctive and leading industry of each town, and then we improve and optimize some original industries and introduce new high-end industries, thus we achieve the multi-level and diversified industrial system, as shown in Fig. 4.7.

The mineral resources and historic culture resources in Shipai Town is rich. The mineral resources include red sandstone, white mud, black mud and river sand; the historic culture resources include the Yanling ancient quarry, Tangwei ancient villages from the Ming and Qing dynasties and Baotang Kangwang Temple. In 2006, Shipai Town has established the city brand of "Chinese town with Lingnan characteristic" (http://baike.baidu.com/view/183018.htm). The industry types in Shipai include electronic information industry, apparel industry and print industry. Driven by the high-end industry in the Eco-industrial Park, the electronic and information industry has been upgrading and some robotics industry has developed, and then leisure and tourism can be developed according to the town's resources advantages.

Chashan Town is China's historic and cultural village and Chinese National Key Cultural Relics Protection Units. The tourism resources are rich and the town built the Chashan international apparel wholesale market, a major tea market, plant market and a large-scale professional logistics market. The main industry types in Chashan include electronic industry and apparel industry, at the same time Chashan Town is known for its foodstuff (http://www.baike.com/wiki/%E8%8C%B6%E5%B1%B1%E9%95%87). Driven by the high-end industry in the Eco-industrial Park, Chashan has developed logistics and its supporting industry and upgraded the apparel wholesale and manufacturing industry, and then we will develop catering according to the town's own resources advantage.

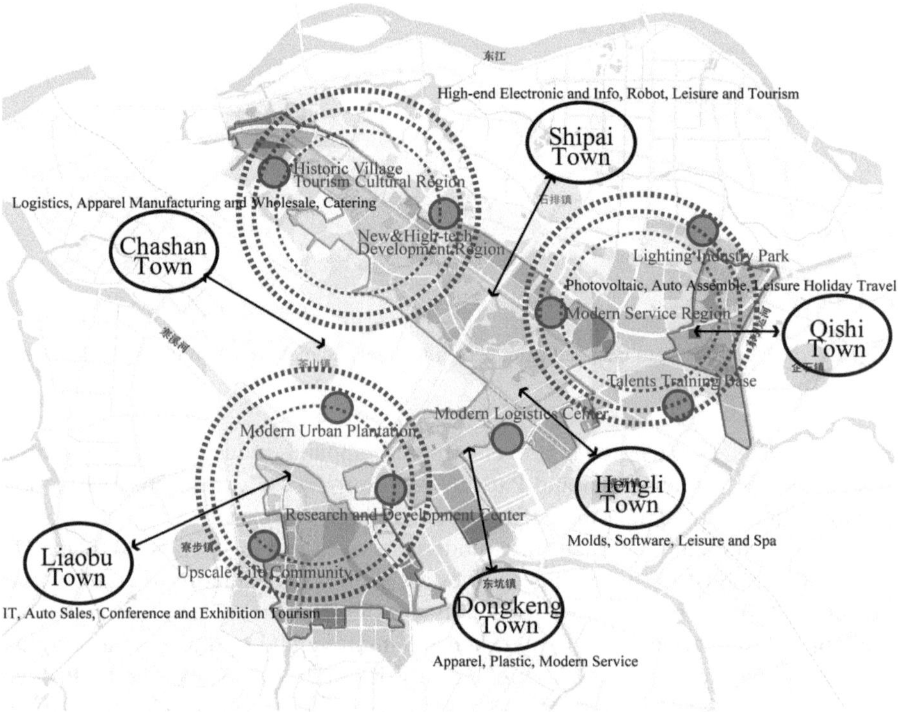

Fig. 4.7 Industry driven model in Dongguan Eco-industrial Park

The Xiang City in Liaobu Town is rated as the non-material cultural heritage town of the Guangdong province. The main industry types in Liaobu Town include electronic information industry, auto sales industry and mechanical equipment manufacturing industry (http://baike.baidu.com/view/182924.htm?fr = wordsearch). Driven by the high-end industry in the Eco-industrial Park, the auto sales industry has been kept and the IT industries has been upgraded and develop on the basis of electronic information industry, and then we will develop conference and exhibition tourism according to the town's own resources and advantages.

Dongkeng town has formed its industrial system which has electronic, apparel, and plastic industries as its pillars (http://baike.baidu.com/subview/183010/5767139.htm). Driven by the high-end industry in the Eco-industrial Park, the apparel and plastic industry has been preserved and upgraded while actively developing a modern service industry.

Hengli Town is a beautiful ecological town, the green area in town center is 439.5 ha and its coverage rate is 31.7 %. It is the Provincial Health Town and is famous for its molds. The main industry types in Hengli Town include mold, footwear industry and electronic industry (http://baike.baidu.com/subview/183004/6401317.htm?fromtitle=%E6%A8%AA%E6%B2%A5&fromid=4737085&type=syn). Driven by the high-end industry in the Eco-industrial Park, Hengli Town actively develops and upgrades the molds industry and developed software industry on the basis of workforce training in the Eco-industrial Park, and then we will develop leisure and spa industry according to the town's own resources and advantages.

The landscape in Qishi Town is rich and it has ecological advantages. It can provide high quality water across one-third of the city's area. The main industry types in Qishi Town include photovoltaic industry, electronic and information industry and automobile assembly (http://baike.baidu.com/view/183016.htm?fr=wordsearc). Driven by the high-end industry in the Eco-industrial Park, the photovoltaic and automobile assembly industries have been upgraded and optimized, and then we will develop a leisure holiday travel industry according to the town's own resources and advantages.

3. The transformation of the economic development model

The upgrading and transformation of industry means the transformation of the economic development model. The above research proves that, the construction of Dongguan National Wetland Park system achieves the upgrading and transformation of industry by its environmental attractiveness and industrial driving effects. By this way, it can transform the economic development model from relying on manufacturing, which is high-input, low-output and high-emission to relying on modern energy conservation and environmental protection industries, which are high-end, diverse, multilevel, low-input, high-output and low-emission. Specifically, we can talk about the transformation of the economic development model from two aspects. Firstly, water system improvements will preserve water resources in the wetland park, improve tourism and create a society which is resource-saving and environmentally friendly. Secondly, the high-end industries in the Eco-industrial Park lead to diversity and multilevel of the industries in surrounding towns. The development of high tech and high value-added industries such as service, tourism, conference, and exhibition industries could reduce the low-end manufacturing which is high-consumption and high-pollution. In this way, we can form a diversified and multilevel industry structure, these industries can coordinate and compete with each other and the industrial resources between them can be recycled, regenerated and reused. Thus the regional economy can achieve sustainable development.

Since the construction of the Dongguan National Wetland Park system, the amount of waste water caused by every billion GDP has dropped about 1,300 thousand tons, the amount of waste residues caused by every billion GDP dropped from about 2.5 thousand tons (Dongguan City Statistical Yearbook 1978–2012). Thus it can be seen that, the construction of Dongguan National Wetland Park system could promote the transformation of regional the economic development model from extensive to intensive and it will drive the transformation of the economic development model of whole Dongguan City.

4.9 The Dongguan National Wetland Park System Promotes the Transformation of Society

The city's social transformation means that the society will transform from one of urban-rural conflict to a harmonious society, transform from an unstable society with many conflicts to a harmonious society with different ranks and classes, from an inharmonious and incomplete society to people oriented and an all-round harmonious and sustainable development society. The main challenge for Dongguan's social transformation is the problem of how to achieve harmonious and people oriented society and how to promote the transformation of people in Dongguan.

Firstly, the transformation of industry in Dongguan promotes the transformation of society. The construction of Dongguan National Wetland Park system promotes the transformation of industry from a single, low-end manufacturing to a diversified, multilevel industry. The changing of the industrial structure promotes the changing of demographic, economic and cultural structures, thus, the social operating mechanism and interest relationship also has been changed.

Secondly, the transformation of economic development model promotes social transformation to some extent. The Dongguan National Wetland Park System promotes an economic development driven by resource-saving and environmentally friendly industries. This transformation must result in the changing of people's idea and behavior and thus people's way of production and lifestyle.

Thirdly, as regional ecological base, the Dongguan National Wetland Park System formed a relatively equal and high-quality environment in this area and it will expand the ecological base in whole Dongguan City in the near future. It will definitely promote social equity and reduce social conflicts.

Fourthly, as green infrastructure, Dongguan National Wetland Park system can promote sustainability. This sustainable natural environment contributes to mental health and thus promotes the transformation of society from uncoordinated, incomplete and unhealthy to people oriented, all-round, harmonious and sustainable.

At last, as green infrastructure, the Dongguan National Wetland Park System provides a high-quality ecological network. Therefore, it can provide space support for people's diversity and healthy life and also the spiritual and cultural needs, and then improve the quality of city's culture.

Hence, one can see that the green infrastructure is very important for the promotion of social transformation. Since the construction of the Dongguan National Wetland Park System, the number of catering services in surrounding the area has increased from 132 in 2007 to 354 in 2012 and the social crimes in the surrounding area has reduced by 200 (Dongguan city statistical yearbook 1978–2012). But the social transformation also needs the guidance of relevant policies. In recent years, in order to promote the development of a harmonious society, Dongguan has made great efforts to make relevant policies, such as the "New Dongguanese" policy. This is an important action to promote Dongguan social transformation and city upgrading. It can eliminate the psychological obstacle of the immigrant population and let many foreign workers realize their authentication, thus it could reduce crime rates among immigrant populations and build a harmonious Dongguan.

4.10 Conclusion

As the city's green infrastructure, the Dongguan National Wetland Park System improves the city's ecological environment and promotes Dongguan's transformation of economy and society. Firstly, the Dongguan National Wetland Park System forms a regional ecological base and restores the region's natural environment. Thus it changes waste into treasure and ties together a large area. And then the creative class will come here and the Eco-industrial Park will be constructed. The high-end industry and creative class will lead to the upgrading and transformation of industry in the larger area, including the surrounding towns, so the economic development model will be transformed from a resource- based economy to innovation-based economy. At last, on the basis of the improvement of ecological environment and the transformation of the economic development model, green infrastructure gives full play to its social service effect to promote the construction of a harmonious society by changing the way people live and producing a spiritual and cultural life supported by the ecological network. In this way, Dongguan will change form, form low-level urbanization toward advanced urbanization.

China's cities generally have the problem of large populations, resource shortages, urban sprawl, heavy pollution, low-end industries, social conflicts, and so on. These problems are connected and they have reciprocal causation. Low-end industry and urban sprawl lead to waste of resources and environmental pollution; such a large population that lives in a bad environment must produce many social problems. Thus, in order to achieve urban transition in China we should pay more attention to industry upgrading and transformation and city intensified development. Green infrastructure has shown its advantages to cope with these problems. On one hand, green infrastructure develops a big ecological network in the city and improves the ecological environment to form regional cohesion and achieves an intensification of land use. On the other hand, it forms a regional center that attracts high-end industry and a talented workforce, in this way it promotes industry upgrading and transformation and then changes the economic development model to fundamentally improve the natural environment. Thus, this is a virtuous circle which can achieve the city's social transformation.

References

Accenture and Chinese Academic of Sciences (CAS). (2013). *Towards a new resource economy: To promote China's harmonious urban transformation*. http://www.accenture.com/cn-zh/Pages/insight-new-resource-economy-city-transition.aspx

Baidu Encyclopedia. *Dongkeng Town*. http://baike.baidu.com/subview/183010/5767139.htm

Baidu Encyclopedia. *Hengli Town*. http://baike.baidu.com/subview/183004/6401317.htm?fromtitle=%E6%A8%AA%E6%B2%A5&fromid=4737085&type=syn

Baidu Encyclopedia. *Liaobu Town*. http://baike.baidu.com/view/182924.htm?fr=wordsearch

Baidu Encyclopedia. *Qishi Town*. http://baike.baidu.com/view/183016.htm?fr=wordsearc

Baidu Encyclopedia. *Shipai Town*. http://baike.baidu.com/view/183018.htm

Benedict, M., & McMahon, E. (2006). *Green infrastructure: Linking communities and landscapes* [M] (pp. 1–3). Washington: Island Press.

Chen Guiming, & Li Ganqiu. (2005). *Apply the scientific outlook and vigorously develop the circular economy*. http://www.gdei.gov.cn/zwgk/jmzk/gdjm/200506/200507/t20050719_42515.html

Dongguan City Bureau of Statistics. (2009). *The work and living conditions of the new Dongguanese had to be improved*. Dongguan developments.

Dongguan City Statistical Yearbook. (1978–2010). http://tjj.dg.gov.cn/website/web2/art_list.jsp?columnId=10000

Gu Chaolin. (2011, November). Transitional development and future urban forms. *City Planning Review*, *35*(11):23–34.

Li Kairan. (2009). Green infrastructure: Concept, theory and practice. *Chinese Garden*. 1000-6664 (2009)10-0088-03.

Li Qingyu, & Zhou Guiyin. (1994). *A new approach in the study of American history and modernization* (pp. 103–104, p. 121). Beijing: People's Publishing House.

Liang Liangfeng, & Chen Zhongnuan. (2007). *Analysis of Dongguan rapid population urbanization, take Dongcheng district for example*. 1001-7852(2007)05—0025—05.

Lin Yudong. (2014). *Add green for "Green Dongguan" There are thousands trees in Dongguan Eco-industrial Park*. SUN0769.COM. http://news.sun0769.com/dg/sh/201402/t20140228_3547042.shtml

Mo Anda, Wang Siyu, & Tang Yuansong. (2009). *Accelerate social management problems in the industrialization process: Dongguan case*. Dongguan social science. 1000-114X(2009)03-0171-05.

Qian Wei. (2011). *An innovative city development road, the experience and enlightenment of American typical city transformation*. 1004-7069(2011)-02-0016-04.

Ruchelman, L. I. (2007). *Cities in the third wave: The technological transformation of Urban America* [M]. Lanham: Rowman & Littlefield Publishers Inc.

Shuai Ping, & Zhou Lijun. (2013). The way of the Chicago transition. *Oriental Morning Post*. 2013-08-06.

Toffler, A., & Toffler, H. (1994). *Creating a new civilization: The politics of the third wave* [M]. Atlanta: Turner.

Wang Dan, & Wang Shijun. (2007). Understandings on development view of new urbanism and smart growth of the USA. *Urban Planning International*, *22*(2):61–66.

Wang Lixin. (2011). Los Angeles city transformation enlightenment to Shenzhen. *China Opening Herald*. http://city.ifeng.com/special/Hollywood/20110304/43789.shtml

Wikipedia. *Chashan Town*. http://www.baike.com/wiki/%E8%8C%B6%E5%B1%B1%E9%95%87

Chapter 5
The Symbiotic Strategies Study of Low-Carbon Eco-City Based on Multi-symbiosis Theory

Xia Zhu and Yangyue Zhou

Under the dual drive of industrialization and urbanization, countries have pursued city development, following the western model, which originated during the western industrial revolution. While cities accumulate wealth and realize dreams for human beings, they also require use of 85 % the earth's resources and energy, and produce similar percentages of greenhouse gases, and helping cause global warming, ozone reduction, carbon balance disorder, a decrease in biodiversity and a series of ecological environmental problems. Therefore, facing the unprecedented challenge of global climate change and the increasingly serious global environmental and urban crisis, human beings have to start reexamining the old urban development model in regards to the relationship between humanity and nature. Due to accelerated period of urbanized development, China has been facing a "4p" crisis: resource poverty, environmental pollution, population explosion, and social problems. A transformation of China's cities into multi-symbiotic low carbon eco-cities is inevitable, in order to transform the development model, help reduce contradictions between development and survival, and optimize the relationship between urbanization and the ecological environment.

5.1 The Academic Research

5.1.1 Symbiosis Theory

"Symbiosis" is the most common phenomenon in nature, the concept of which stems from biology. In order to describe the mutual relationship between algae and

X. Zhu (✉) • Y. Zhou
School of Architecture and Urban Planning, Huazhong University of Science & Technology, 1037 Luoyu Road, Hongshan District, Wuhan, Hubei Province, People's Republic of China
e-mail: 2003888@vip.163.com; zyy1981022@sina.com

Q. Pan, J. Cao (eds.), *Recent Developments in Chinese Urban Planning*, GeoJournal Library 114, DOI 10.1007/978-3-319-18470-8_5

fungi, De Barry, a botanist of Germany, firstly proposed the concept of symbiosis in 1879 (Zhang and He 2003). "Symbiosis" refers to different species living together through some sort of physical contact, depending on each other to adapt to complex and changeable environment in a long-term evolutionary process, which is beneficial to both organisms (Li 2000). Since the 1950s, symbiosis theory has gradually been used in anthropology, sociology, economics, management science, politics, and other exploration. With deepening research and practice of symbiosis theory, the theory's connotation and denotation continue to expand. The generalized concept of "symbiosis" can be understood as coexistence, communion and harmony. Symbiosis refers to the relationship between symbiotic units in a symbiotic environment according to some co-occurrence patterns, which generally includes three elements: a symbiotic unit, a symbiotic model and a symbiotic environment. Symbiotic unit refers to basic symbionts or symbiotic energy production and exchange units, which characteristically lie in the complex properties of symbionts. A symbiotic model (or symbiotic relationship) refers to the interaction or combination of symbiotic units, which reflects the characteristics of how material, energy and information are exchanged between symbiotic units. Symbiotic system state changes, mainly reflect on the symbiotic mode changes. The symbiotic environment refers to all influential factors outside of symbiotic unit. Symbiotic units, the symbiotic model and the symbiotic environment all influence and interact with one another, reflecting on the dynamic direction change of symbiotic system. Regarding the relationship of the three elements, the symbiotic model is key, the symbiotic unit is a foundation, and symbiotic environment is an important external condition.

5.1.2 Low-Carbon Eco-City

The concept of a "low-carbon eco-city" is derived and developed from the concept of the "eco-city". In 1971, the United Nations Educational, Scientific and Cultural Organization (UNESCO) formally proposed the concept of an eco-city during the program of Man and Biosphere (MAB). After more than 40 years, the eco-city became the leading development direction for city construction in the world, in terms of theoretical discussion. In 2003, the British government originally proposed the concept of a "low-carbon economy" by publishing the essay of "THE FUTURE OF ENERGY: CREATING A LOW CARBON ECONOMY", this paper caught the attention of the international community. Based on the concepts and theory in the paper, Chinese scholar Qiu B.X. combined the ideas of the eco-city and low carbon economy and presented the concept of a "low-carbon eco-city" in 2009. The low-carbon eco-city is a type of energy saving and environmental protection city, marked by low energy consumption, low pollution, and low emission; which, is a new urban development model emphasizing the comprehensive balance of ecological environment (Shen 2013). In essence, a low-carbon eco-city belongs in the category of an eco-city, as a primary process towards eco-cities, and a type of eco-city with a main focus of reducing carbon emissions.

5.1.3 Multi-symbiotic Low-Carbon Eco-City

At present, there's no clear definition of a multi-symbiotic low-carbon eco-city domestically or internationally. The low-carbon eco-city from new urbanization in China should be multi-symbiotic; namely, human elements should harmoniously coexist with nature, science and technology, economy, culture, and other diverse elements in cities (Register 2006). From different interfaces or dimensions, the basic compositions and characteristics of symbiotic units are different. Based on the current development of low carbon eco-cities, a set of guiding indices should be established at the national level, helping to identify development goals for the low carbon eco-city. In short, the indices for the multi-symbiotic low carbon eco-city shall at least include four dimensions: resource saving, environment friendliness, sustainable economy, and social harmony (Mcharg 1981).

5.2 The Connotation and Characteristics of Multi-symbiotic Low-Carbon Eco-City

5.2.1 The Connotation of Multi-symbiotic Low-Carbon Eco-City

The basis for understanding the basic concepts of a multi-symbiotic low-carbon eco-city can be acknowledged from eight different levels: function, economy, philosophy, society, ecology, efficiency, space, and culture. A detail connotation system can also be constructed (Table 5.1). Table 5.1 shows that the connotation of a multi-symbiotic low-carbon eco-city is reflected in the following eight aspects: (1) equality in philosophy; (2) coordination in economy; (3) circulation in function; (4) harmony in society; (5) coexistence in ecology; (6) synthesis in benefit; (7) compact in space; (8) inheritance in culture.

5.2.2 The Characteristic of Multi-symbiotic Low-Carbon Eco-City

A substance's characteristics are unique, differing from the basic signs and flags of other substances (Roseland 1997). The multi-symbiotic low-carbon eco-city has five basic characteristics: symbiosis, diversity, efficiency, circulation, harmony (Fig. 5.1; Table 5.2).

Table 5.1 The connotation system of multi-symbiotic low-carbon eco-city

Aspect	Connotation	Key words
Philosophy	Achieve equality in human and natural interaction from perspective of symbiosis	Equality
Economy	Develop a circular and low-carbon economy, promote urbanization, industrialization and agricultural modernization and information	Coordination
Function	Promote the circulation of material, energy and information smoothly; also form a self-organization symbiosis system between urban and nature	Circulation
Society	Advocate ecological civilization, realize harmony of the social system and natural system	Harmony
Ecology	Improve the quality of the living environment and optimize the ecologic environment from the angle of symbiosis, coexistence and co-prosperity	Coexistence
Benefit	Symbiotic space model rich in diversity, connotation, compact and compound	Synthesis
Space	Symbiotic space model marked by diversity, connotation, compactness and compound	Compact
Culture	Fuse local culture and heterogeneous culture, symbiosis of history and future	Inheritance

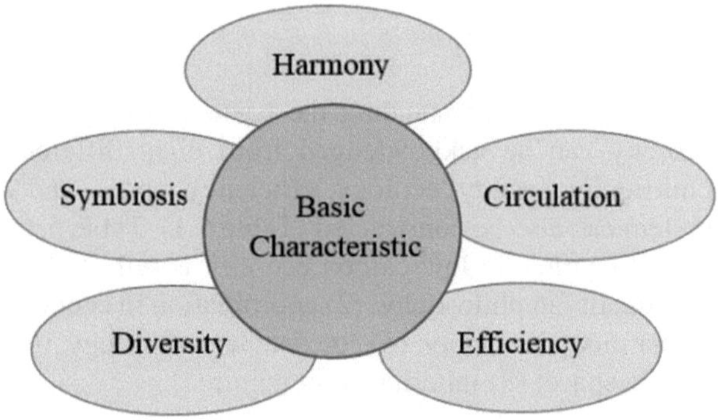

Fig. 5.1 The characteristic of multi-symbiotic low-carbon eco-city

Table 5.2 An analysis of basic characteristics of a multi-symbiotic low-carbon eco-city

Characteristic	Main idea	Status
Symbiosis	Achieve multi-symbiosis of ecology, economy, energy and the sustainable development of human settlements	Core factor
Diversity	Urban gene-diversity, urban species diversity, urban system diversity, urban landscape diversity	Fundamental power
Efficiency	High efficiency of urban energy system, high benefit of urban transformation system, high efficiency of urban circulation system	Basic guarantee
Cyclicity	Virtuous cycle of system, material and elements	Foundation
Harmony	Realize the harmony of human and nature	Bond

5.2.2.1 The Characteristic of Symbiosis

The ecologic theory believes that the various life levels, the overall characteristics and system functions of a variety of eco-systems both are the coevolution products of biological and environmental long-term symbiosis. Symbiosis and coevolution is a common phenomenon in the ecosystem, the symbiotic relationship is the basis of self-organizing structure within certain functions of the ecosystem (Phdungsilp 2010), and the characteristic of symbiosis is one of the most important features of a self-organizing system. It is through the symbiosis of multi-systems that a low-carbon eco-city realizes the sustainability of ecological environment, energy using, society development, and human life, thereby improving operational efficiency and benefits of various urban systems, reducing internal friction and damage to urban environment, which ultimately achieves symbiosis between human beings and nature. Therefore, symbiosis planning and design should be served as the core of low-carbon eco-city planning.

5.2.2.2 The Characteristic of Diversity

Acting as a complex system involving multiple dimensions, multi-symbiotic low-carbon eco-city has diversities just similar with ecosystem, including genetic diversity, species diversity, ecosystem diversity and landscape diversity (Chen and Zhu 2010), which are the fundamental motivation of benign development of urban. Genetic diversity determines species diversity, characterized by the complexity of the society, economy, culture and so on. Species diversity is mainly manifested in a variety of functions on mixed land use and composite space. The highly networked internal links of urban energy, industrial, transportation, spatial, and other systems, reflect the diversity of urban ecosystem. The landscape diversity of a low-carbon eco-city can be highlighted through blending city into the surrounding natural landscape elements, such as mountains, water, and farmland.

5.2.2.3 The Characteristic of Efficiency

The high efficiency of a symbiotic low-carbon eco-city is reflected through four aspects: (1) the efficient utilization of urban energy and resources; (2) the high efficiency of urban conversion systems, namely lower input, higher output and less pollution during the conversion process of the natural substance, the economic substance, and emissions of the low-carbon eco-city; (3) the high efficiency of urban transfer systems, specifically the low-carbon eco-city based on the ecological infrastructure, accelerating the flow of material, energy, information, value, and the stream of people, minimizing the economic loss, carbon emissions as well as environmental pollution; (4) reducing the energy consumption during the production, operation, and maintenance of the city.

5.2.2.4 The Characteristic of Circulation

The characteristic of circulation of a multi-symbiotic low-carbon eco-city is reflected in all aspects within a city. It refers to recycling of the society, economy, and nature from a macro scale; includes circulation of material, nutrition, waste, etc. at middle scale; from micro scope, namely, water cycle, carbon cycle, nitrogen cycle and so on. The positive role of circulation lies in improving eco-efficiency of material resources utilization and promoting urban independence, then greatly reducing energy and material input from outside system (Kei et al. 2010).

5.2.2.5 The Characteristic of Harmony

The characteristic of harmony in a multi-symbiotic low-carbon eco-city is reflecting urban development degree in the higher values, both containing few symbiotic characteristics, but also reflecting symbiotic state achieved through harmonious development, which is the core connotation in a low-carbon eco-city. The characteristic of concordance is reflected in two aspects: (1) It is one of the key bonds to realize the coordinated development of human, urban and nature from the relationship between human beings and nature; (2) It is a harmonious society with trustful interpersonal relationships from the relationship between people and people.

5.3 The Symbiotic Interface of Multi-symbiotic Low-Carbon Eco-City

The paper takes journals published in the China Journal Full-text Database (CJFD) since 1980–2012 as a data source, adopting keywords fuzzy retrieval methods to get theses regarding "low-carbon" &"eco-city" in high impact journals. It retrieved a total of 2077 theses which is called as sample documents. By analysing the sample documents using qualitative and quantitative research methods, we can know main research direction and subject distribution of low-carbon eco-city, and then summarize key symbiotic interface. The research, firstly, indicates that "eco-city" has been becoming popular research based on the number of papers published in the last 10 years and the growth trend (Fig. 5.2). Secondly, within the scope of the top 20 high impact journals published regarding low-carbon eco-cities (Table 5.3), 1469 papers has been published and accounted for 70.72 % of the total samples, and the top three journals are "Chinese Garden", "Planners" and "Urban Development Research". According to discipline division standard, all papers related to low carbon eco-city are mainly from seven disciplines or research field: urban and rural planning, ecology, architecture, environment, geography, resources, economics and so on. Correspondingly, the symbiotic model or symbiotic relationship of the low-carbon eco-city is mainly embodied in the eight key symbiotic interface

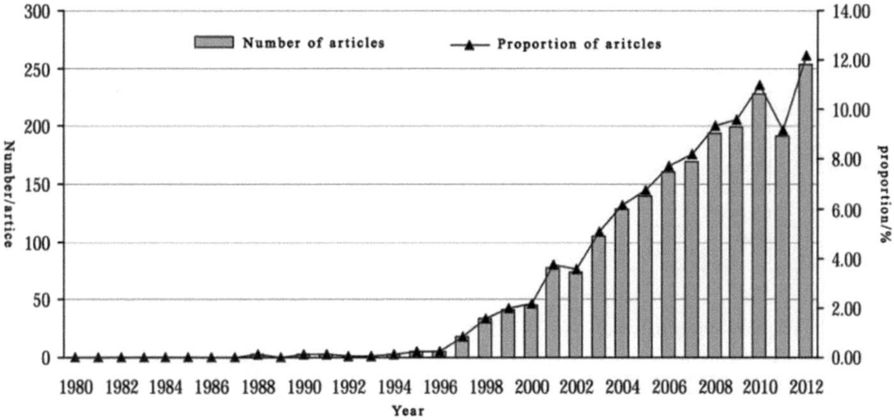

Fig. 5.2 Distribution of documents from 1980 to 2014 in CJFD journals with composite impact factor ≥ 0.9

Table 5.3 List of the journal according to number of papers on eco-city (top 20)

Rank	Journal title	Journal impact factor (JIF)	Number of journal articles(article)	Proportion (%)
1	Chinese Garden	1.343	176	8.47
2	Planners	1.475	157	7.56
3	Urban Development Research	1.822	152	7.32
4	Journal of Ecology	2.542	115	5.54
5	Modern Urban Research	1.295	90	4.33
6	Urban Planning	2.039	85	4.09
7	Urban Planning Forum	3.034	79	3.80
8	China Population Resources & Environment	2.768	76	3.66
9	Urban problems	1.212	67	3.23
10	Resource and environment	0.986	62	2.99
11	Research of Soil & water	0.905	62	2.99
12	Journal of Architecture	0.943	52	2.50
13	Chinese Journal of Ecology	1.804	52	2.50
14	Environmental Science & Technology	1.167	48	2.31
15	Economic Geography	1.978	37	1.78
16	Chinese Journal of Mechanical Engineering	1.738	35	1.69
17	Chin J App Ecol	2.350	31	1.49
18	Areal Research & Development	1.346	30	1.44
19	Progress in Geography	2.239	29	1.40
20	Tropical Geography	0.900	28	1.35

Note: The proportion refers to the number of eco-city papers in the total of samples

Table 5.4 The corresponding relationship between disciplines and key symbiotic interface

Disciplines	Urban and rural planning	Ecology	Architecture	Environics	Geography	Resources	Economics
Key symbiotic interface	Land use, Space layout, Green transport ation	Eco-community	Green architecture	Green infrastructure	Space layout	Energy utilization	Low-carbon industry

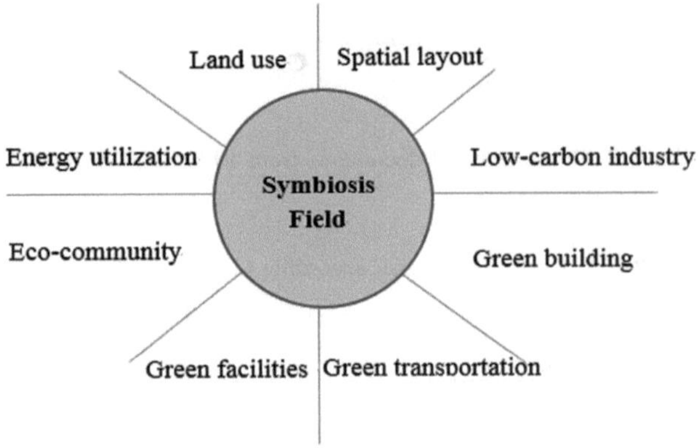

Fig. 5.3 The key symbiosis interface of low-carbon eco-city

(Table 5.4): land use, space layout, green transportation, low-carbon industry, eco-community, green architecture, green infrastructure and energy utilization (Fig. 5.3).

5.4 The Symbiotic Strategies of Multi-symbiotic Low-Carbon Eco-City

The corresponding strategies or models of a multi-symbiotic low-carbon eco-city have been put forward based on the understanding of the symbiotic interface of a low-carbon eco-city.

5.4.1 Intensive Land Strategy

A city will become a symbiotic city by combination with a variety of land uses and functions in the twenty-first century. First of all, the "booth pie" expansion model of a traditional urban should be turned into a compact development model in order to promote mixed land use (Fig. 5.4), creating a compact symbiotic space model.

Fig. 5.4 Multi-mixed pattern of land use

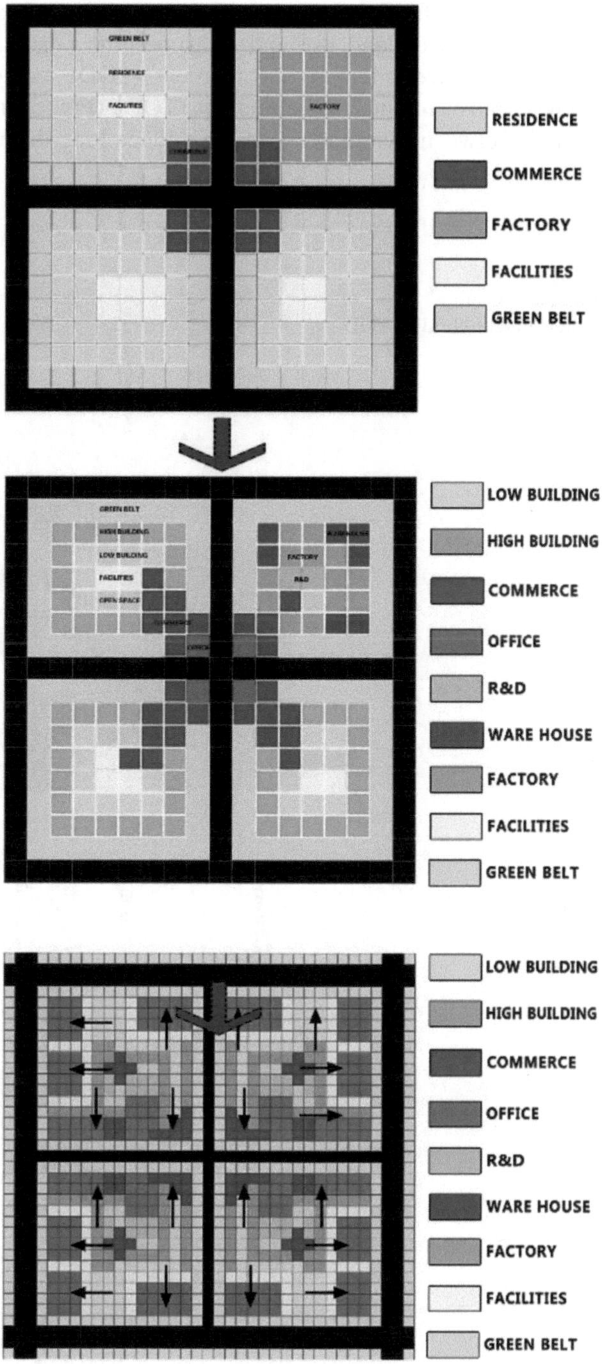

Secondly, the SET development pattern should be adopted in planning to promote the integration of land use and transportation (Zhang et al. 2011); namely, we advocate service oriented development (SOD), ecology oriented development (EOD), transit oriented development (TOD) (Fig. 5.5) model in order to achieve balance between living and working, reduce long-distance commuting and ensure safety in blocks.

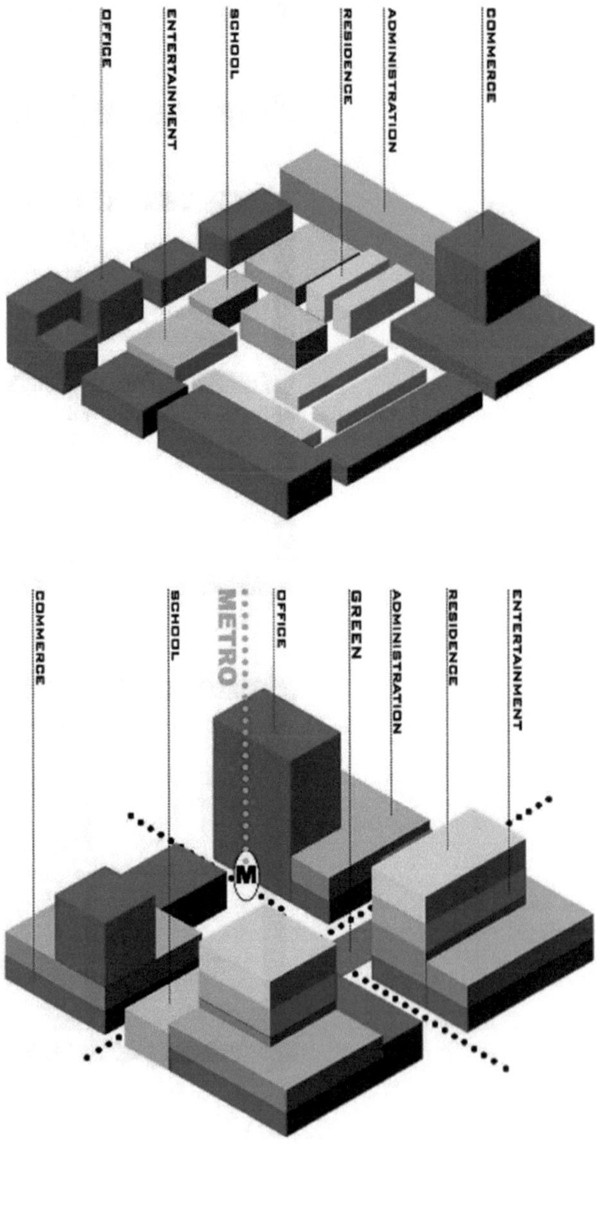

Fig. 5.5 A three-dimensional hybrid of TOD

5.4.2 Compact Space Strategy

The biggest problem of urban expansion is the urban growth boundary (UGB) tends to be blurred (Shen et al. 2010). Defining a clear urban growth boundary by determining a rigid limitation of urban expansion can limit disorder of urban space, while maintaining regional ecological landscape and keeping balance between urban construction land, and ecological control area. Besides, a compact city can greatly reduce reliance on private cars, especially by alleviating traffic pressure and decreasing the consumption of oil and other resources, and reducing air pollution. Simultaneously, a compact city can improve relative density of urban space, function combination, and physical form. These benefits are advantageous for sharing resources, service, and infrastructure, while reducing duplicate construction on the land, lessening the cost of energy and resources, and finally, improving the sustainable development of city.

5.4.3 Low-Carbon Industry Strategy

Along with industrialization, the transformation of urban development in China is the product of industrialization in other developed countries (Shen and XU 2010). Therefore, urban development must be combined with the building of low-carbon industry. Firstly, it is necessary to upgrade traditional industry of high energy consumption, high pollution and low efficiency, and encourage low-carbon environmental industry. Then, the eco-industrial parks should manage to achieve "zero carbon emissions" during material production cycle through forming "end to end closed chains" between industrial enterprises symbiosis (Fig. 5.6).

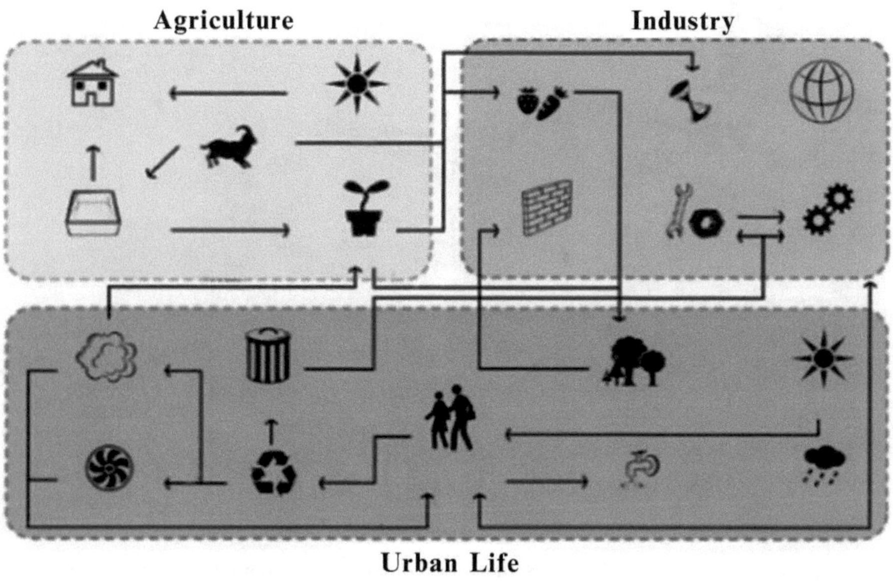

Fig. 5.6 Closed cycle industrial chain

5.4.4 Green Building Strategy

Green building (Fig. 5.7) is a basic component unit of a multi-symbiotic low-carbon eco-city. Therefore, a green building should be vigorously developed through an optimized energy system to reduce energy consumption. Within the whole life of a green building, resource conservation including energy, land, water, and materials should be maximized to protect the environment and reduce pollution, and also provide human beings with healthy, applicable and efficient living space, and harmonious coexistence with nature.

5.4.5 Green Transportation Strategy

The core concept of green transportation is to choose a healthy, eco-friendly transportation. Traffic organization should emphasize systems, prioritizing buses and slow systems, which include pedestrian system and non-motor vehicle transport system. Firstly, the development of integrated green public transportation system and a non-motor vehicle transportation network shall be encouraged and supported at all levels by the government. Secondly, the proportion of green transportation should be raised (Fig. 5.8) by reducing cars in the urban traffic structure, gradually the green traffic pattern, "subway + bus + public bicycle" (Fig. 5.9) and "zero transfer system", achieved to create a suitable pedestrian system and a slow environment.

Fig. 5.7 Green building (Note: Source http://image.baidu.com/i?tn=baiduimage&ps=1&ct=201326592&lm)

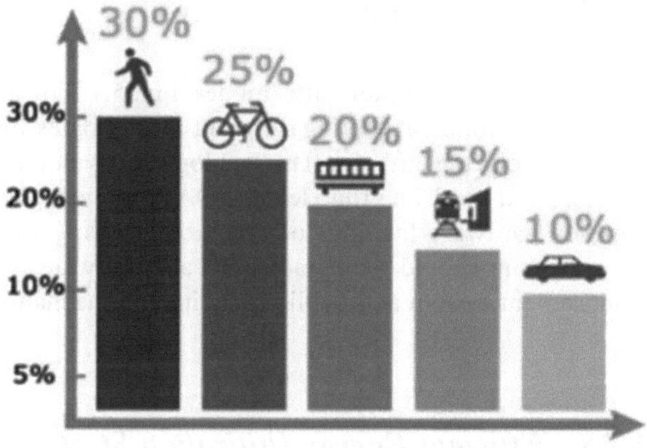

Fig. 5.8 The proportion of green transportation

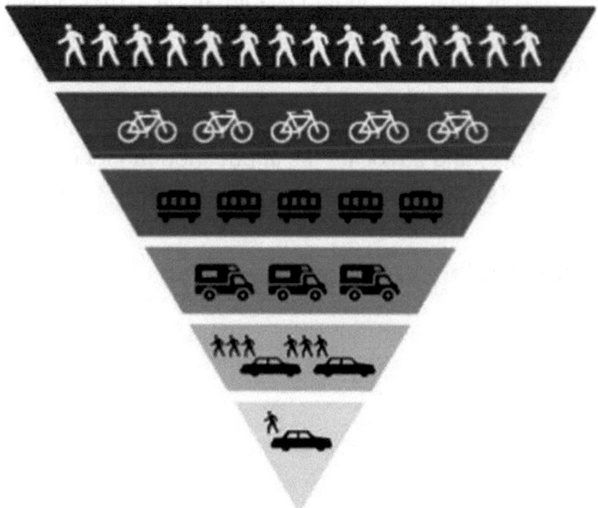

Fig. 5.9 The green traffic pattern

5.4.6 *Green Infrastructure Strategy*

By using the eco-elements of original water system, wetland, forest shelter belt and park, ect., forming the basic spatial pattern on the basis of the green infrastructure, building the optimal ecological security pattern of "matrix- corridor- patch", exerting the function of ecosystem carbon sinks. At the same time, based on the evolution mechanism of urban ecosystem, a set of complete indices for the ecological and environmental system should be established. In other words, for subsystems of transportation, new energy development, water use, public facilities, garbage disposal, ect, a set of specific and operational ecological indicators should be listed (Qiu 2009).

5.4.7 Eco-Community Strategy

Applying the "eco-community" model (Fig. 5.10) by adjusting measures according to local conditions. Adopting different technologies for ecological symbiotic units in three levels: the basic level of "ecological development unit" (EDU), "ecological development area" on group level (EDA) and the highest level "ECOCITY". Then, not only hardware facilities of green buildings, green communities, and garbage classifications should be upgraded, but also software systems such as environmental management system and public participation mechanism.

5.4.8 Efficient Energy Utilization Strategy

On the one hand, we should pay attention to diversification of energy utilization and clean production, optimize energy structure, develop renewable energy technology, improve renewable energy utilization system for city. On the other hand, from the perspective of integrated energy planning, it is necessary to consider efficient use and configuration of the urban municipal facilities energy system (heating, gas, electricity) and advocate the application model of conventional energy and renewable energy to ensure the safety of the urban energy supply (Tan 2011).

Fig. 5.10 The eco-community model

5.5 Conclusion and Prospect

Since the concept of a low carbon eco-city put forward in 2009, Chinese urban construction has experienced a development history from a single spark to a prairie fire, but Chinese current development of low-carbon eco-city still has many problems, for example, ignoring Chinese national conditions and objective requirements when developing low carbon eco-city; lacking a unified index system of authority and orientation in national level; keen on single technology application while ignoring construction cost; more slogan but less practice, existing bubble phenomenon of "pseudo low carbon" or "pseudo ecology" and so on. Based on the perspective of "multi-symbiosis", the key development of low-carbon eco-city is put forward from "amount ascension" to "qualitative leap". Its fundamental purpose is to realize urban transformation, while the transformation seriously achieves four shifts "pattern of economic growth from brown to green, growth model from linear to circular, spatial pattern from extensive, disorderly, unbalanced to intensive, orderly and balanced, development model from sublation of A and B model to exploration of new C model". Under four symbiotic dimensions- resource saving, environment friendliness, sustainable economy and social harmony- and eight key symbiotic interfaces, applying different symbiotic strategies to realize sustainable development and coexistence.

References

Chen, F., & Zhu, D. J. (2010). Research on the content, models and strategies of low carbon cities. *Urban Planning Forum, 9*(4), 125–128.

Kei, G., Kou, J. S., & Yu, M. (2010). A low-carbon scenario creation method for a local-scale economy and its application in Kyoto city. *Energy Policy, 9*, 237–241.

Li, B. (2000). *Ecology* (pp. 110–111). Beijing: Higher Education Press.

Mcharg, I. L. (1981). Human planning at Pennsylvania. *Landscape Planning, 8*(2), 109–220.

Phdungsilp, A. (2010). Integrated energy and carbon modeling with a decision support system. Policy scenarios for low-carbon city development in Bangkok. *Energy Policy, 9*(4), 167–172.

Qiu, B. X. (2009). The transformation trend of urban development model – Low carbon ecological city. *Urban Studies, 16*(8), 232–236.

Register, R. (2006). *Ecocities: Rebuilding cities in balance with nature* (pp. 25–27). Gabriola Island: New Society Publishers.

Roseland, M. (1997). *Dimensions of the future: An eco-city overview, eco-city dimensions* (pp. 1–12). New York: New Society Publishers.

Shen, Q. J. (2013). A new type of urbanization theory based on ecological civilization. *Urban Planning Forum, 1*(30), 25–36.

Shen, Q. J., & Xu, S. Y. (2010). State of urban diversity and compactness: Characterization and relationship. *Urban Planning, 10*(3), 45–49.

Shen, Q. J., Yan, C., & Liu, C. S. (2010). The discussion of connotation, characteristics and the basic principles of low-carbon eco-city. *Urban Planning Forum, 5*(12), 276–279.

Tan, X. M. (2011). Low carbon culture and its fundamental role in the development of low carbon. *Natural Dialectics Research, 4*, 122–126.

Zhang, X., & He, X. L. (2003). *Phytology* (pp. 406–407). Beijing: China Agriculture Press.

Zhang, J., Zhao, C. K., & Miao, B. T. (2011). The urban space planning and of enlightenment Chinese low-carbon eco-city. *Shanxi Architecture, 37*(33), 29–33.

Chapter 6
The Paradoxes of Land Resources Allocation's Game Playing: From Zhoukou Grave Events

Ze Zhang and Ying Gu

The Henan Dig Graves Movement, which aimed to increase the area of arable land and balance urban construction land, lasted from roughly 2012 to 2013 and took place in Henan province, mainly in Nanyang, Luoyang, Shangqiu, and Zhoukou. Based on its scale and influence, the movement was most prominent in Henan province. The "Dig Graves Movement" became one the most discussed topics in Zhoukou in 2012 and 2013. Although the purpose of the movement was to reclaim land for farming, the policy of reclaiming land was controversial among the public. After much news media exposure, which contributed to the controversy, supporters of the movement described it as "typical and a good experience", while opponents said "What a foul action it was!" (Wikipedia 2012). According to the official statement, the "Dig Graves Movement" was motivated by the funeral reform and rehabilitation of graves. However, some scholars argued that government's true target was more construct land resources (Ping Chen 2012). In addition, other scholars believe that the Henan government pushed for the "Dig Graves Movement" eagerly not because it wanted to rehabilitate the graves, but because it wanted more access to land for construction (Henan Daily 2014). As evidence for this argument, only 2 months after the "Dig Graves Movement" had begun, the Henan Development and Reform Commission (Feng Wang 2012) approved a 7 km^2 western expansion project for its industrial park. This action increased suspicions that the "Dig Graves Movement" was pushed to allow for "land sales" (Fig. 6.1).

Based on analysis by relevant scholars, the local government of Zhoukou's process under the "Dig Graves Movement" consisted of digging graves, flattening a considerable amount of land, and turning it into arable land, as shown in Fig. 6.2. According to the "Arable Land's Pothook of Increase or Decrease" policy, the local government can legally use a corresponding amount of land for construction as the new arable land obtained through digging graves. As such, by freeing up land

Z. Zhang (✉) • Y. Gu
Department of Urban Planning, Tongji University, Shanghai, China
e-mail: zhangze47@126.com

Q. Pan, J. Cao (eds.), *Recent Developments in Chinese Urban Planning*,
GeoJournal Library 114, DOI 10.1007/978-3-319-18470-8_6

Fig. 6.1 A flat grave and its gravestone in arable land, Zhoukou (Reproduced from 21st century business herald, 2012)

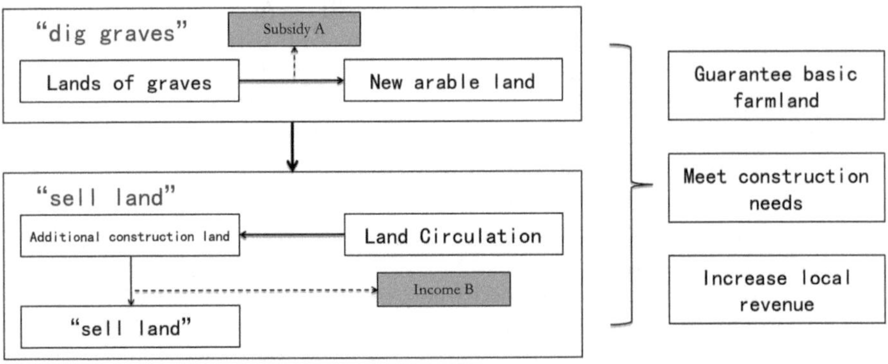

Fig. 6.2 Diagram of the "Dig Graves Movement"

through digging graves, local governments gain tremendous economic benefits because they can sell land for construction. In doing so, the governments also satisfy the large demand for land, which is needed to increase urbanization and industrialization. Through this process, the local government gains additional construction land and balances the total amount of basic farmland. Meanwhile, local governments also receive additional economic income because the income from "selling constructional land" is far more than the subsidy they allocate for "digging graves".

The "Dig Graves Movement" raised many concerns, but the core problem is related to the game playing of the arable land resources, especially for traditional agricultural areas, such as Zhoukou. The case study of Zhoukou discussed later in this paper will further explore this issue.

6.1 Object of Study

As for the allocation of land resources, contradictory choices between urban construction and farmland protection are problems faced by the urbanization and industrialization of China. This contradiction is more prominent in districts that mainly export rather than import grain products. This paper aims to study the allocation of land resources in the grain-export districts under the current development situation. Generally speaking, these grain-exporting districts have terrains consisting mostly of plains with undulations, and their traditional agricultural economy accounts for a large proportion of their GDP. These regions usually lag behind the eastern coastal areas in their levels of urbanization. The districts are just beginning industrialization, and their economy and society still show some typical characteristics of a traditional agricultural society (Liu Bin 2010).

One of the grain-exporting districts is Zhoukou in the Henan province. Zhoukou is located on the Huang-Huai plain, and its terrain is entirely flat. In 2012, Zhoukou city's total population was 12.38 million, and total area of arable land was over 12 million mu. The first industry in Zhoukou accounted for 28.1 % of the GDP, which is far more than the national average. The urbanization rate of Zhoukou is 31.29 %, which is far lower than the national average. The large proportion of the GDP held by the first industry and the low rate of urbanization both indicate that Zhoukou is still a traditional agricultural district. Combining the specific case of Zhoukou, this article will focus on the paradox of land resource allocation's game playing in districts that export grain.

6.2 Paradoxes of Land Resource Allocation in Zhoukou

6.2.1 Stage Characteristics of Zhoukou

In 2013, the GDP of Zhoukou city reached 179.07 billion yuan, which was a growth rate of 9.4 percentage points over the previous year. As well, the retail sales of social consumer goods achieved 75.68 billion yuan, which was a large increase. Accordingly, the Zhoukou local government's public finance income reached 7.6 billion yuan, its urban per capita disposable income reached 18,046 yuan, and its rural per capital disposable income reached 6950 yuan. These economic figures demonstrate that Zhoukou is at a turning point moving from being an agricultural society to an industrial society, and its economy will likely continue to grow rapidly. The agricultural industry of Zhoukou consists of two main development characteristics that have shaped its economic growth, which are discussed next.

6.2.1.1 Large Rural Population, But Not Used as an Advantage

Zhoukou City, with a population of more than ten million, has a Children Dependency Ratio as high as 33 %, 2009 (Office of Zhoukou Local Chronicles 2009), placing it first in Henan province. The high Children Dependency Ratio means that Zhoukou city has advantages over other districts in terms of both the total population and the supply of population of working age. The urbanization rate of Zhoukou city is only 33.44 %, which means nearly 70 % of the population live in rural areas, and large portions of the rural population are agricultural (Fig. 6.3).

According to the sixth nationwide census of the population, the emigrations from Zhoukou to other places had been increasing every year. Nevertheless, industry, which requires a high-quality labor force, can only absorb small numbers of rural workers. As Fig. 6.4 shows, from 2005 to 2009, the total percentage of secondary industry workers rose only by 0.19 percentage points (Office of Henan Local Chronicles 2013). Therefore, although Zhoukou has a large population, much of the population is from rural areas and industry alone cannot merely absorb the surplus rural labor. As a result, Zhoukou faces a dilemma of having both a "labor shortage" and "unemployment". Therefore, the population advantage of Zhoukou has not yet been fully realized.

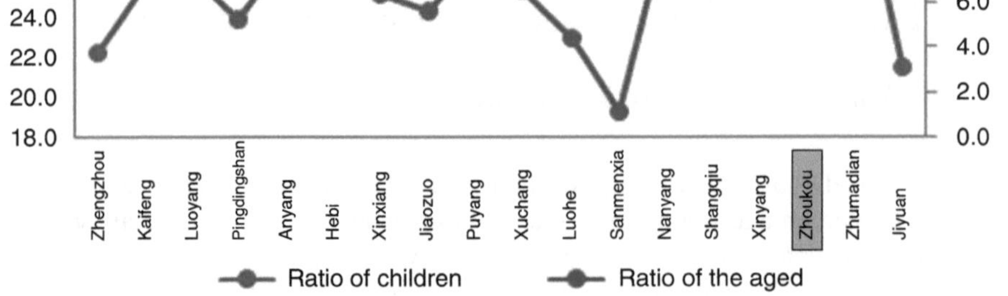

Fig. 6.3 Ratio of children and the aged for cities in Henan province, 2009

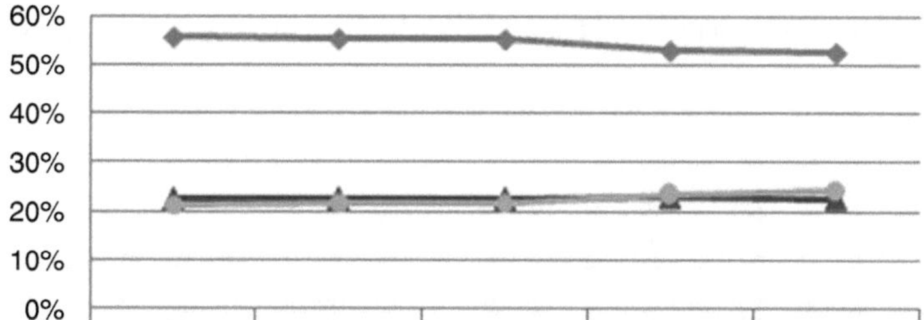

	2005	2006	2007	2008	2009	Growth rate
1st industry	55.87%	55.54%	55.46%	53.21%	52.78%	−3.10%
2st industry	22.71%	22.77%	22.81%	23.11%	22.90%	0.19%
3st industry	21.42%	21.70%	21.73%	23.68%	24.32%	2.90%

Fig. 6.4 Evolution of the proportion of labor in Zhoukou's three industries, 2005–2009

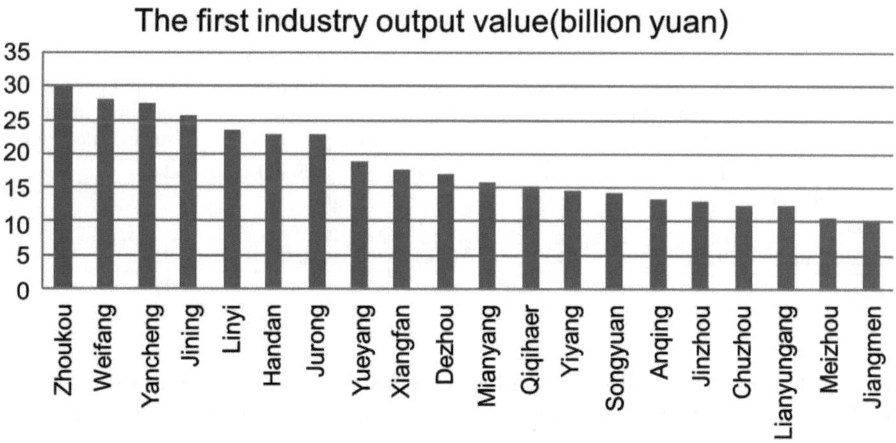

Fig. 6.5 First industry output value of important agriculture cities in 2008

6.2.1.2 First Industry Accounts for a Large Portion But Has Low Productivity

Although Zhoukou city has advanced the process of industrialization, the first industrial output has declined from 1 year to the next and still consists of a larger portion of the regional economy. In 2010, the first industrial output of Zhoukou accounted for 29.3 % of GDP, while the first industrial employment population accounted for 53.5 % of total employment.

As shown in Fig. 6.5, in 2008, the first industry output value of Zhoukou city reached 29.82 billion yuan, making it the largest agricultural city in the nation (Xiaolong Chen 2010). However, Fig. 6.6 shows that the first industry's production efficiency of Zhoukou was merely 8372 yuan per labor, and Zhoukou ranked at the

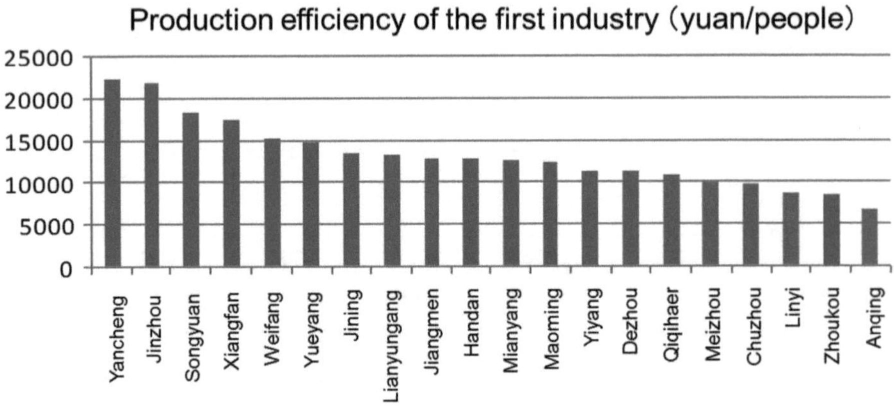

Fig. 6.6 Important agriculture cities' production efficiency of the first industry in 2008

Table 6.1 Comparison between construction land and arable land in Zhoukou, 2009

	Output value (billion yuan)	Labor (thousand people)	Land area (square kilometers)	Output per people (yuan/people)	Output per land (million yuan/square kilometers)
Urban construction land	11.06	436.4	44.4	**25,300**	**249**
Arable land	29.82	6430	8541	**4637**	**3.5**
Ratio	–	–	–	**5.5**	**71.2**

bottom for production efficiency among the important agricultural cities. Having inefficient production in Zhoukou is not conducive to the development of agricultural industrialization because it results in higher agricultural product costs.

6.2.2 *The Paradoxes of Land Resource Allocation*

Land resources mean that the land can be used now or in the foreseeable future to facilitate the development of society. From the perspective of labor productivity, land resources have economic value and allow for the means for production.

Zhoukou City owns land of 120 million square kilometers. The land is one of the most important development resources for Zhoukou. Zhoukou city's land resources are mainly used for two types of activities: urban construction and agriculture. To illustrate, in 2009, 44.4 km^2 land was used for urban construction and 8541 km^2 of land was used as arable land. Table 6.1 compares the output of urban construction land and rural arable land. Surprisingly, the labor of the arable land was 5.5 times the labor of urban constructional land and the output of rural arable land per land is only 1/70 of urban construction land.

Because the marginal economic benefits of construction land are much greater than the marginal economic benefits of farmland, the local government tends to allot more land resources to construction rather than farming. Meanwhile, according to *Function Zoning of the Major Function Development-optimized County*, Zhoukou is located in the "huang-huai plain", which is one of the main grain-producing areas. It provides an important national base for grain and food production, and it needs to provide more than seven billion catties of commodity grain for the entire nation every year. The national policy established by the "main agricultural regional division of labor" requires that Zhoukou promote the industrialization and urbanization while maintaining its responsibility of guaranteeing food security for the country. Therefore, promoting industrialization and urbanization by sacrificing agriculture, as was done by the eastern coast cities when their industrialization and urbanization began, will not work in Zhoukou (Fig. 6.7).

Thus, the Zhoukou government faces two contradictory forces in the process of land resource allocation. On the one hand, it wants to promote the local economy industry development by allocating more land resources toward urban construction because its marginal economic benefit is much higher than that of rural arable land. On the other hand, it is constrained by the central government's farmland protection policy that requires it retain a defined and a large amount of land resources as arable land, which in turn, increases the opportunity cost of the local government's land resource allocation.

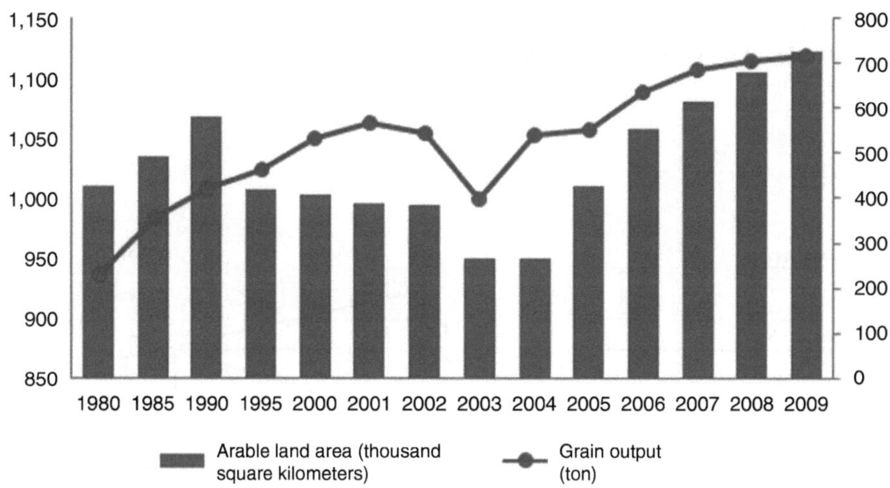

Fig. 6.7 The arable land area and grain output of Zhoukou from 1980 to 2009

6.3 Cause Analysis of Paradoxes

One of the reasons for the paradox facing Zhoukou regarding its land resources is that the positive externalities generated by the process of protecting farmland do not receive appropriate compensation even though the external effects related to food security is a type of public good.

6.3.1 Food Security as a Type of Public Good

Food is a special commodity that has both general merchandise and public good attributes. Food sold on the market is purely a private good and has the characteristics of private goods. For example, food as a private good may be "split". However, other food acquired by the government in case of famine or food reserves kept for natural disasters has some of the "indivisible" features of public goods. For example, in the case of famine, the state grain reserves would provide a basic food supply for every citizen. As a result, food security has both "non-exclusive" and "non-competitive consumption" features (Xinhua Zhu 2008) (Fig. 6.8).

For many countries, the national food policy historically has proven that food security has the characteristics of a public good. When it acts as a private good, food supply can be automatically adjusted by the market mechanism without the need for government intervention. However, the market mechanism could fail in regards to arable land protection leading to externalities, such as lack of key types of food, and therefore, for food safety, the supply of food requires a government guarantee to ensure the protection of farmland. As a result, food security acts as a type of public good.

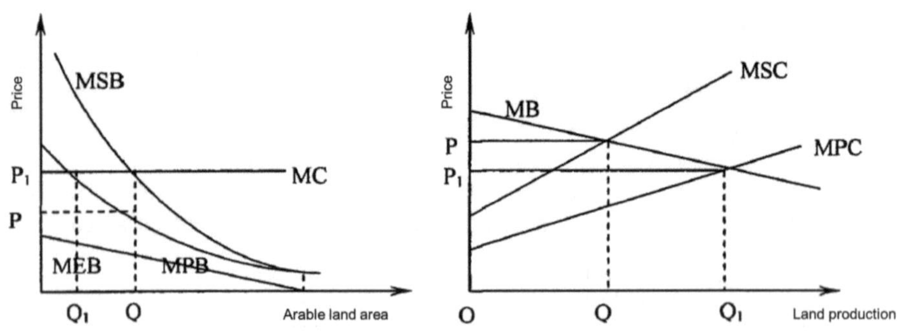

Fig. 6.8 Externalities of protecting arable land (Reproduced from Galloway 2001)

6.3.2 *Loss of Compensation Mechanism for Land Protection*

Arable land protection is an important means to ensure food security remains a public good. Under the current pricing mode, grain-planting income cannot reflect the intrinsic value of arable land because food is inexpensive when priced as a private good (Galloway 2001).

Because the marginal revenue of farmland protection is divided from the marginal social benefits, revenues or costs of individual land-use behavior are shared or assumed by other members of society. This structure is not conducive to the efficient allocation of land resources and has led local governments of traditional agricultural areas to pay for the country's food security. As a result, the local governments pay for the corresponding opportunity cost of land resources, but are unable to obtain sufficient external compensation.

6.3.3 *Verification of Paradox*

Table 6.2 shows a comparison between the national average and Zhoukou city data. Compared with the proportion of urban construction land nationwide, the proportion of urban construction land in Zhoukou is only about half. Meanwhile, the proportion of rural arable land is five times the national average. Therefore, with such large portions of traditional agricultural areas in Zhoukou, the district pays a large opportunity cost to protect arable land, and benefits spillover when dealing with the allocation of land resource in this district.

The imbalance between grain production districts and grain consumption districts exacerbates the imbalance of economic development and lessen the interests of grain production districts. Grain production districts have sacrificed opportunities to develop its industry by instead adjusting their rural districts to maintain national stability and grain acreage to ensure a steady supply of food. Steps should be taken to better address the unequal sacrifices in these districts resulting from this phenomenon. The example presented next section offers some insights into mechanisms that deserve consideration.

6.4 Learning from the CAP Case

6.4.1 *Case Introduction*

The Common Agricultural Policy (CAP) is the agricultural policy of the European Union (EU). It implements a system of agricultural subsidies and other programs. It was introduced in 1962 and has undergone several changes since then (Wikipedia

Table 6.2 Comparison of land resources' allocation between national average and Zhoukou, 2009

	Total area (thousand square kilometer)	Population (million)	Construction land (square kilometer)	Arable land (million mu)	Construction land per people (square kilometer/ thousand people)	Percentage of construction land	Arable land per people (mu/thousand people)	Percentage of arable land
Zhoukou	11.9	10.91	429.3	11.7	**3.9**	**3.61‰**	**612,450**	**65.54 %**
Nationwide	9600	1353	70,000	1826	**5.2**	**7.29‰**	**59,930**	**12.68 %**

2014). The CAP budget as a percentage of the total EU budget has declined over the years. In 1984 it was 71 %, in 2010 it was 42 %, and in 2013 it was 39 %.

6.4.2 *Mechanism of CAP*

CAP recognized the need to account for the social structure of agriculture and for the structural and natural disparities between the various agricultural regions to make appropriate adjustments. It is an integrated system of measures that works by maintaining commodity price levels within the EU and by subsidizing production.

CAP consists of a number of mechanisms. First, farmers can receive a subsidy for farm production. Each country can choose whether the payment will be established at the farm level or at the regional level. Farmers receiving the subsidy have the flexibility to produce any commodity on their land except for fruit, vegetables, and table potatoes. In addition, they are obliged to keep their land in good agricultural and environmental condition. Farmers have to respect environmental, food safety, sanitary, and animal welfare standards. Penalties exist for noncompliance. Specifically, if farmers do not respect these standards, their payment will be reduced.

The direct subsidies for farmers were originally intended to encourage farmers to choose to grow crops with subsidies and to maintain homegrown supplies. Subsidies were generally paid based on the area of land growing a particular crop rather than on the total amount of crop produced. Reforms implemented starting in 2005 have been phasing out specific subsidies in favor of flat-rate payments based only on the area of land in cultivation and on the requirement that farmers adopt environmentally beneficial farming methods. The change is intended to give farmers more freedom to choose to grow the crops most in demand and therefore, reduce the economic incentive to overproduce (Fig. 6.9).

Another mechanism of CAP is that import levies and quotas are applied to specified goods coming into the EU. The levies are set to raise the world market price to the EU target price. The target price is chosen as the maximum desirable price for goods within the EU (Grant 1997). Import quotas are a means of restricting the amount of food being imported into the EU. Some non-member countries have negotiated quotas that allow them to sell particular goods in the EU without tariffs. This notably applies to countries that had a traditional trade link with a member country.

CAP also establishes an internal intervention price. If the internal market price falls below the intervention level then the EU will buy up goods to raise the price to the intervention level. The intervention price is set lower than the target price. The internal market price can only vary within the range between the intervention price and target price (Fig. 6.10).

CAP also introduced production quotas and "set-aside" payments in an effort to prevent overproduction of some foods (for example, milk, grain, and wine) that attracted subsidies well in excess of market prices. However, a secondary market

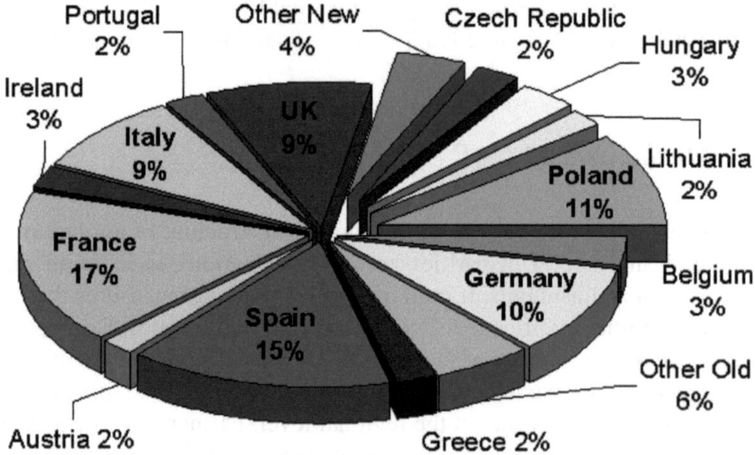

Fig. 6.9 Percentage of EU farmland by country (Wikipedia 2014)

Fig. 6.10 CAP beneficiaries 2004 (Wikipedia 2014)

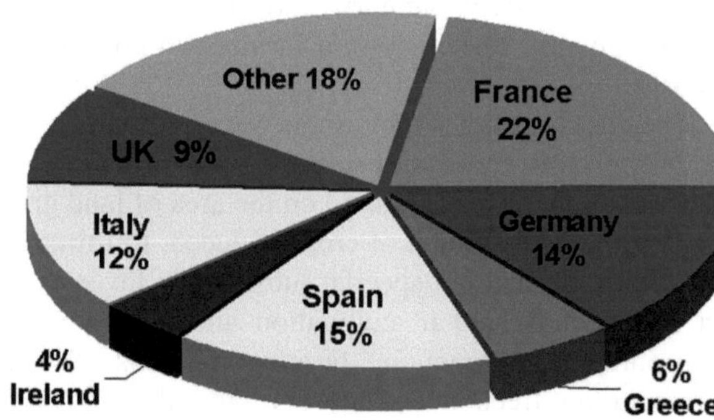

has evolved, especially in the sale of milk quotas, and some farmers have made imaginative use of "set-aside" payments, for example, setting aside land that was difficult to farm. Currently the set-aside payments have been suspended as decisions about their future are being made amidst a context in which the prices for some commodities rise and interest in growing biofuels increases.

6.4.3 Case Summary

Although CAP has been highly controversial since its inception, the interests of EU food production and marketing, in general, have improved considerably. CAP has prompted the coordination of regional economic disparities and stabilization of the agricultural markets, which has benefitted EU food production and marketing. Also, CAP has successfully balanced both competition in European agriculture and the world market.

A key concern in China around food production is the existence of regional production and imbalance in demand. The EU Common Agricultural Policy Coordination Mechanism production provides a useful example for how to achieve a basic balance between grain production districts and grain consumption districts. In China, a feasible externality benefit compensation mechanism for farmland protection is needed to reduce the "free rider" of food security, solve the "efficiency" spillover problem and balance regional development resources.

6.5 Strategies and Prospects

This paper uses the case study of Zhoukou to illustrate the land resource allocation paradoxes faced by grain producing areas. The dilemma has arisen, in part, because grain-producing areas pay a huge opportunity cost of land resources to maintain supplies needed for food security, but they were not getting adequate compensation. A compensation mechanism of spillover benefits is missing.

Two suggestions are proposed. The first solution is to (1) prepare for the establishment of regional land resource allocation and coordination mechanisms, (2) formulate the external compensation standard of internal effects, and (3) promote the balance of regional development while improving the overall benefits. The second is to (1) seize the opportunity of system reforms, (2) promote the development of system innovations, (3) improve the benefit of arable land resources.

References

Feng Wang. (2012). *Henan Dig Graves movements: Land reform or funeral? 21st century business herald* (Vol. 2015). http://finance.ifeng.com/roll/20121129/7362118.shtml

Galloway, J. N. (2001). Acidification of the world: Nature and anthropogenic. *Water, Air and Soil Pollution, 130*, 17–24.

Grant, W. (1997). *The common agricultural policy*. New York: Macmillan Press Ltd.

Henan Daily. (2014). *Arable land's pothook of increase or decrease in Henan province* (Vol. 2014). http://newpaper.dahe.cn/hnsb/html/2012-11/29/content_818085.htm?div=−1

Liu Bin. (2010). *Comparative study on agricultural areas industrialization*. Harbin: Heilongjiang University.

Office of Henan Local Chronicles. (2013). *Henan Province, Henan Statistical Yearbook.*

Office of Zhoukou Local Chronicles. (2009). *Zhoukou City, Henan province, Zhoukou yearbook.*

Ping Chen. (2012). Game playing of the land –Zhoukou dig graves movement. *Chizi, 2013*(12), 34–35.

Wikipedia. (2012). *Henan Grave Removal Project – Wikipedia, the free encyclopedia* (Vol. 2012). http://en.wikipedia.org/wiki/Henan_Dig_Graves_Movement

Wikipedia. (2014). *The common agricultural policy – Wikipedia, the free encyclopedia* (Vol. 2014). http://en.wikipedia.org/wiki/Common_Agricultural_Policy

Xiaolong Chen. (2010). *The editorial board and the editorial department China City Statistical Yearbook*. China Statistics Press, Beijing.

Xinhua Zhu. (2008). *Studies on external compensation of arable land protection for food security*. Nanjing: Nanjing Agricultural University.

Chapter 7
Conflicts in the Urban Renewal of the Historic Preservation Area—Based on the Investigation of Nanbuting Community in Nanjing

Meicheng Wang and Tian Ruan

7.1 Introduction

With the transformation of urbanization in China, the focus of urban development has gradually transferred from the growth of total construction land to the quality improvement of built-up land (Zhang et al. 2008). Urban planning in this context needs to focus on the renewal of old communities, especially those with historic preservation. The deprivation of old communities in Chinese cities is widely noticeable, but neither community-led renovations nor government-led renewal favored by local residents are often observed. On the one hand, when local governments plan to upgrade infrastructure and buildings in a deprived block, they have to address complicated issues of property rights due to the transformation of China's land system in the past century (Zhao et al. 1998; Wu 2002).

Both approaches involve redistribution of spatial resources among stakeholders. Local residents, governmental agencies and real estate developers all seek to maximize their individual benefits during the transaction of property rights (Zhang 2004). In international practices, scholars frequently report and criticize forced evictions conducted by "demolition and relocation offices/companies" (Hooper and Ortolano 2012). Local residents choose to react against demolition by "petitioning" or violent resistance, which draws extensive attention from the public and the government at various levels. Complicated conflicts among stakeholders require urban planners to look beyond the chaos and refer to the nature of the planning process: what are stakeholder's interests? Where do their interests

M. Wang (✉) • T. Ruan
Nanjing University, School of Architecture and Urban Planning,
Nanjing, China
e-mail: meichengwang09@gmail.com; ruantian2010@gmail.com

Q. Pan, J. Cao (eds.), *Recent Developments in Chinese Urban Planning*,
GeoJournal Library 114, DOI 10.1007/978-3-319-18470-8_7

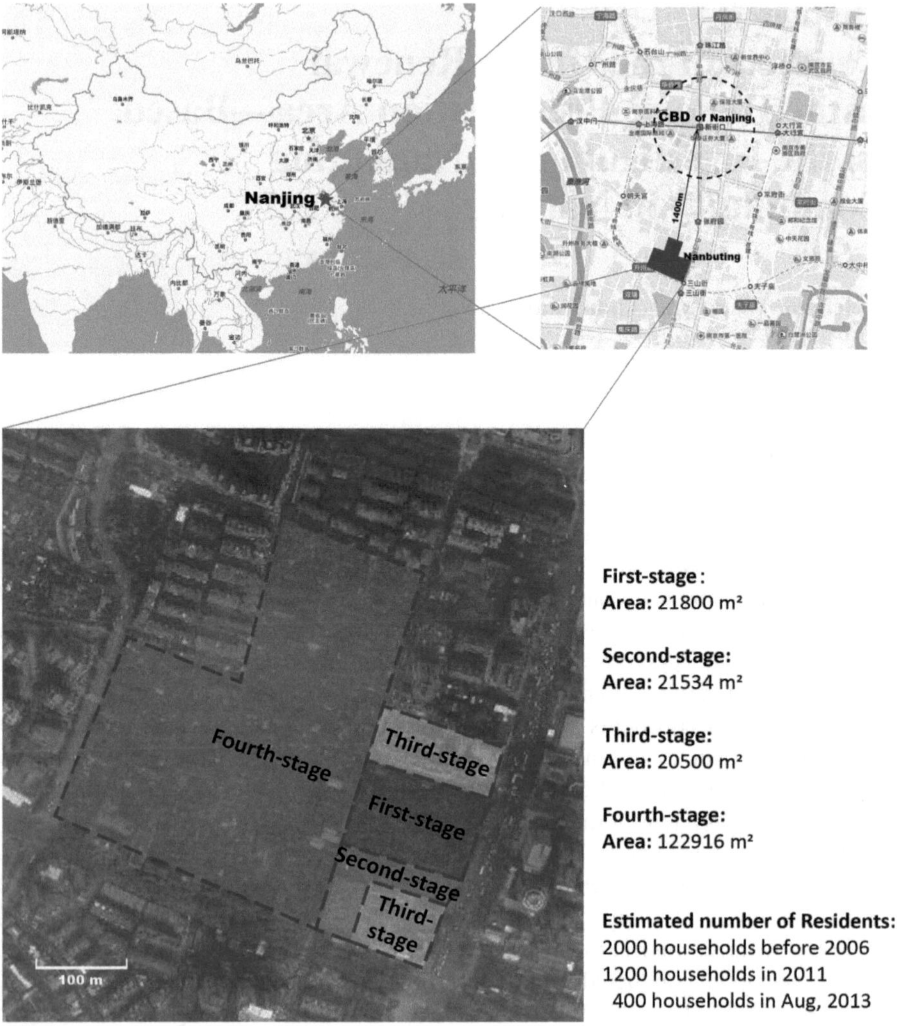

Fig. 7.1 Information of Nanbuting in Nanjing, China

intersect and conflict? In addition what actions should planners introduce to the mediation process (Forester 1987; Healey 1997; Susskind et al. 1999)?

This article analyzes these issues by investigating a community renewal case of a deprived historic block named Nanbuting, located 1.4 km away from the CBD of Nanjing (Fig. 7.1). In its 10-year demolition process, the number of households in this area dropped from more than 2000 (before 2006) to approximately 400 (August, 2013) (Fig. 7.2). This block is located in the traditional residential area of Nanjing since the early Ming Dynasty, and has maintained high commercial vitality during the late Qing Dynasty and the Republic of China (Zhou and Zhang 2010).

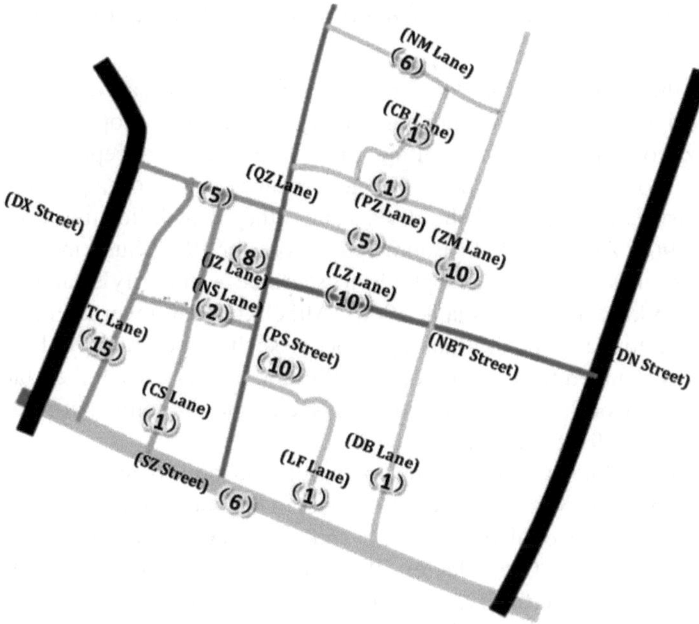

Fig. 7.2 Map of the questionnaire surverying (with the ground map derived from Municipality of Nanjing 2012c; and name of streets derived from Lv 1991)

Compared with other historic residential blocks in Nanjing, Nanbuting is the most adjacent to the CBD of Nanjing, thus is regarded as an important target for real estate development. Meanwhile, it owns the most satisfying traditional texture, thus a large number of scholars advocate for protecting it from for-profit development.

7.2 Conceptual Frameworks

This article builds a conceptual framework on three bodies of literature. The first part reviews an evolving discussion on urban planning and public policy-making process, which provides this case with a general cognition of its movement in every stage. Then to understand core values in this urban renewal case with historic preservation appeals, this article introduces the concept of land development rights and its international practices. Finally, to define and analyze multi-stakeholder conflicts, this article draws on lessons from conflict assessment, mediation, and participatory planning theory. These three parts of literature jointly provide perspectives to explore conflicts in this case and define the significance of conflict study in the urban planning process.

7.2.1 Urban Planning and Public Policy-Making Process

Observing the failures of top-down planning strategies, urban development projects, especially at the local and community level, have adopted lessons from decentralization of the policy-making process and the concept of governance (Innes and Booher 1999). Under the governance framework, public decision-making must value every stakeholder's voice other than following any authority (Davidoff 1965; Arnstein 1969; Baum 1997; Forester 1987; Susskind et al. 1999). Furthermore, it is no longer acceptable to sacrifice a minority's interests for the majority (Susskind and Cruikshank 2006). Although local governments hold most responsibility to direct policy-making process such as community development, their role has changed from regulator to facilitator, and as a result they face a more comprehensive workload with more diverse social and economic participators than before (Peterson 1995; Thatcher 1998).

Forced by problems in practicing, planning and policy-making theories have accordingly evolved. Many scholars contribute to the knowledge of why and how to build consensus, recognize conflicts and facilitate collaboration (Susskind et al. 1999; Forester 2006; Healey 1997; Innes and Booher 1999; Hajer and Wagenaar 2003). Specifically, Susskind et al. (1999) identifies critical steps throughout consensus building, in which conflict assessment is placed as an initial procedure for further convening and mediation. He defines interests as "participants' underlying values and needs" (Susskind et al. 1999, p. 6). Practically, Forester (2009) contributed substantial records of how planners deal with conflict in their practices, in which he extracted a notion that indirect approaches to mediate participation works better than merely narrowing discussion to the conflict itself.

Enlightened by the lessons above, it is necessary for a conflict study to first identify stakeholder's interests (their values and needs); but for planners and policy-makers, the more practical task is to propose accessible ways of moving forward, building consensus and achieving development beyond conflict.

7.2.2 Core Value Behind Urban Renewal Conflict: Land Development Right (LDR)

The concept of land development right (LDR) comes from the density transfer mechanism for protecting landmarks in New York City, 1968 (Mills 1980). Chinese scholars pay attention to this because the Chinese institution of land leasing is gradually not able to adapt to the radical change in urban development. In China, the land right contains many parts, but only the land use right is able to be traded in the land market. Developers need to acquire the land use right through a land-leasing progress, which indicates that the land use right consists of "rights of use" and "rights of development" (Zhu 2004). Hence the land-leasing progress is featured by high costs, especially in the central areas of cities. Land developers need to improve the intensity of development or reduce the compensation to the former land use right owners to balance their cost and revenue. This causes many problems such as conflicts of demolition and destruction of historic areas.

Land development right, defined by Chinese scholars, is a part of land ownership that reflects the incremental earnings due to the dynamic use of land (Zhang 2004 and Zhao et al. 1998). Guo (2007) interpreted the land development right with three situations: (1) changing land-use types, e.g., to convert residential lands into commercial lands; (2) raising land development intensity, e.g., to convert low-rise buildings into high-rise ones; (3) investing lands without changing their use or development intensity, e.g., to convert run-down houses into upscale villas.

Many scholars regard the separation and transfer of the LDR as an efficient method to manage land use, which introduces advantageous application for urban development and became commonly known as transferable development right (TDR) (Mills 1980; Timothy 1998; Kaplowitz et al. 2008; Wang et al. 2010). Timothy (1998) believes that the TDR is an effective method to protect farmland. Wang et al. (2010) used a case in Zhejiang to demonstrate that the transfer of LDR is able to improve the flexibility of land use, and thus liberates the local development from regulatory constraints.

One essential function (or advantage) of TDR is historic preservation which has attracted scholars' attention since the beginning (Niemann 1978; Allison 1996; McConnell and Walls 2009). TDR is able to overcome the above shortages of the land leasing. For example, Niemann indicated that TDR is not only effective for the urban landmarks' preservation, but it also liberates the burden from the landmark owner. McConnell and Walls (2009) also described the practices of TDR in the U. S.: development rights binding with acknowledged historic preservation appeals are able to be transferred to other areas, where building restriction is less intense and thus more suitable for new development.

7.2.3 Beyond Conflict Assessment: Participation in Planning

As early as five decades ago, Davidoff and Reiner (1962) highlighted the diversification in modern society. They regarded the planning process as a series of choices driven by different values. Many scholars started to explore the possibility of reaching consensus and rectifying notions through practices. Arnstein (1969) warned that the participation process must witness the redistribution of power; otherwise there is no evidence of its significance. Healey (1997) also argued that planners should achieve "enabling" more than "controlling" when referring to the notion of involvement. Thus, it is not the gesture of convening that helps reach consensus, but meaningful interactions that facilitate communications among stakeholders.

Besides contents and meaning, scholars also stressed the significance of time dimension during participation processes. Jürgen (1985) pointed out that any current consent is incomplete and temporary because no one can go further than the time flows, thus continuous community participation is always required to reach a better vision, other than a perfect state. His "communicative theory" supports Harvey's view that community is not fixed but constantly interacting with its temporal background (Harvey 1997). These insights require a sustainable system for flexible, long-term and robust involvement. From another aspect, this continuousness actually accelerates the planning process in the long-term because it initially establishes a constructive framework and therefore saves money as well as time and finally helps to formulate a virtuous circle for planning as a whole in the local area (Baker et al. 2007).

7.3 The Four-Stage "Renovation Project of Nanbuting" in Nanjing

This case is representative because demolitions in this community drew the attention of China's previous Premier twice in the last decade. The demolitions resulted from a four-stage "Renovation Project of Nanbuting" (Fig. 7.1), conducted by a state-owned company named Nanjing Urban Construction Investment Holding (Group) Co., Ltd. (NUCIH). NUCIH was established in 2002 for the purpose of investing and managing the infrastructure and public utilities in Nanjing, and was in charge of the Nanbuting project on behalf of the municipality.[1] In the same year, the municipality revised the "Preservation Planning of Historic and Cultural Famous City of Nanjing," identifying several blocks in the traditional south area of Nanjing as "historic preservation areas" including Nanbuting (Municipality of Nanjing 2012a).[2] The 1st-stage Nanbuting Project was completed in 2002, renovating the historic "Ganxi Mansion" and opened it to the public as a historic museum.

Soon in 2003, the municipality appointed Southeast University (in Nanjing) to make a detailed preservation and renewal plan for Nanbuting. The plan was submitted in 2004 but was not implemented. Two year later, NUCIH conducted a 2nd-stage Nanbuting Project in the name of preservative development, but in fact the nature of the 2nd-stage project was reconstructing an area to the south of Ganxi Mansion into a commercial block. Upon the reconstruction of this commercial block, the municipality announced that the large areas near the 2nd-stage project were to be demolished, which caused strong opposition from sixteen scholars and experts in China, including two prestigious academicians. These scholars appealed to Premier Wen Jiabao to stop the commercial development in this historic block. The Premier responded to their letters and asked the municipality to conduct a detailed research before further actions (Xie 2011). As a result, the Nanbuting

[1] NUCIH was a product of "government-enterprise separation", more details at http://www.njcjjt.com.cn/

[2] The official plans, unless specifically noted, can be reviewed at the website of Nanjing Urban Planning Bureau: http://www.njghj.gov.cn/ngweb/page/index.aspx

Table 7.1 Timeline of the plans and demolitions regarding four areas

	Publicity of the plan	Action of demolition been conducted	Decision of demolition been made
1st-stage project	2003	–	–
2nd-stage project	2006	2003	2003
3rd-stage project	Aug 2012	Aug 2006	June 2006
4th-stage project	Jan 2013	Jan 2009	Mar 2009

Project was set aside for 3 years, while demolitions for other development projects were under implementation in similar historic blocks in Nanjing.

In early 2009, the municipality suddenly announced a new project named the "Renovation of Dilapidated Buildings," in which all of the 3rd- and 4th-stage areas of Nanbuting (Fig. 7.1) were included.[3] Although more than a hundred local residents of Nanbuting immediately appealed to the local government to stop this substandard project, the "demolition team" went to work on the block, causing violent forced evictions in March (Chen 2011). Soon 29 experts appealed to the central government in April. In May the Premier sent an investigation team to Nanjing. Although the project was under examination, forced evictions did not stop; in July, a real estate developer published that there would be top-level villas in Nanbuting block, with an estimated value of 40,000 yuan/m^2.

From 2010 to 2013, a large number of houses in Nanbuting were gradually demolished. Meanwhile multiple preservation plans of Nanbuting or related to Nanbuting were made and published. At the beginning of 2013, the 4th-stage project plan was announced (Municipality of Nanjing 2013; see also Liu 2013), followed by a new round of demolitions in May.

After reviewing the planning and implementation of the four-stage renovation project, we found that demolitions were announced before stakeholders were informed of the plans for renovation (Table 7.1). The renovation was implemented without sufficient negotiations and proper stakeholder involvement, and severe conflicts were observed among local residents, real estate developers, governmental sectors and scholars. The purpose of this article is to analyze the key issues underlying the conflicts among stakeholders, and to discuss why similar conflicts related to community renewal have become pervasive in China's urbanization.

[3] The Baixia district government did announced strict regulation in early 2009 (See http://www.njpf.gov.cn/Web/Info/2009-03/261555068593.html), but still the incidents in 2009 was pernicious. Baixia district has been merged into Qinhuai district in 2013.

7.4 Data and Methodology

This article applies both qualitative and quantitative analytical techniques over 18 months of investigation. From March 2012 to June 2012, the research team mapped, evaluated and analyzed historic buildings and streets and interviewed households mainly by questionnaire survey. Then from March 2013 to September 2013, the team conducted in-depth interviews with local residents, governmental officials, real estate developers, and critical scholars who were directly involved in the 10-year renewal process, in order to discover their respective interests and demands.

Quantitatively, the research team mapped and evaluated 133 listed historic buildings in the 4th-stage project area and classified its 16 streets using cluster analysis, combined with qualitative ways. Four featured areas were identified.

Qualitatively, questionnaire surveys of local residents were conducted twice. The first round of surveys were conducted between March and June of 2012, with a sample of 94 households (Fig. 7.2); the second round of surveys were gathered from 22 households in August 2013. Only one individual in each household was interviewed, but samples in two rounds of surveys overlapped. In the first round of surveys, housing conditions were recorded including areas, years and generations of living, and properties of houses. Then concerning their housing conditions, the team evaluated residents' attitudes towards their community situations, the development project and corresponding demolitions. In the second round of surveys, the research team focused on residents' feedback on the demolitions starting in May 2013.

During the first-round of surveys, the team also used questionnaires to investigate the cognition of Nanbuting among citizens in Nanjing, with a sample of 109 individuals in the Xinjiekou CBD, in comparison with that of local residents. In-depth interviews were conducted with two voluntary leaders of residents, four scholars who were directly involved in Nanbuting Renovation Project, three governmental officials in charge of the project, one manager of NUCIH.[4] Interviewees were not given any financial incentive for participation. The transcripts reflected their opinions on the project and the demolitions, and were regarded as an important resource of timelines and events.

[4] Because of the sensibility of the conflicts, all interviewees, unless specifically noted, are anonymous in this article. The research team keeps all transcripts.

7.5 Features of the Historic Block and Residents

7.5.1 Location and Development Restrictions

The location of Nanbuting determines its high land value in central Nanjing. Housing price around Nanbuting had risen from 3000 yuan/m^2 in 2003 to 25,000–30,000 yuan/m^2 in 2013 (Chen 2011). However, due to various historic preservation plans and regulations, development of the block was strictly restricted with very limited floor area ratio (FAR).[5] Specifically, Nanbuting was classified into historic preservation areas in the "Preservation Planning of Historic and Cultural Famous City of Nanjing (2002)," and had been planned in detail three times since 2003.[6]

7.5.2 Deprivation and Segregation

7.5.2.1 Living Conditions

Nanbuting is a low-income community. In 2012, the average urban residents' per capita disposable income was 3,026.8 yuan per month (Nanjing Municipal Bureau of Statistics 2013), while more than half of the interviewees (57 %) reported their household monthly income below 3000 yuan (Fig. 7.3).

Local residents were noticeably unsatisfied with the sanitation and landscape of their community (scoring in a 1–7 scale). Participants reported being less unsatisfied with other infrastructure, but highly satisfied with their neighborhood relations. In particular, they were proud of the cultural influence of Nanbuting (Fig. 7.4).

7.5.2.2 Building Conditions

All of the 133 listed historic buildings were mapped and evaluated (Yao 2009). The research team recorded the street name and number of each building, status of occupancy, score of quality (1 – unbroken structure and facades, 2 – unbroken structure with broken facades, 3 – broken structure with existing facades, 4 – torn

[5] Floor area ratio (FAR) is flexibly regulated in China, compared with other institutions that designates fixed FAR for various land-use categories. FAR for blocks can be reviewed in corresponding regulatory plans.

[6] The three plans are (1) 2003 Preservation planning of historic and cultural famous city of Nanjing; (2) 2010 Historic preservation plan and urban design of Nanjing Old South District; (3) The four-stage "Renovation Project of Nanbuting".

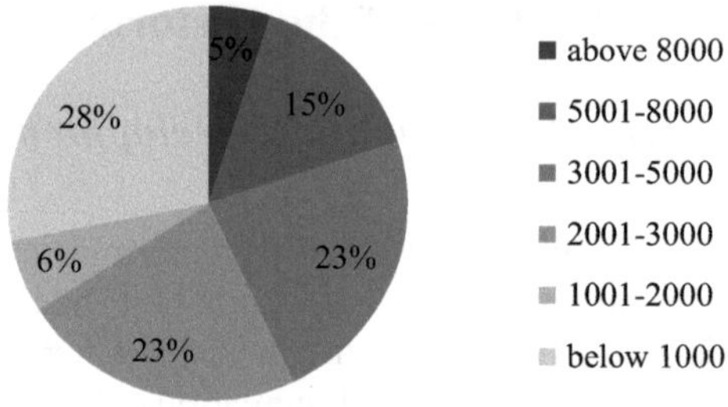

Fig. 7.3 Monthly income (yuan) per household

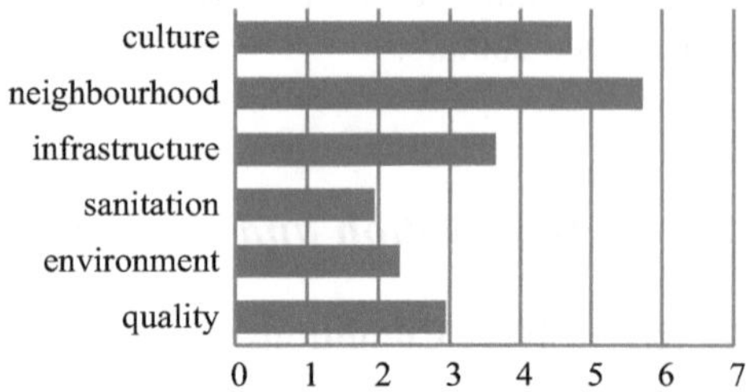

Fig. 7.4 Scores of the living quality by residents

down). The general quality of these listed buildings was inferior in June 2012, with the occupancy rate fewer than 70 % (Fig. 7.5).

7.5.2.3 Segregated Cognition and Social Activities

Although local residents showed confidence in the cultural influence of their community, the concept of "Nanbuting" was unfamiliar to other citizens in Nanjing. In contrast, 70 % of residents perceived that their community ranked "very high" or "high" popularity among citizens in Nanjing, while the actual number reflected in the survey was less than one third (Fig. 7.6). Meanwhile, local residents strongly held that their community has unique cultural values which attract tourists and cultural activities, but only 8.3 % of the interviewees visited Nanbuting for tourism. Instead, more than half of them reported that they came to Nanbuting for restaurants and shopping (54.5 %) (Fig. 7.7).

During the investigation, local residents demonstrated strong negative emotions against the surveys. A large number of interviewees questioned if the team was sent by governmental agencies to negotiate for demolitions. However, when they were informed that the team only consisted of undergraduate students and was for the sole purpose of research, they reminded team members to be careful about personal safety around the site. It was clear that after several rounds of demolitions, residents became very sensitive to outsiders entering their community.

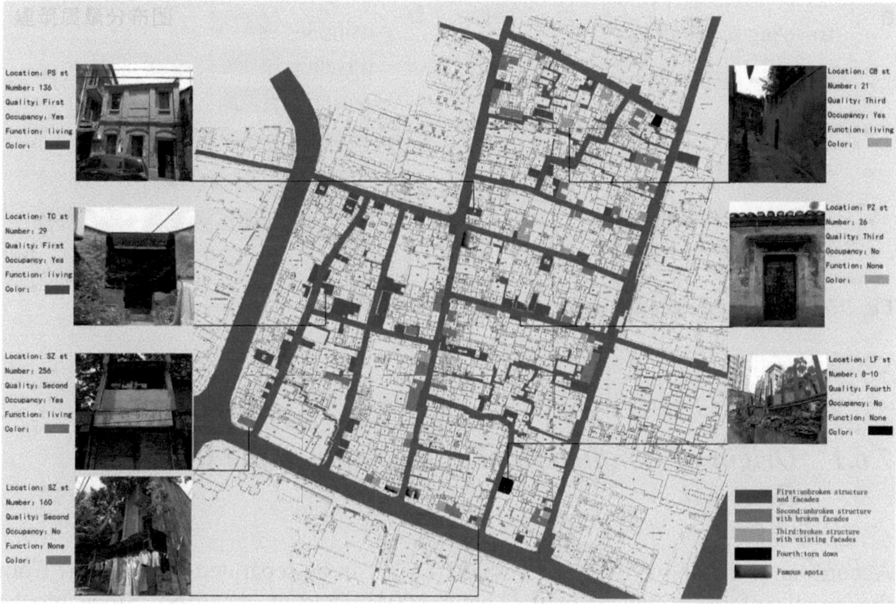

Fig. 7.5 Map of the quality of listed historic buildings (with the ground map derived from Municipality of Nanjing 2012c)

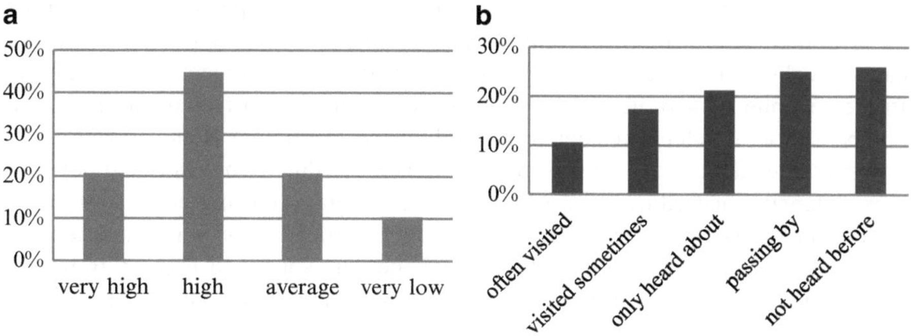

Fig. 7.6 (**a**) Evaluation of the block by local residents, (**b**) Evaluation of the block by outside citizens

7.6 Analysis of Multi-stakeholder Conflicts

The origin of the conflicts lied in the considerable potential benefits of community renewal. Conflicts were formed among different subsystems of all stakeholders. Common multi-stakeholder conflicts were exacerbated by voluntary community participation in the renewal process.

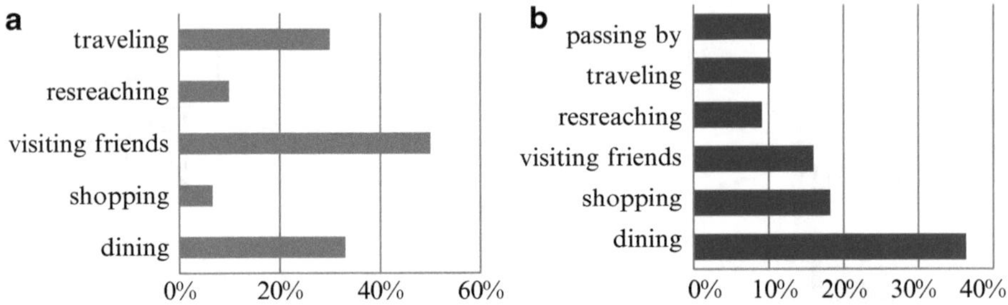

Fig. 7.7 (**a**) Local residents guessing visiting purposes, (**b**) Actual purposes of visitors

7.6.1 Origin of Conflicts: The Core Value in the Urban Renewal

According to the above literature review, LDR can be separated and sold off from the bundle of the rights of land (Zhang 2004; Zhao et al. 1998). Since blocks adjacent to Nanbuting were developed with high intensity, stakeholders held high expectations of potential earnings in redeveloping Nanbuting. However, there is a gap between expectations and the development intensity (e.g., FAR) allowed by historic preservation plans and regulations (Fig. 7.8). This gap is sometimes defined as "hypothetical development right" (Zhang 2004). In most Chinese urban renewal cases, local governments require capital balance in renewal projects, indicating that future revenue of a project should be no less than the cost of demolitions and compensations for local residents. Regarding the Nanbuting Renovation Project, however, the capital balance was difficult to achieve due to the restricted development intensity, according to an official in the Nanjing Housing and Construction Bureau and a manager of NUCIH. Given that no compensation was guaranteed for the hypothetical development right, all stakeholders sought to break through the restrictions set by preservation plans and to enhance their earnings: more revenue of land leasing for local governments, more compensations for local residents, and more commercial returns for real estate developers. On the contrary to their expectations, scholars and the general public suggested that although the development restriction seems to hinder the immediate earnings of each stakeholder, it must be applied since historic preservation will enhance the overall image of a city, which brings uncountable benefits (Zhou and Zhang 2010; Wang 2012). Therefore, unrealistic expectations from each stakeholder became the origin of conflict during the renewal process.

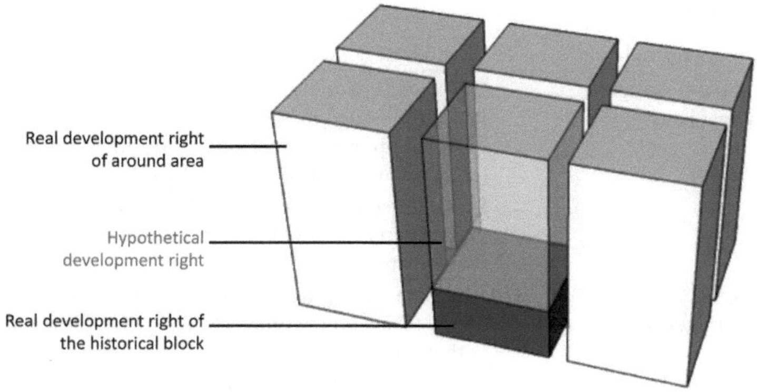

Fig. 7.8 Conceptual diagram of the development right

7.6.2 Formation of Conflicts: Different Subsystems of Each Stakeholder

The multi-stakeholder conflicts were then formed among different subsystems of all stakeholders. The investigation found that the complicated internal structure of each stakeholder greatly impeded the demolition negotiations and compensation.

For local residents, subsystems were classified according to different conditions of their housing property rights, which led to various appeals in the community renewal. Combining narrative accounts of scholars and local residents, two types of property rights were identified with various methods of acquisition (Table 7.2).

Property rights in Nanbuting were highly mixed. Only 17 % of the interviewees claimed owning complete property rights of their home. Nearly half (44 %) of the interviewees were renters, and many were not informed of who the owners were (Fig. 7.9). Residents with different types of property rights demonstrated various appeals (Table 7.3). The unified policy of demolitions and compensations implemented by the municipality could not meet their demands. Some private homeowner noted: "Since my house has been owned by several generations of my family, why can't I keep my own property? Or why should I accept the same compensation standard as those public house dwellers?" Even worse, the municipality chose to negotiate with each household in a non-transparent manner, which caused suspicion and mistrust among different groups of residents. Interviewees were not willing to reveal the offer that they received from the government, but many of them assumed that their neighbors might negotiate for a better offer.

Besides local residents, subsystems were also identified within other stakeholders. For local government, both land development revenues and public interests were concerned during the renovation project; various governmental agencies had

Table 7.2 Types of property rights' historical formation

Types of property rights	Methods of acquisition
"Private houses" (property owned by residents)	I. Acquired by inheritance – the house owner once held the land deed issued by the Republic of China, and thus received property ownership certificate issued by the new government of P.R.C at around 1951.
	II. Acquired by transaction – the house owner bought the property in housing market or was gifted the house by previous owner.
"Public houses" (property owned by the government)	III. Some houses were abandoned during the wars before 1949.[a] The new government then took over those houses without identified owners.
	IV. During the Socialist Transformation (in the early 1950s), some houses were confiscated by the new government.
	V. During the Great Leap Forward Movement (from 1958 to 1960), private houses larger than 225 m^2 were partly taken over by the government. Although nominal property rights still belonged to the private owners, actual ownerships became controversial since multiple users occupied these houses in the following decades.
	VI. During the Cultural Revolution (1966–1976), nearly all private houses were taken over by the Housing Authority. Most of these houses were successively required to be returned since the 1990s, but many of them are still owned by the Housing Authority.

[a]The People's Republic of China was established as 1949, marked as a turning point of land system reforms

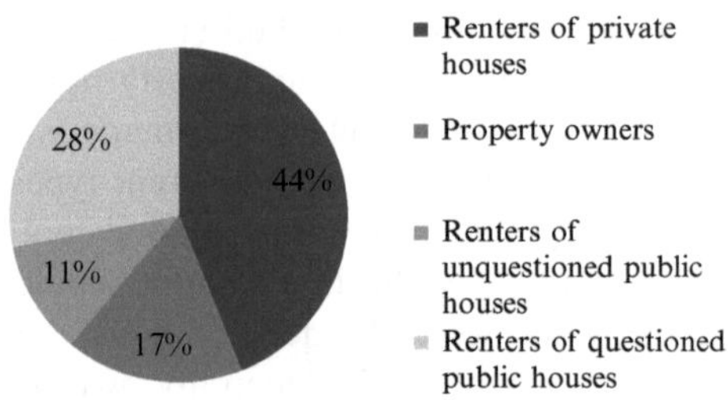

Fig. 7.9 Rate of different types of property right

disputes in renewal methods and implementation. The timeline (Table 7.1) reflects the incapability of planning regulations, which was respectively confirmed by an official of Nanjing Land Resource Center, a scholar in charge of the 2003 preservation plan of Nanbuting, and a scholar advocating for the 2006 petition to the central government. The interviewee working for Nanjing Urban Planning Bureau did not respond to the questioned timeline regarding the plan for publicity and project implementation.

Table 7.3 Feature of different types of property rights

Types of residents	Types of property rights	Features of residents	Appeals
Property owners	Type I & II houses (used by owners)	Their family has lived in the house for generations; they held a strong emotional connection with the community.	Keeping living in the block; protecting historic buildings and cultures.
Renters of unquestioned public houses	Type III & IV houses	They mostly moved into the community in the early 1950s with poor financial circumstances, and demonstrated an emotional connection with the community.	Meeting basic housing needs, preferably keeping living in the block; improving housing conditions.
Renters of questioned public houses	Type V & VI houses	They mostly moved into the community after 1958 with poor financial circumstances.	Meeting basic housing needs; improving housing conditions.
Renters of private houses	Type I & II houses (used by renters)	They were highly mobile and were not much affected by the renewal.	No particular appeals

To understand this situation, it is important to recall that the four-stage Renovation Project of Nanbuting was conducted by a state-owned company (NUCIH) running on behalf of the municipality. When the development decision was made, NUCIH could immediately run their business with real estate developers; corresponding decision of demolitions would be announced by the district government (lower level of the municipality). Since Nanjing Urban Planning Bureau was also subordinated to the municipality, it provided a plan when there "should have been a plan". Looking into this process, it is easier to understand that although all households in the 4th-stage renovation area were asked to move out since early 2009, a corresponding plan for this area was not published until January 2013, with the "vision" that 650 households should be kept in the block.

Scholars and experts influenced the Nanbuting project with different aims and expectations. On the one hand, they were able to communicate with local residents in field studies and enhance public opinions on official hearings; on the other hand, they were also capable of lobbying for commercial development by NUCIH. The team identified three types of scholars and experts during the investigations: (1) a senior urban planner who advocated for detailed surveys and planning for a rational project; (2) a policy analyst and a historian that strongly advocated for preserving the block and its residents as a whole; (3) a senior sociologist who advocated for commercial development. Interviews with scholars revealed that they believed they held the best solution for the project, which caused disputes among themselves, and consequently affected positions of governmental agencies as well as those of local residents.

In summary, conflicts in community renewal projects have long been seen as disagreements among multi-stakeholders, but our analysis found that the subject units of comprehensive conflicts are subsystems of stakeholders. Disputes are generated within each of them, and are formed based on unrealistic expectations held by all.

7.6.3 Exacerbation of Conflicts: Effect of Voluntary Community Participation

Three groups of voluntary participators were identified during the Nanbuting Renovation Project. In the absence of recognized participation platforms, each group brought more negative and uncontrolled effects than positive contributions.

The most noticeable group of voluntary participators were local residents, with violent resistance incidents reported in 2006. Residents were the most directly affected stakeholders but they were not officially involved in the decision-making processes of the renovation project. Two voluntary leaders were recognized by most residents who advocated for preservation and anti-demolition: Mr. X in Taicang Street (TC) and Mr. Y in Qianzhang Street (QZ).[7] Although X was a property owner while Y lived in an unquestioned public house, both of them had lived in Nanbuting for decades and were highly trusted by their neighbors. They frequently quoted Article 10 of Constitution, laws/regulations on land and housing administration of both P.R.C. and Nanjing (The Central People's Government of the P.R.C. 2011; Municipality of Nanjing 2012b), insisting that buildings in Nanbuting were entitled to be well-protected according to the laws of historic preservation and should not be demolished. These laws and relative newspapers were pervasively posted on the doors and walls of local households, arousing anti-demolition sentiments throughout the community, and raising unrealistic hope among residents that their dwellings could eventually be saved. When residents were informed of demolition decisions, allied groups voluntarily opposed the demolition in extreme ways, such as writing petition letters with blood fingerprints.

The second important group of participators were scholars and experts. Many of them believed that Nanbuting possessed such irreplaceable cultural significance that they must be preserved. However, when there was no platform for their participation and their voices were ignored by the local government, they voluntarily participated in the process in a very unusual way by appealing to the central government (twice respectively in 2006 and 2009). Their efforts drew the attention of China's previous Premier, and they successfully achieved their immediate goal of stopping demolitions. In the long run, however, conflicts among subsystems of

[7] To protect human subjects, the personal information of Mr. X and Mr. Y has been redacted.

government and local residents were exacerbated, given the lack of regular participation platforms and accumulated mistrust.

The media played a key role in creating tension between local residents and the government by frequently making announcements that were inaccurate or unconfirmed. For example, on September 8th, 2011, the Yangtze Evening Post reported that over a thousand households were planned to remain in Nanbuting without demolition (Zhao and Qiu 2011). The news was inspiring to the residents but the government quickly clarified the following day that the news was a rumor (Liu 2011). This conflicting news caused oppositions among a large number of residents, leading to an assembly of infuriated residents in front of the Nanjing Urban Planning Bureau, which, ironically, is a governmental subsystem that advocated for the preservation of houses and not for demolitions.

In summary, the conflicts in Nanbuting grew from the gap between stakeholders' high expectation of financial gains during the renewal project. Then the progress was significantly complicated by the subsystems of different stakeholders. Moreover, voluntary participation by local residents, scholars and the media jointly impeded the problematic four-stage project. Because of a lack of normalized participation platforms, their actions lead to uncontrolled consequences, which exacerbated the conflicts and deviated from their initial intent.

7.7 Discussion on Urban Renewal Patterns

This sections will discuss the interaction between two renewal patterns of the historic blocks and the resulting conflict. There are obvious distinctions between the two renewal patterns in revenue and cost. Also, the characteristics from above section (differentiated subsystem of each stakeholder and voluntary community participation) influenced the results of the two patterns distinctively.

7.7.1 The Real-Estate Renewal Pattern for Historic Blocks

In China, deprived urban blocks are often demolished as a whole and reconstructed by real estate developers who also pay for the demolition. In this real-estate renewal pattern, reconstruction will not start until all the residents and old buildings have been removed, which requires considerable cost and time (Zhang et al. 2008).

In fact, the cost of real-estate renewal progress in deprived historic blocks is often very high because of their central locations. According to an official from the Nanjing Housing and Construction Bureau, the extra expense that the municipality paid for the delay of Nanbuting Project is considerable. On the other hand, revenue from regular land development of historic preservation areas are extremely limited if real estate developers comply with the restricted FAR of the block; but if the restrictions are violated, residents are likely to demonstrate strong discontent. In the case of Nanbuting, a large number of local residents quoted that "the price of villas built on the 2nd-stage area was more than 40,000/m^2 in 2009. Then why should they accept the compensation of 8000/m^2 in 2013, which is the same as the offer that they received in 2009?" Poorly defined land development rights, combined with imprudent choice of the renewal patterns, jointly led to misunderstandings and accumulated conflict among multi-stakeholders.

7.7.2 *Self-Renewal Pattern for Historic Blocks*

In recent years, however, more and more municipalities have applied another method of urban renewal, especially in respect to historic preservation areas. By improving aging infrastructure, municipalities encouraged local residents to renovate their own homes and to make money through commercial activities. Both the Yandaixie Street case (Beijing) and the Wenhuali (Yangzhou) case proved that this pattern could considerably reduce the risk of severe conflict between residents and the government, without involving real estate developers (Wang 2012). Without any demolition, this self-renewal pattern is able to protect the quality of the traditional blocks at a lower cost.

Under the self-renewal pattern, confirming housing property rights is a crucial strategy to solve the problems caused by residents' differentiated subsystems (Guo 2007). The aforementioned investigation illustrated that residents with different types of housing property rights had different requests. Many residents in Nanbuting reported that their property rights were disregarded during the renovation project. Thus they lost the motivation of self-renewal, as they were afraid that their investment would eventually be demolished without proper compensation. The research found that 85 % of residents also agreed that the property owners had the right to stay on the block. These residents were interested in staying in the block because of their family heritage, as well as established relationships with neighbors and the local community.

7.8 Strategies for Both Patterns

Given the unexpected social cost it is essential to identify effective strategies for facilitating the urban renewal in a peaceful and sustainable manner. Two major goals include: (1) identifying the interests of each stakeholder to mediate any potential conflict; (2) balancing the needs of both the historic preservation and the economic development in the historic block renewal processes.

7.8.1 Establishing Participating Platforms

By analyzing multi-stakeholder conflicts, this case study calls for immediate attention to apply legitimate community participation as a strategy for both two renewal patterns. Residents from Nanbuting and scholars have demonstrated the negative effects of voluntary participation under the lack of a platform for stakeholders to voice their concerns.

Under the real-estate development pattern, the main task of consensus building is financial distribution between developers and local residents, with additional attention to their internal subsystems. Under the self-renewal pattern, significant mediation needs to focus on governmental guidelines and incentives for residents. As a robust community requires constant efforts of consensus building and interactions, each pattern needs fundamental support from long-term and flexible participation platforms, on which further proposal can be presented and discussed.

7.8.2 Mixed-Use Spatial Plan

According to the aforementioned evaluations of 133 listed historic buildings, the team continued to analyze and classify the 16 streets where these buildings were located, and eventually classified them into four feature classes. Due to the nature of land use planning and the large quantity of data analysis, these contents are better reserved for another article, however they still provide salient implications.

The team found that the subsystems of local residents demonstrated spatial distribution, so the renewal plan should use participation platforms to encourage them to make voluntary decisions. In this way, the block would be organically divided into subareas. For example, since property owners have constantly maintained their living environment, their streets, occupying only 1/4 of the whole block, remained the most satisfying preservation area, and meanwhile they were exactly the ones who strongly opposed demolition. The strategic recommendation for NUCIH is to create a mixed-use spatial plan, preserving this subarea for traditional residence while working with real estate developers to reconstruct other subareas for commercial development. By looking closer into subareas, each subsystem among stakeholders will fit their interests into the mixed-use plan.

To verify the feasibility of this strategy, the team investigated the visions of local residents in September 2013. The results showed that residents who were willing to stay and did not agree with demolition in the last decade were around 350–450 households, which was less than the envisioned 650 households to stay in the plan for the 4th-stage area. Their subarea was sufficient to accommodate this number of households, and nearly all of them agreed to renovate their homes with a unified style, paying the bill themselves and even paying for improving the neighborhood environment. Once the government could invest and manage the community infrastructure, these residents were inspired to voluntarily renovate their block. In conclusion, a reasonable number of the residents had the ability and the aspiration to renew their own subarea, leaving adequate land for commercial development.

7.9 Conclusion

This article investigated the urban renewal process of a historic preservation area in Nanjing in the Nanbuting community. It drew on a framework from the planning and policy-making process, linking multi-stakeholder conflict assessment to potential suggestions for participatory planning.

By extensively interviewing the local residents, governmental officials, project managers and scholars directly involved in this process, this article demonstrated the origin, formation and exacerbation of the multi-stakeholder conflicts. The origin of the conflicts lies in the considerable potential benefits of community renewal. Conflicts were formed among differentiated subsystems of all stakeholders. Common multi-stakeholder conflicts were exacerbated by voluntary community participation in the renewal process.

The article then identifies two patterns of urban renewal for historic preservation areas, featured by whether to involve real estate developers. Given the complicated conflicts in community renewal with historic preservation appeals, either the real-estate developer pattern or the self-renewal pattern requires a long-term and flexible participation platform for consensus building.

Further research may discuss how to establish participation mechanism among stakeholders with complicated internal subsystems, and how to facilitate plan implementation in a peaceful and sustainable manner.

Acknowledgments The research team members are grateful to all the interviewees who provided insights and first-hand information related to the Nanbuting Renovation Project.

References

Allison, E. W. (1996). historic preservation in a development-dominated city: The passage of New York City's landmark preservation legislation. *Journal of Urban History, 22*(3), 350–376. doi:10.1177/009614429602200304.

Arnstein, S. (1969). A ladder of citizen participation in the USA. *Journal of the American Institute of Planners, 35*(4), 216–224. doi:10.1080/01944366908977225.

Baker, M., Coaffee, J., & Sherriff, G. (2007). Achieving successful participation in the new UK spatial planning system. *Planning, Practice & Research, 22*(1), 79–93. doi:10.1080/02697450601173371.

Baum, H. (1997). *The organization of hope: Communities planning themselves*. Albany: State University of New York Press.

Chen, H. (2011). *The mechanism and its countermeasures of imbalanced politics in urban residential redevelopment projects during the period of institutional transition*. PhD dissertation, Nanjing University, Nanjing.

Davidoff, P. (1965). Advocacy and pluralism in planning. *Journal of the American Planning Association, 31*(4), 331–338. doi:10.1080/01944366508978187.

Davidoff, P., & Reiner, T. (1962). A choice theory of planning. *Journal of the American Institute of Planners, 28*(2), 103–115. doi:10.1080/01944366208979427.

Forester, J. (1987). Planning in the face of conflict: Negotiation and mediation strategies in local land use regulation. *Journal of the American Planning Association, 53*(3), 303–314.

Forester, J. (2006). Making participation work when interests conflict. *Journal of the American Planning Association, 72*(4), 447–465. doi:10.1080/01944368708976450.

Forester, J. (2009). *Dealing with differences: Dramas of mediating public disputes*. New York: Oxford University Press. doi:10.1080/19460171.2011.555685.

Guo, X. (2007). Inner city conservation and revitalization planning in Beijing: A perspective from land development right. 城市规划(*City Planning Review*), *31*(12), 66–72.

Hajer, M. A., & Wagenaar, H. (2003). *Deliberative policy analysis: Understanding governance in the network society*. Cambridge: Cambridge University Press. doi:10.1007/s11077-005-1765-z.

Harvey, D. (1997). Contested cities: Spatial process and spatial form. In N. Jewson & S. MacGregor (Eds.), *Transforming cites*. London/New York: Routledge.

Healey, P. (1997). *Collaborative planning: Shaping places in fragmented societies*. London: Macmillan. doi:10.1016/S0743-0167(97)00052-1.

Hooper, M., & Ortolano, L. (2012). Confronting urban displacement: Social movement participation and post-eviction resettlement success in Dar es Salaam, Tanzania. *Journal of Planning Education and Research, 32*(3), 278–288. doi:10.1177/0739456X12439066.

Innes, J. E., & Booher, D. E. (1999). Consensus building as role playing and bricolage: Toward a theory of collaborative planning. *Journal of the American Planning Association, 65*(1), 9–26. doi:10.1080/01944369908976031.

Jürgen, H. (1985). *The theory of communicative action*. Boston: Beacon.

Kaplowitz, M. D., Machemer, P., & Pruetz, R. (2008). Planners' experiences in managing growth using transferable development rights (TDR) in the United States. *Land Use Policy, 25*(3), 378–387. doi:10.1016/j.landusepol.2007.07.004.

Liu, H. (2011, September 9). Nanjing Urban Planning Bureau refuted the rumour of "Fourth stage Plan of Nanbuting". 南京晨报*(Nanjing Morning)*. http://njcb.jschina.com.cn/mp3/html/2011-09/09/content_425053.htm

Liu, J. (2013, March 17). Jiangnan historic cultural resources to be extended by the "Jiangnan seventy-two workshops". 文汇报(*Wen Wei Po*).

Lv, W. (1991). 南京地方志 *(The origin of toponymy in Nanjing)*. Nanjing: Jiangsu Science and Technology Press.

McConnell, V., & Walls, M. (2009). Policy monitor: U.S. experience with transferable development rights. *Review of Environmental Economics and Policy, 3*(2), 288–303. doi:10.1093/reep/rep008.

Mills, D. (1980). Transferable development rights market. *Journal of Urban Economics, 7*(1), 63–74. doi:10.1016/0094-1190(80)90026-1.

Municipality of Nanjing. (2012a). 南京历史文化名城保护规划 *(Historical and cultural city of Nanjing protection planning) (2010–2020)*. http://www.nanjing.gov.cn/njszf/szf/201209/t20120926_1176953.html

Municipality of Nanjing. (2012b). 南京市土地征收补偿条例 *(Nanjing House Acquisition from State-owned Land and Compensation Ordinance)*. http://www.nanjing.gov.cn/njszf/szf/201205/t20120511_1183232.html

Municipality of Nanjing. (2012c). 南京老城南历史保护规划与城市设计(征询意见稿) *(South part of old Nanjing city historic preservative plan and urban design (for public opinions))*. http://www.njghj.gov.cn/ngweb/Page/Detail.aspx?InfoGuid=228a1197-23c4-4cda-a425-f06059a9ff89

Municipality of Nanjing. (2013). 评事街历史风貌保护规划(征询意见稿) *(Pingshi Street historic featured area preservative plan (for public opinions))*. http://www.njghj.gov.cn/ngweb/Page/Detail.aspx?CategoryID=633fd117-b833-410f-8bb5-4b6ba311a366&InfoGuid=fa7d4c07-b885-4d9a-8ee1-8323c04aed61

Nanjing Municipal Bureau of Statistics. (2013). *2013 年南京市统计年鉴* (2013 statistical yearbook of Nanjing). http://221.226.86.104/file/nj2004/2013/renmin/4-2.htm

Niemann, G. A. (1978). Historic preservation and transferable development rights. *The University of Illinois Law Forum, 1978*(4), 927–951.

Peterson, J. (1995). Decision-making in the European Union: Towards a framework for analysis. *Journal of European Public Policy, 2*(1), 69–93. doi:10.1080/13501769508406975.

Susskind, L., & Cruikshank, J. L. (2006). *Breaking Robert's rules: The new way to run your meeting, build consensus, and get results*. Oxford: Oxford University Press.

Susskind, L., McKearnan, S., & Thomas-Larmer, J. (1999). *The consensus building handbook: A comprehensive guide to reaching agreement*. Thousand Oaks: Sage Publications.

Thatcher, M. (1998). The development of policy network analyses: From modest origins to overarching frameworks. *Journal of Theoretical Politics, 10*(4), 389–416. doi:10.1177/0951692898010004002.

The Central People's Government of the P.R.C. (2011). 国务院590号令"国有土地上房屋征收与补偿条例" *(State Council Order No. 590 "House Acquisition from State-owned Land and Compensation Ordinance")*. http://www.gov.cn/flfg/2011-01/21/content_1791480.htm

Timothy, L. (1998). *Transfer of development rights*. Ohio State University Fact Sheet. Columbus: Ohio State University. http://ohioline.osu.edu/cd-fact/1264.html

Wang, H., Tao, R., Wang, L., & Su, F. (2010). Farmland preservation and land development rights trading in Zhejiang, China. *Habitat International, 34*(2010), 454–463. doi:10.1016/j.habitatint.2009.12.004.

Wang, J. (2012). *Ten years*. Beijing: SDX Joint Publishing Company.

Wu, F. (2002). China's changing urban governance in the transition towards a more market-oriented economy. *Urban Studies, 39*(7), 1071–1093.

Xie, W. (2011). *Historic districts in urban regeneration: A research of social changes in the community of Nanbuting*. Master dissertation, Nanjing University, Nanjing.

Yao, Y. (2009). *Application on affirming immovable cultural relics in Nanbuting*. http://blog.sina.com.cn/s/blog_4d9d92790100fzc7.html

Zhang, H. (2004). 中国土地发展权研究:土地开发与资源保护的新视角*(Research on Chinese land development rights: A new perspective of land development and resource preservation)*. Beijing: Renmin University of China Press.

Zhang, J., Wu, F., & Laurence, J. M. (2008). *Institutional transition and reconstruction of china's urban space: Establishing an institutional analysis structure for spatial evolution*. 城市规划 *(City Planning Review), 32*(6), 55–56.

Zhao, F., & Qiu, H. (2011, September 8). Nanbuting Renovation Project will preserve a thousand indigenous residents. 扬子晚报*(Yangtze Evening Post)*. http://news.sohu.com/20110908/n318701662.shtml

Zhao, M., Bao, G., & Hou, L. (1998). 土地使用制度改革与城乡发展 *(Reforms of land use system and development of cities and the country)*. Shanghai: Tongji University Press.

Zhou, X., & Zhang, W. (2010). *The cultural value of the historic street blocks in the South part of old Nanjing city*. 建筑创作 *(Archicreation)* (2), 158–159.

Zhu, J. (2004). From land use right to land development right: Institutional change in China's urban development. *Urban Studies, 41*(7), 1249–1267. doi:10.1080/0042098042000214770.

Chapter 8
Planning for Emerging Megacity Regions in China: Preliminary Research Within a Socioeconomic Framework

Li Hao and Qian Zhu

8.1 Introduction

In this era of globalization, the megacity region is an emerging phenomenon in recent urban development, a trend of large-scale space transformation, occurring in both developed and developing countries. Unlike conventional metropolis and city regions, megacity regions are distinctly new for theorists and planning practitioners. The concept of the megacity region (MCR) was defined by Hall P. and Pain K. (2006) in the book *The Polycentric Metropolis: Learning from Mega-City Regions in Europe*:

> Its recent rediscovery has been in Eastern Asia……A new form: series of anything between 10 and 50 cities and towns, physically separate but functionally networked, clustered around one or more larger central cities, and drawing enormous economic strength from a new functional division of labor. These places exist both as separate entities, in which most residents, and as part of a wider functional urban region (FUR) connected by dense flows of people and information carried along motorways, high- speed rail lines and telecommunications cables: the "space of flows" (Castells 1996, pp. 376–428) with major implications for sustainable development (Blowers and Pain 1999). It is no exaggeration to say that this is the emerging urban form at the start of the 21st century. (p. 3)

Although MCR is a relatively new concept in urban study, it has its roots in several urban theories developed several decades ago, such as "world cities" by

L. Hao (✉)
Research Center for Master Planning, Beijing Tsinghua Tongheng Urban Planning & Design Institute, 17th Floor East Tower, Building Jia1, East Qinghe Jiayuan, Haidian District, Beijing 100085, China
e-mail: interhow@hotmail.com

Q. Zhu
School of Planning, University of Waterloo, 200 University Avenue West, Waterloo, ON N2L 3G1, Canada

Q. Pan, J. Cao (eds.), *Recent Developments in Chinese Urban Planning*, GeoJournal Library 114, DOI 10.1007/978-3-319-18470-8_8

Friedman and "global cities" by Sassen. In order to gain an understanding of this urban concept, a literature review of the topic becomes necessary.

8.2 Theoretical Background

8.2.1 World City and Global City

For decades, large cities have attracted the attention of many scholars. One example is Peter Hall, who first published *The World Cities* in 1966 (Hall 1966). In the book, Hall identified seven world cities, including two polycentric metropolises: Rhine-Ruhr and Randstad. Since then, there has been a trove of literature concentrating on the phenomenon of "world cities" and/or "global cities", such as Friedmann and Wolff (1982), Sassen (1991, 1994), Castells (1996), and Short et al. (1996). The main focus of these studies has been on metropolitan cities on the global level, with the authors proposing collections of indicators to measure and define the idea of world city.

8.2.2 From Megacity to Megacity Region

Recently, some scholars have attempted to extend this notion to a wider metropolitan region through the term "global city-region" (Scott et al. 2001). Scott et al. (2001) found that in the contemporary world, global city-regions are emerging from large metropolitan areas, or contiguous sets of metropolitan areas. He further states that today it is city regions, not central cities or even nation states, which appear to be the boundaries for global economy competition (Scott et al. 2001). In addition, Hall and Pain (2006) noticed a similar phenomenon: the polycentric mega-city region, which is emerging in highly urbanized areas. Sassen (2006) also noticed the development trend from the node of global city to a wider scope region related to global city. In fact, it was Gottmann (1961) who originally identified the phenomenon as a megalopolis, decades before the current academic analysis. However, until recently, this idea had not attracted many scholars' attention. Hall and Pain (2006) give an updated definition for a megacity region (MCR). Though relatively similar to Gottmann's megalopolis, MCR is a recent urban form that is more complicated, through a high degree of interconnectivity. Hall and Pain (2006) further emphasize the fact that an MCR is highly codependent, more so than previous urban clusters. An MCR is based on Castells (1996) concept of "space of flows", which connects the individual urban elements as a relatively dependent spatial entity, and this makes it fundamentally different from the previous geographical concept of a megalopolis. In other words, an MCR is more complex, dynamic, functional and spatially active in the context of the present global economy. A popular empirical methodology was first devised by an academic organization, Globalization and World Cities (GaWC), on global urban research (Taylor 2004). This interlocking network model is utilized to measure regional polycentricity and network connectivity in a megacity region. Via similar quantitative research on intra-regional spatial flow, Hall and Pain (2006) lead a POLYNET project to study eight emerging megacity regions in northwest Europe. Thierstein and Forster (2008) conducted similar research on several megacity regions in Europe, in an attempt to visualize this urban phenomenon.

8.2.3 Megacity Region in Asia and China

Since 1980s, Asia has attracted many researchers' interest on the topic of megacity regions. T.G. McGee (1989) researched the urban rural fringe in Southeast Asia, and creatively came up to the concept of *desakota*: the mixing zone of city and countryside. Thereafter, he raised the notion of mega-urban region in East Asia (McGee 1991). The new spatial form of big city clusters has been illustrated in several regions of East Asia, such as the Pearl River Delta and Yangtze River Delta in China, the Tokyo-Osaka corridor in Japan, the growth triangle lead by Singapore, etc. (Lo and Yeung 1996; Hall 1999).

There have also been some recent theoretical innovations with regards to the metropolises of China in the context of globalization. These discussions have concentrated on single mega cities such as Shanghai, Beijing and Guangzhou (Cai and Sit 2003; Lin 2004; Wei and Yu 2006). Generally, the current international literature about world cities primarily focuses on the experiences of advanced economies in Europe and North America (Wei and Leung 2005), while the research on China is relatively limited. Although there are some studies of megacities in China, they are focused on the regional scope. More specifically, large cities in China, such as Beijing and Shanghai, are considered very low level in the world urban hierarchy within some studies, even though there are some deviations to the real situation. The deviations have mainly been attributed to the continued focus on single city analysis, with little consideration on the whole megacity region. In recent years, more Chinese literature from mainland China and Hong Kong have discussed Chinese city region cases (Xie and Ning 2005; Cai and Sit 2003). However, these studies focus on different urban clusters, especially the comparison between several megacity regions within China. Few academics consider neither the globalization context nor the international perspective of such analysis.

8.3 China's Megacity Regions

8.3.1 MCRs' Rising in Post-reform China

Under the current trend of globalization, the centre of world industry and economy continues to move toward the Asian Pacific region. China, especially the coastal area, has already become a focus of this international shift of industry and capital. The continual investment of foreign capital has been a leading factor in Chinese urban development, especially in the economically developed coastal areas (Gu 1999).

During the past 30 years, with the complexity of urbanization, industrialization, and marketization, many big cities and city clusters have been rising in China. In the globalization era, metropolises and city clusters became a new form of economic agglomeration. Megalopolises are, in fact, centers of both economic growth and destinations for foreign capital investments, with their vibrant economies which act as a core growth component of the national urban system. Recently, the new urban clusters have been in the form of MCRs within China. Those MCRs are essentially different from conventional cities and regions, with different socio-economic structures, as well as spatial formation.

8.3.2 An Economic Perspective: MCR and Economic Agglomeration

China's MCR are fundamentally driven by economic agglomeration. In the context of globalization, international trade and transnational industry shifts contribute to the development of MCRs in China. Many small cities and towns within MCRs have fostered their own local industrial clusters by agglomeration economy. Cities specializing in one or several types of production are networked, linked together via industry correlation and organization. An intense network of investment, manufacturing technology and human resources has formed in these city regions, creating a new kind of spatial form. This transformation could be construed as a capital accumulation process (Harvey 2001) from a political perspective, or an agglomeration economy (Fujita and Thisse 2002) from the New Geography Economy perspective, as it is mainly driven by market forces involving foreign and private firms.

In the dynamic process, metropolises and their hinterlands are rising together through industry division and linkage. For instance, since China launched its reform and open policy in the late 1970s, Hong Kong has shifted nearly all of its manufacturing industry to its adjacent hinterland, the Pearl River Delta. This region has become the industry base for Hong Kong, and Hong Kong accepts the role of marketing, financing, designing and managing those industries. A similar trend has occurred in the Yangtze River Delta with its central metropolis of Shanghai. The world cities, large cities, and their hinterlands are connected linked together at a vast spatial scale through a complex urban system.

8.3.3 A Spatial Perspective: MCR and Uneven Spatial Development

When it comes to spatial vision at a macro level, we could find why megacity regions are rising in certain parts in China. Generally, the socioeconomic strength of China has risen rapidly, and the comprehensive national competence has enhanced unceasingly. However, this development has been quite uneven, both socially and spatially. China's current socioeconomic status, with a free market system combined with powerful state intervention is conceptualized by Harvey (2007) as neo-liberalism with Chinese characteristics. Under this kind of neo-liberalism, the uneven geographical implementation concentrates mainly in coastal regions. In these regions, where infrastructure and geographic connections already exist, industrialization and international trade all contribute to the formation and development of metropolis and megalopolis regions. Thus, the MCRs could be perceived as a result of this uneven spatial development. For example, rich Pearl River Delta (Fig. 8.1), part of Guangdong Province, accounted for 23.2 % of land within the province, 50 % of the population, and 85 % of the GDP of the whole province in 2010 (Statistics Bureau of Guangdong Province 2011). This region

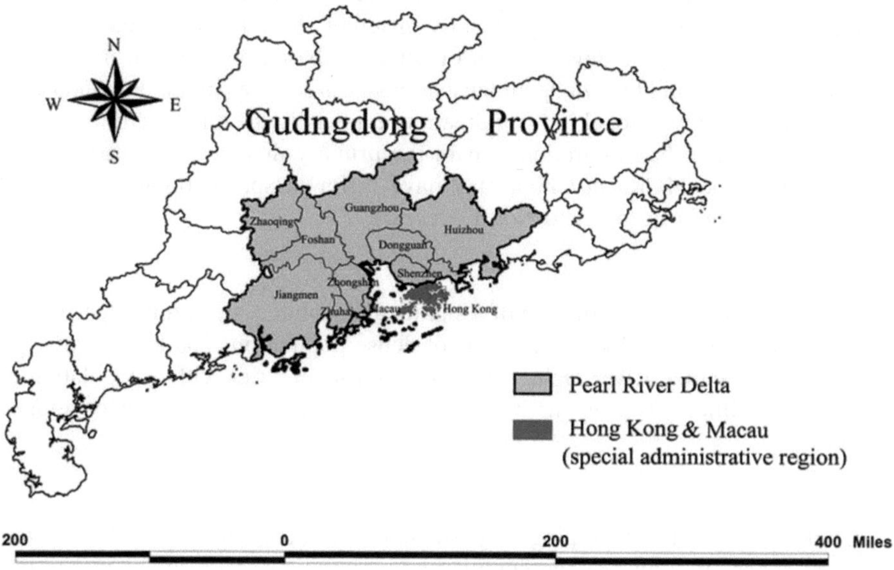

Fig. 8.1 Pearl River Delta within Guangdong Province (Source: the author)

even accounts for 9.4 % of China's GDP and 27.4 % of national exports. Under these circumstances, not only do megacities (regions) have high population and high economic growth rates, they have also been the main factor of national economic growth and urbanization.

8.3.4 MCR with Chinese Characteristics

Although MCR is a phenomenon in a wide range of nations, the MCRs in Asia, especially in China, have distinct socioeconomic characteristics. These characteristics are categorized into four aspects:

1. Mixture of urban and rural

   Although China has a high urbanization rate, its overall ratio of urbanization is low, i.e., there are still many people living in rural regions. Thus a Chinese MCR is more likely a combination of urban and rural; in the developed world an MCR would be only a cluster of cities. Morphologically, a MCR in China (especially MCRs in southern China) has a development pattern similar to McGee's (1989) *desakota* model, which is a continuous space of scattered urban cores and their rural hinterlands. The town and village enterprise (TVE) in southern coastal China also lead the urbanization and industrialization in rural areas around cities. A contiguous set of villages, towns and cities have formed into an MCR.

2. Driving Forces

   Though megacity region is a worldwide phenomenon, the driving forces behind China's megacity regions are significantly different from developed countries. In the developed world, the development of MCRs is based on a spatial network of high-tech industry and knowledge intensive business services in the post-industrial era; in contrast conventional industrialization, especially the manufacturing industry and the rural–urban migration remain the primary driving forces for China's MCRs. Thus, the MCRs in China may have different spatial organizations and forms from those in Europe and North America.

3. Larger scale

   The MCR is an urban structure of a significant spatial scale; China's MCRs are even larger, as they are located in areas of dense population. The largest MCR in China is the Yangtze River Delta: 16 prefectural level cities with a population of nearly 100 million. This is far beyond the scale of most MCRs in developed countries.

4. Complex degree of development

   China's MCRs are developing very irregularly, in both intra-regional and inter-regional scope. In addition, the complex administration arrangement makes this more problematic. Thus, the analysis of MCRs under Chinese context becomes quite necessary.

8.4 Planning and Governance

8.4.1 Background: Decentralization, Pro-growth Policy and Entrepreneurial Governance

China's reform is essentially to introduce market mechanism under a decentralized administrative framework. By opening the economy to the outside world, and adhering to international standards, the central state releases its power through reform of administrative decentralization, in order to carry out economic liberalization and marketization.

This reform leads cities to greater freedom, with independent authority in development. In this process, especially after the tax distribution system introduced in 1994, local governments became the dominating power and key stakeholder in urban development (Vogel et al. 2010). This change has had some profound influence. For instance, urban vigor strengthens, as the city and region have direct international connections. In addition the Chinese cities have been promoting urban entrepreneurialism (Wu and Zhang 2007).

China's urban development has become a 'state-centered political bureaucratic' process, as China is a kind of developmental state (Hill and Kim 2000). However, the simple performance assessment for officials results in the GDPism, and the 'growth first' mentality set by the central government becomes a key parameter for fostering an ideological foundation for local governments' entrepreneurial governance (Vogel et al. 2010). Economic growth becomes the first target for different levels of government, and local officials are enthusiastic but single-mindedly focused on economic growth. In this case, the large city in fact becomes "the super enterprise" controlled by the local authority or elite (Wu 2000). The local authority uses the control of financial resources as a tool to stimulate economic growth. Thus, the cities within one region, or from different regions, begin a fierce competition for growth, similar to what would occur between competing enterprises. To a certain extent, the pro-growth policy and entrepreneurial governance, under the socialist-market regime, contribute to vicious competition and disorderly development in the mega city regions. However, under isolated development policy, the cities lack regional coordination and cooperation, both of which negatively affect the cities' globalization and urban competitive strength making.

8.4.2 Administration Arrangement

The administrative system in China is very different from many western countries (Fig. 8.2) due to its hierarchal, and hence complex nature, in addition to a dual structure of for urban-rural areas (Fig. 8.2). There are several different kinds of cities: Municipality Directly under Central Government, prefectural level cities, and county level cities. An MCR often consists of these different kinds of cities, as

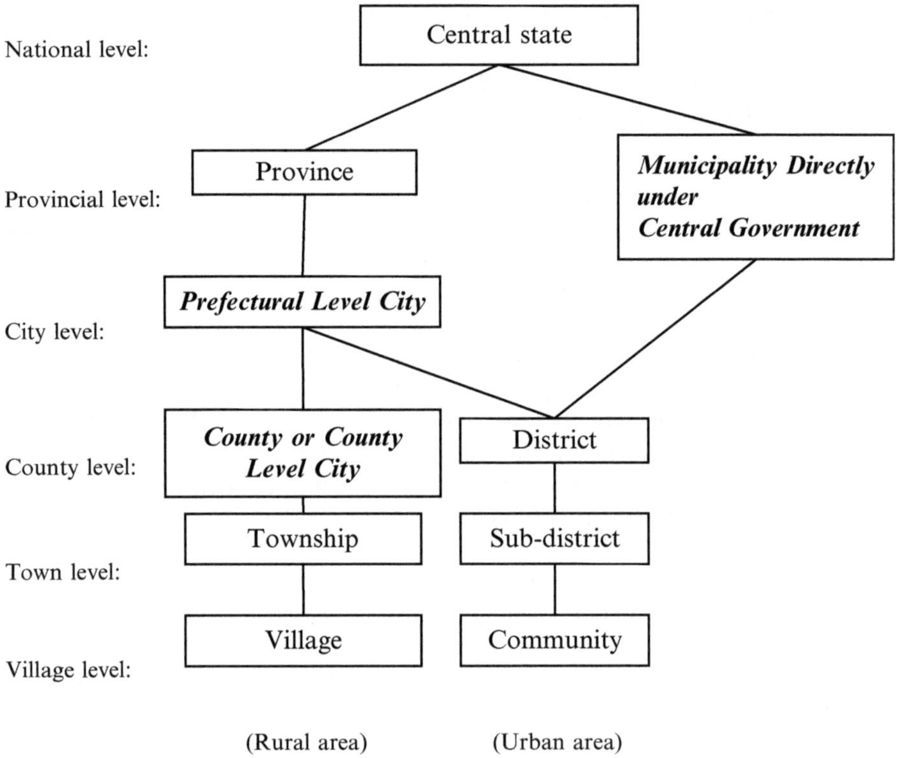

Fig. 8.2 The Jurisdiction hierarchy and MCR's components (cities in italic) (Source: the author)

well as some rural counties. There are also some sub-prefecture and sub-provincial cities in the administration domain, which makes the multi-level governance even more complicated.

Thus, the governance of MCR involves not only cross boundary issues but also cross administration level issues. The administration system has been the same for decades, a legacy of China's centralized socialism. In the post-reform era, the political system remains rigid, at odds with the significant economic reform. For instance, there is no particular administration arrangement for the MCRs (Xu 2008); for the largest MCR, the Yangtze River Delta, the only governing body is the annual joint meeting of mayors from the prefectural level cities and Shanghai (the only Municipality Directly under Central Government in this region). This meeting aims to enhance regional competitiveness and facilitate regionalization. However, it is quite informal. It lasts only for 2 days and focuses on general policy. In the Jing-Jin-Ji Region (the MCR around Beijing), these coordination meetings do not even occur.

Without cooperation and collaboration, fierce intra-regional competition results in numerous issues, such as duplication in constructions, cross boundary pollution, and redundant investment projects. A typical example can be found in the Pearl

River Delta Region, as nearly all the major cities are building their own international airports. This destructive competition results in an oversupply of infrastructure and development, with a waste of resources and capital funds.

From an economic perspective, this kind of governance approach has weakened the scale economy and resulted in regional diseconomies. Thus, the functional advantage of MCRs is in doubt for the long term.

MCRs constitute an emerging urban form in China; however, no corresponding political jurisdiction arrangement has been established to govern and plan this new scale of territory. Until recently, institutional initiatives occurred sluggishly. For instance, in the last 10 years no administration mechanism or structure has been established to augment the annual mayors' meeting for the Yangtze River Delta. Moreover, the household registration system (Hukou in Chinese) which was based on administrative boundaries hinders the population fluctuating between different cities within one MCR.

8.4.3 Fragmented Planning

Of note, China's planning system is fragmented (Fig. 8.3), with various kinds of parallel planning legislation within the statutory planning framework, and each kind of planning is overseen by a different ministry. The primary focus of this system is socioeconomic development planning, urban planning, and land use planning.

Urban planning, considered the most important, is influenced by the socialist planning in a planned economy from the former Soviet Union; change has been slow. Currently, the only statutory planning related to MCRs is the national urban system planning (Fig. 8.4), which defines the border, population scale and urban functions of various urban clusters. However, this planning is too macro, with little detail and; thus its use in implementation is questionable. Generally, local authorities pay the most attention to more detailed planning such as land use planning and construction planning. Furthermore, all the provinces and cities have their own urban system planning. Thus, the national urban system planning works largely at a conceptual level, with little implementation capacity. More importantly, urban system planning covers only the urban area, excluding rural areas, due to the urban-rural dual administration structure in China. Since a MCR is an integrated region composed of cities and their rural hinterland, this planning is ineffective for a city region.

Also, the planning system is consistent with the administrative system; different level authorities have their own planning departments. The system lacks cooperation mechanisms to deal with multi-level and cross boundary issues.

In a MCR, the regional planning functions are usually divided between different government departments, and local authorities mainly make legislation regulations that favor their own interests in prompting local economic growth due to GDPism. Planning is designed to make cities into growth machines for GDP in the short term,

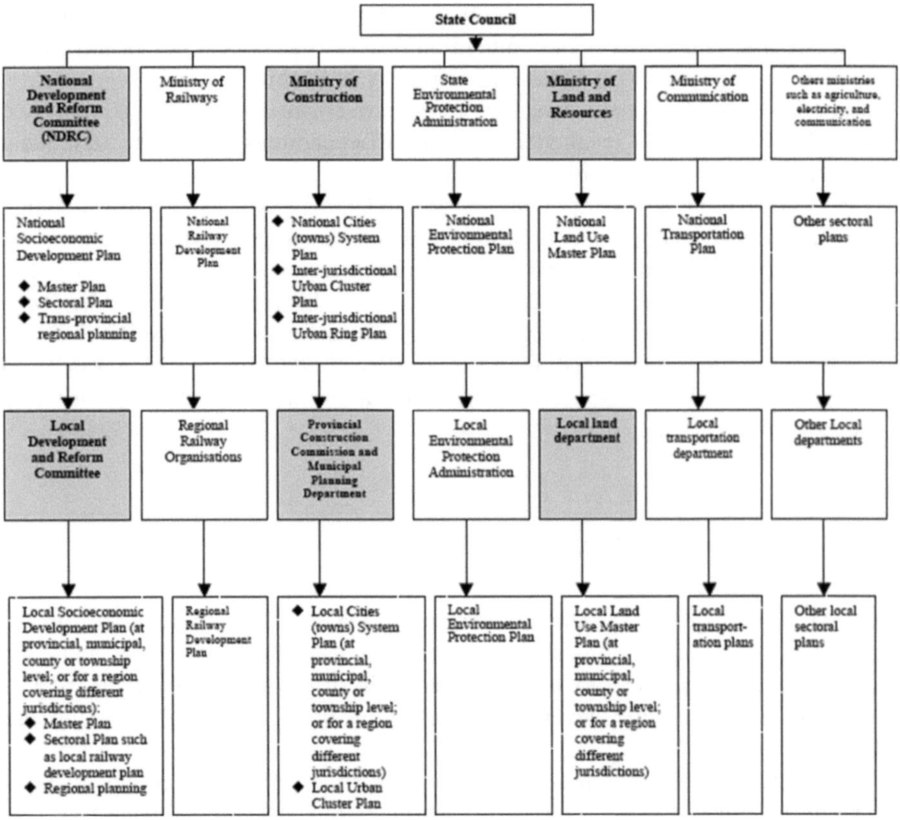

Fig. 8.3 Fragmented and dispersed planning system in China (Source: Xu 2008)

especially during local officials' tenure in office. Contradictions often exist between planning and regulations at various levels of government (Xu 2008).

The dispersed planning structure contributes to the disordered spatial development. Sometimes the urban-rural hybrid landscape (the so called "city-not-city" and "countryside-not-countryside" landscape) has emerged in Chinese cities (Zhang 2008), and this especially phenomenal in MCRs due to the fragmented jurisdiction and disordered regional development. The disordered regional development also results in severe environment problems, such as significant loss of farmland, urban sprawl and environment deterioration.

8.4.4 Current Regional Planning: Awkward Status

City regionalism is a new trend in China, and it is prompted by the central government's desire to achieve regional equality.

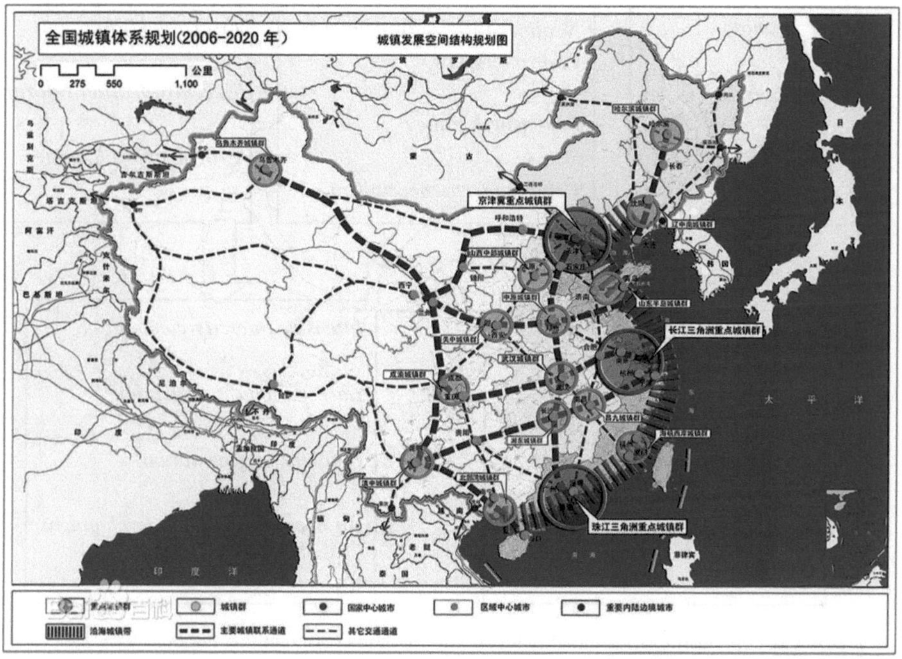

Fig. 8.4 The national urban system planning (2006–2020) (Source: Ministry of Construction of China (2006))

At a local level, many city-region planning policies are intended to prompt regionalization and enhance urban cluster competitiveness (Luo and Shen 2008). However, these plans are ineffective. Although lots of "metropolitan region planning", "urban cluster planning", "regional planning" and such occur, these types of planning are not a part of any statutory planning system. Moreover, not all the existing and proposal regional planning is integrated in one comprehensive system, and so has little effect on the current chaotic development situation.

For a city region within a single province, such as the Pearl River Delta, the regional planning comes occurs within the provincial government. For cross-boundary city regions, such as Jing-Jin-Ji region or the Yangtze River Delta, the central government (Luo and Shen 2008) mainly controls the regional planning. The planning process can extend for long periods, with many conflicts, when dealing with cross boundary and cross level cooperation within the process. In most cases, this typical top-down planning is effective; the local development tends to be led by local officials' and political influence, with limited consideration for the superior government.

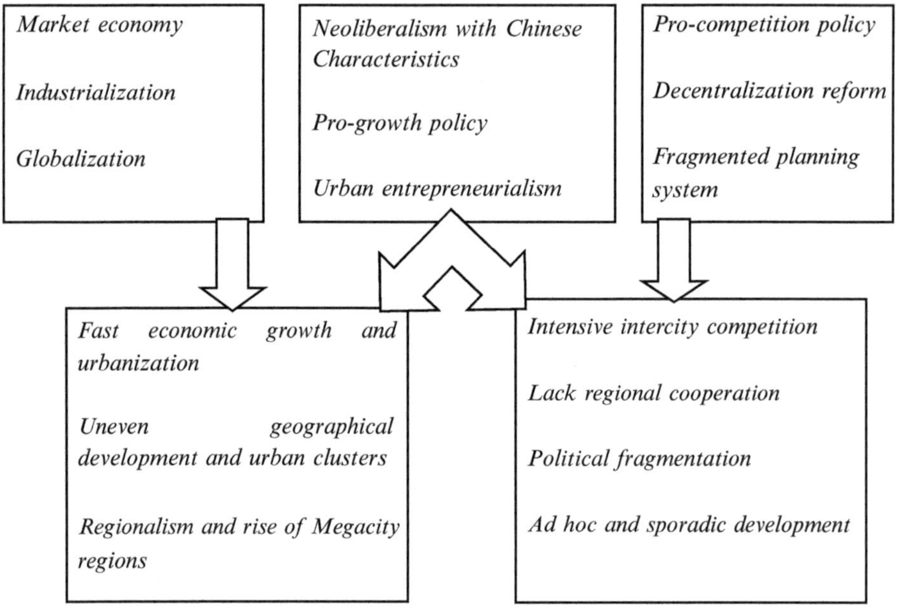

Fig. 8.5 Conceptual framework of MCRs' rising in China (Source: the author)

8.4.5 Central Dilemma: Regionalism vs Localism

As demonstrated by Fig. 8.5, the problematic development of MCRs is rooted in the socioeconomic transformation process during the reform period.

Market forces are increasingly dominating the driving forces of MCR's development; however, changes within the administrative and planning framework move at a much slower pace. The regional integration is restrained by the rigidity of the planning system. For instance, the economic sector calls for territorially integrated growth, yet the hierarchal administration and fragmented planning framework limit economic integration. Therefore, the current formation and organization of MCRs is far from mature and perfect. What could be considered the most integrated MCR, the Pearl River Delta, which emerged in the 1980s, is yet to be considered a single functional region (Yeh 2001; Yang 2004).

The central consideration of local officials is the economic growth within their jurisdiction. Fragmented administration power within individual administrative units creates invisible walls along jurisdictional boundaries, which curbs intra-regional flow and suppresses the formation of city regions. Local officials, in coalition with private-sector developers, contribute to ad hoc urban development through this process. The core contradiction in MCR development is the uniformity of city clusters' rising, against fragmented regional governance and planning. From another perspective, this could also be seen as regionalism driven by market forces versus localism led by local authorities.

8.5 Conclusion and Implication

The emergence of megacity regions has various implications in numerous aspects of city life, such as transportation, economic development, administration, land use, environment, etc. This paper focuses on its implication for regional planning and governance.

From the previous discussion, we conclude that the current governance and planning for MCRs face several severe challenges:

1. Under the laissez-faire economic mechanism, the economic impetus functions mainly in a bottom-up approach and cities within urban clusters are highly correlated and networked within the economic growth process, yet the planning system and regional governance is quite fragmented, with a top-down approach. Furthermore, the cities often lack interaction and communication under this kind of planning mechanism (Luo and Shen 2008). Thus, this planning practice results in the imperfect and disorderly urban development currently observed in the megacity regions.
2. There is no official governance body or institution of MCRs. The current administrative hierarchies are bureaucratic in its approach to inter-city and intra-city economic competition. The current communication methods between cities within one MCR are often accomplished through unofficial meeting and conferences. Coordination in MCR remains a challenge for different level's authorities, with efficient administration on a city-regional level still lagging behind. More powerful and efficient regional institutions are needed for MCRs.
3. Though there is "metropolitan region planning", "urban cluster planning", "regional planning" within government, these kinds of planning are not a part of the statutory planning system. Also, most third-sector or market actors are not included in the regional governance process (Heeg et al. 2003). However, lacking public participation, the market's needs still could not be reflected within the current planning framework. Plans are generally made by high level government as an expression of official's ideas, similar to catchwords and propaganda policies. Without statutory and comprehensive spatial planning in MCRs, the existing planning framework of implementation and its overall effectiveness is questionable.

Nowadays administration and planning arrangement for MCRs is quite problematic. MCRs are a new urban form for a rapidly urbanizing country such as China. These regions drive China's economy and have become one of the central tenets of China's urban development. Yet, this new territory form requires specific regional location policies. Compared with rapid urban transformation, current planning and governance on this issue trail the issue. Different from conventional city and regional planning, the MCRs are where attention should be focused. The current ill-defined planning arrangement for MCRs fails to achieve effective cross-border and multi-level governance in the city regions, thus it will be unlikely to secure sustainable urban development in the long term.

Under these circumstances, new comprehensive and practical planning legislation for MCRs should be established. This legislation could work as a coordinator and integrator, to achieve spatial integration, as well as sustainable economic and spatial development in megacity regions. Regional planning should guide the development of MCRs within a rational and scientific framework. A comprehensive framework for regional cooperation and sustainable development is need urgently. Planners and policy makers should look forward to the future, and have a broader vision. In addition, the public should be included in this regional development process. Moreover, China could learn from other developed countries about regional cooperation and urban networks, such as the European Union's spatial planning ESDP (European Spatial Development Perspective).

In conclusion, in the near future, initiatives for city-region planning should be established to accommodate the changing urban scale. This calls for an urgent need for comprehensive, inclusive, coordinative and statutory spatial planning on MCRs. There is a lot of work to be done in this field.

References

Blowers, A., & Pain, K. (1999). The unsustainable city? In S. Pile, C. Brook, & G. Mooney (Eds.), *Unruly cities? Order/disorder* (pp. 247–298). London: Routledge.

Cai, J., & Sit, V. F. S. (2003). Measuring world city formation – The case of Shanghai. *The Annals of Regional Science, 37*, 435–446.

Castells, M. (1996). *The rise of the network society*. Oxford: Blackwell.

Friedmann, J., & Wolff, G. (1982). World city formation: An agenda for research and action. *International Journal of Urban and Regional Research, 6*, 309–344.

Fujita, M., & Thisse, J. F. (2002). *Economics of agglomeration: Cities, industrial location, and regional growth*. Cambridge: Cambridge University Press.

Gottman, J. (1961). *Megalopolis*. Cambridge, MA: MIT Press.

Gu, Z. L. (1999). *Chinese urban geography*. Beijing: The Commercial Press (in Chinese).

Hall, P. (1966). *World cities*. London: Wiedenfeld & Nicholson.

Hall, P. (1999). Planning for the mega-city: A new Eastern Asian urban form? In J. Brotchie, P. Newton, P. Hall, & J. Dickey (Eds.), *East-west perspectives on 21st century urban development: Sustainable eastern and western cities in the new millennium*. Hunt/Brookfield: Ashgate Publishing Ltd.

Hall, P., & Pain, K. (2006). *The polycentric metropolis: Learning from mega-city regions in Europe*. London: Earthscan Publications.

Harvey, D. (2001). *Spaces of capital: Towards a critical geography*. Edinburgh: Edinburgh University Press.

Harvey, D. (2007). *A brief history of Neoliberalism*. Oxford: Oxford University Press.

Heeg, S., Klagge, K., & Ossenbruügg, J. (2003). Metropolitan cooperation in Europe: Theoretical issues and perspectives for urban networking. *European Planning Studies, 11*(2), 139–153.

Hill, R., & Kim, J. W. (2000). Global cities and developmental state: New York, Tokyo and Seoul. *Urban Studies, 37*(12), 2167–2195.

Lin, G. C. S. (2004). The Chinese globalizing cities: National centers of globalization and urban transformation. *Progress in Planning, 61*, 143–157.

Lo, F. C., & Yeung, Y. M. (Eds.). (1996). *Emerging world cities in Pacific Asia*. Tokyo: United Nations University Press.

Luo, X., & Shen, J. (2008). Why city-region planning does not work well in China: The case of Suzhou-Wuxi-Changzhou. *Cities, 25*, 207–217.
McGee, T. G. (1989). Urbanisasi or Kotadesasi? Evolving patterns of urbanization in Asia. In F. J. Cost, A. K. Dutt, L. J. C. Ma, & A. G. Nobel (Eds.), *Urbanization in Asia* (pp. 93–108). Honolulu: University of Hawaii Press.
McGee, T. G. (1991). The emergence of desakota regions in Asia: expanding a hypothesis. In N. Ginsburg, B. Koppel, & T. G. McGee (Eds.), *The Extended metropolis: Settlement transition in Asia* (pp. 3–25). Honolulu: University of Hawaii Press.
Ministry of Construction of China. (2005). *National Urban System Planning Outline (2005–2020)*. Beijing (in Chinese).
Sassen, S. (1991). *The global city: New York, London, Tokyo*. Princeton: Princeton University Press.
Sassen, S. (1994). *Cities in a world economy*. Thousand Oaks: Pine Forge/Sage.
Sassen, S. (2006). *Territory, authority, rights: From medieval to global assemblages*. Princeton: Princeton University Press.
Scott, A. J., et al. (2001). Global city-regions. In A. J. Scott (Ed.), *Global city-regions: Trends, theory, policy* (pp. 11–30). Oxford: Oxford University Press.
Short, J. R., Kim, Y., Kuus, M., & Wells, H. (1996). The dirty little secret of world cities research: Data problems in comparative analysis. *International Journal of Urban and Regional Research, 20*(4), 697–717.
Statistics Bureau of Guangdong Province. (2011). *Guangdong statistical yearbook 2010*. Retrieve from, http://www.gdstats.gov.cn/tjnj/ml_c.htm
Taylor, P. J. (2004). *World city network: A global urban analysis*. London: Routledge.
Thierstein, A., & Forster, A. (Eds.). (2008). *The image and the region: Making mega-city regions visible!* Baden: Lars Müller Publishers.
Vogel, R. K., et al. (2010). Governing global city regions in China and the West. *Progress in Planning, 73*(1), 1–75.
Wei, Y. D., & Leung, C. K. (2005). Development zones, foreign investment and global city formation in Shanghai. *Growth and Change, 36*, 16–40.
Wei, Y. H. D., & Yu, D. L. (2006). State policy and the globalization of Beijing: Emerging themes. *Habitat International, 30*(3), 377–395.
Wu, F. (2000). The global and local dimensions of place-making: Remarking Shanghai as a world city. *Urban Studies, 7*(8), 1359–1377.
Wu, F., & Zhang, J. (2007). Planning the competitive city-region: The emergence of strategic development plan in China. *Urban Affairs Review, 42*, 714–740.
Xie, S. H., & Ning, Y. M. (2005). Study on the development of big cities and the formation of metropolitan areas in China. *Urban Problems, 2005*(01), 11–15 (in Chinese).
Xu, J. (2008). Governing city-regions in China: Theoretical issues and perspectives for regional strategic planning. *Town Planning Review, 79*(2–3), 157–185.
Yang, B. J. (2004). Dilemmas and ways out for regional coordinated development in China. *City Planning Review, 81*, 26–34.
Yeh, A. G. O. (2001). Hong Kong and the Pearl River Delta: Competition or cooperation? *Built Environment, 27*, 129–145.
Zhang, L. (2008). Conceptualizing China's urbanization under reforms. *Habitat International, 32*, 452–470.

Chapter 9
Study of and Prospects on the Self-Organizing Evolution Stage of County Urbanization in Hubei

Lingyun Liu

The self-organizing system is a system that can be organized, created, and can evolve by itself without outside specific instructions. It can form an ordered structure from disorder automatically (Wu Tong 2001). By using self-organization theory, this article tries to analyze the historic evolution process of Hubei county's[1] urbanization and forecast future development trend from conditions, performances, and mechanism.

9.1 Self-Organization of Urbanization

9.1.1 Self-Organization Conditions of Urbanization

Self-organization should be formed and maintained under the following four conditions: opened system, disequilibrium, ordered fluctuation, and nonlinear function (under which the self-organization system is activated). Many phenomena in nature and human society are self-organized, including urbanization, which is a process in which a rural population migrates to city continuously. Even with state macro-control, this process is self-organized because settlement, job selection, and house purchases are decided personally (Chen Yanguang 2006).

[1] In this article, county contains counties, county-level cities, and municipal districts far from central urban area.

L. Liu (✉)
Architecture and Urban Planning School, Huazhong University of Science and Technology, Wuhan, China
e-mail: liulingyunphoebe@163.com

Q. Pan, J. Cao (eds.), *Recent Developments in Chinese Urban Planning*,
GeoJournal Library 114, DOI 10.1007/978-3-319-18470-8_9

9.1.2 Self-Organization Performances of Urbanization

There are four performance patterns for self-organization (Sun Zhihai 2004); (Zhang Yongqiang 2006):

9.1.2.1 Self-Creation (Rebirth and Transition)

Self-creation is an ordered process moving from non-organization to organization, or a process moving from a simple organization state with a low degree of ordering to a complex organization state with a high degree of ordering; this process is a system organization development or transition process from nothing; a quantitative change and mutation process; for instance, the initial formation and structural transition update of a urban system.

9.1.2.2 Self-Expansion (Gathering and Connection)

Self-expansion is a process where the organizational complexity is increased without changing the organizational level of a system and an expansion process of the amount of system organization; this process is a gradual change and quantity change process; for instance, the scale expansion of a urban system without changing the structure.

9.1.2.3 Self-Maintenance (Stagnation and Balance)

Self-maintenance is a relatively static process of self-organization behavior that keeps the current situation without changing the organization level and complexity of system; urbanization has a long-term micro-increase state at the early and late periods, which can be approximately considered as the stagnation and balance state (the former is low-level balance state while the latter is high-level balance state; the theoretical stagnation and balance state hardly ever exists in reality).

9.1.2.4 Self-Degeneration (Atrophy and Dissolution)

When the physical energy information of a system cannot satisfy the current structure form operation, the system may be disassembled and degenerated to the lower level state; there are many examples of degenerated cities including, currently, the awkward situation in Detroit.

9.1.3 *Self-Organization Mechanism of Urbanization*

The self-organization system is composed of three parts: external function (power input), internal function (competition and cooperation) and boundary (protection and isolation). Take 'boiling water by erecting boiler' as an example. The boiler is filled with tap water at a normal temperature which is still when not being heated and changed when being heated; the boiler bottom is heated and then the upper part of the water is cold while the lower part is hot; water molecules in the upper part automatically move downwards in order to absorb the heat, thereby forming a macroscopic upper and lower water-circulation phenomenon (Chen Yanguang 2006).

Urbanization is a similar process: some advantages in the area are gathered economically first (the early form of city), then the uniformly distributed population gathers quickly to this place in order to get resources, thereby forming population migration urbanization (Fig. 9.1).

9.1.3.1 External Function: Power Input

Depending on the external situation, the system keeps personnel, material, energy, information, and capital communication with the outside all the time, including input and output; if urbanization is expected to positively run (maintain or develop) continuously, the above-said direction should be a positive input (or, the positive is greater than the negative), otherwise the urbanization will become degenerated or

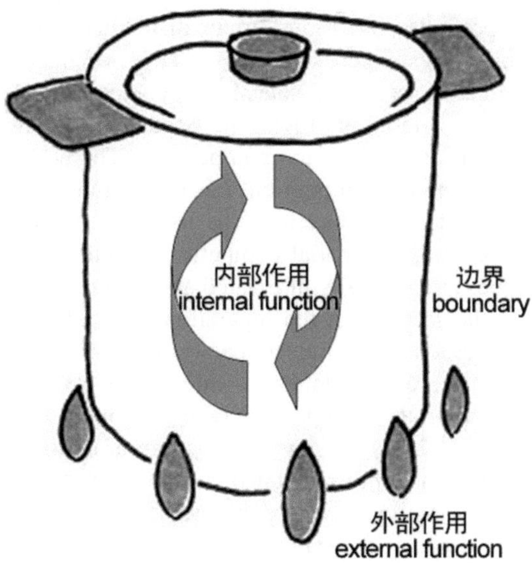

Fig. 9.1 "Boiling water" self-organizing mechanism

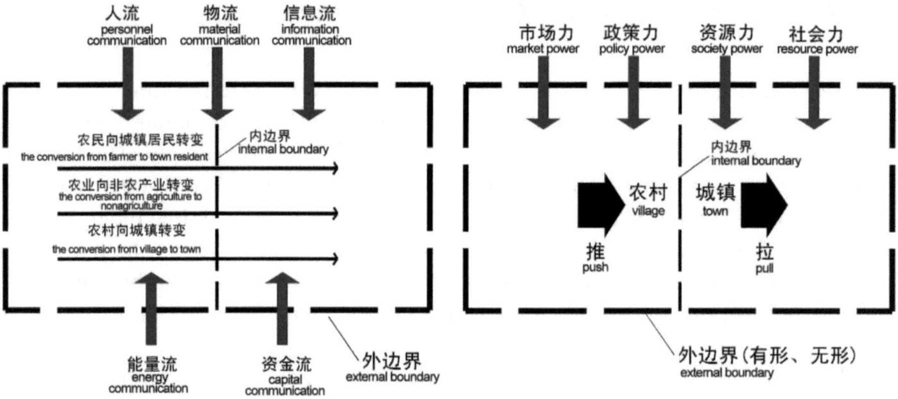

Fig. 9.2 Self-organization mechanism of urbanization

extinct. This input process can be summarized as market power, policy power, society power, and resource power from the perspective of urbanization (Fig. 9.2).

9.1.3.2 Internal Function: Competition and Cooperation

The internal function of a self-organization system is realized by competition and cooperation. Competition in a subsystem makes the system area unbalanced, while cooperation links the movement trends in a system together under unbalanced conditions so as to form order parameters and take advantage of positions supporting or serving the system's integral evolution (Haken 2005).

Urbanization is a spatial competition process getting survival and development opportunity by using certain differences and advantages and interactions under limited resources, selecting the superior and eliminating the inferior on spatial and temporal scales and selecting development. For instance, as the farmer departs the village and enters town, the expansion of town along the river and road reflects spatial occupancy and competition. The ecological niche differentiation resulting from spatial competition and the formation of spatial gradients accelerates area unbalance and decreases the integral benefits. After long-term development and evolution, the spatial competition results in spatial cooperation. Different spatial types are gradually integrated and appear in competition so as to not be eliminated. At the same time, the separation of time and space or the conversion of resource utilization form the cooperation among all types; for instance, the functional structure and scale level structure of a urban system is formed after long-term evolution and competition (Zhang Yongqiang 2006).

In a word, the subsystems in an area make the inside change from disorder to order and from lower level to higher level by mutual competition and cooperation. This is a process where the order parameter forms and reinforces, including three conversions: the conversion from farmer to town resident, the conversion from agriculture to non-agriculture, and the conversion from village to town (further generalized as 'the village push force and the town pull force').

9.1.3.3 Boundary: Protection and Isolation

There are boundaries among systems and within a system (so as to protect the system) as well as links inside and outside, just like the skin and immune system of people. The boundary is universal – the visible boundaries of an area comprise mountains, rivers, borders, geographical position, and traffic while the invisible boundaries comprise administrative boundaries, culture, customs, policies, and systems. The system with open boundaries is well-developed because the external power is easy to digest and absorb by internal functions. The closed boundary is opposite and will resist, defend, and obstruct the system from external power. In the urbanization self-organization process, different types of boundaries result in different urbanization speeds, qualities, forms, and structures.

9.2 Urbanization Self-Organization Evolution in Hubei County

9.2.1 1949–1978: Suspending and Atrophy Period

9.2.1.1 Heavy Industry Priority Strategy and Three-Line Construction Under Catching-up Target

At the initial period of the establishment of modern China, the government selected the catching-up strategy of prioritizing development of heavy industry and took the large-scale industrial construction focused on three-line construction. Hubei province was developed as a nationally important spatial point due to strong industrial and agricultural advantages and central strategic position. Under great national support, Hubei formed an industrial configuration pattern centered on Wuhan, Xiangfan, and Yichang, three traffic hinges linked by the Hubei East, Hubei West and Hubei North economic belts mainly signified by Wuhan Iron and Steel (Group) Corp., Hubei No. 2 Automobile Co., Ltd. and Ke-chou-pa Key Water Control Project during that period. The number of cities increased from two to six (Gu Chaolin et al. 2012), which became the basic pattern of the current Hubei urban system and formed the typical urban-rural dual structure.

9.2.1.2 Town-Commune Transition, Function Degeneration, Quantity Sharp Declination, County Wandering and Shrinking

Big cities grew vigorously while the county development stagnated or shrunk. In 1955, the state council issued *the decision on setting city and organic towns* which canceled and combined small towns and sharply reduced the amount of small towns. In 1963, the state council reissued *the instruction to adjust city and organic towns and reduce city suburbs*, which improved the town formation standard and further reduced the number of small towns. The state also limited the free trade of country business, so many small towns and peddlers' markets further declined and degenerated to villages, which reduced the number of small towns and shrunk the small towns through 1984 (Zhou Zhou 2011). In the population census of 1964, Hubei province had only 189 organic towns (34.1 % of the town population in the whole province), which dropped 14.01 % from 48.37 % in 1949 (Committee for Editing Hubei Chorography 1997). In Yingshan county, for example, there were 13 towns and 164 villages in the Republic of China era, which were changed to communes; in 1964, there was only one town – county seat; there were no towns in 1982.

9.2.1.3 Strong Population Retention of County and Excessive Industrial Output Under Specific Interventions

Under the planned economy, the Hubei urbanization was performed under strong state intervention, the self-organization system was limited, and other-organization ran the leading function. The other-organization is a goal-oriented integral control process, the organization power therefore came from outside, and the internal subsystem just received the power without automatic adjustment and creation. In this period, the government was the prime mover of urbanization; the urbanization requests of a company and people were suppressed and could not form internal support.

This system lacks feedback, adjustment, and cooperation, so the system design will be continuously accumulated and amplified under any little maladaptation until it forms great system unbalance, which is mainly manifested in the following ways:

Firstly, urban-rural division result in seriously stranded population in county. The quick development of big city greatly attracts the farmer; however, the urban-rural segregation policy controls the free flow of farmers to the city. Moreover, the combined towns are canceled so as to improve small town formation, which sharply reduces the number of small towns and a great number of country surplus labor is deposited in country and stops the improvement of population migration and urbanization.

Secondly, industrial separation cause excessive industrial output of county. The heavy industry is hardly related to the resource conditions and industrial basis of a county; the product downstream chain of a state-owned enterprise is arranged by

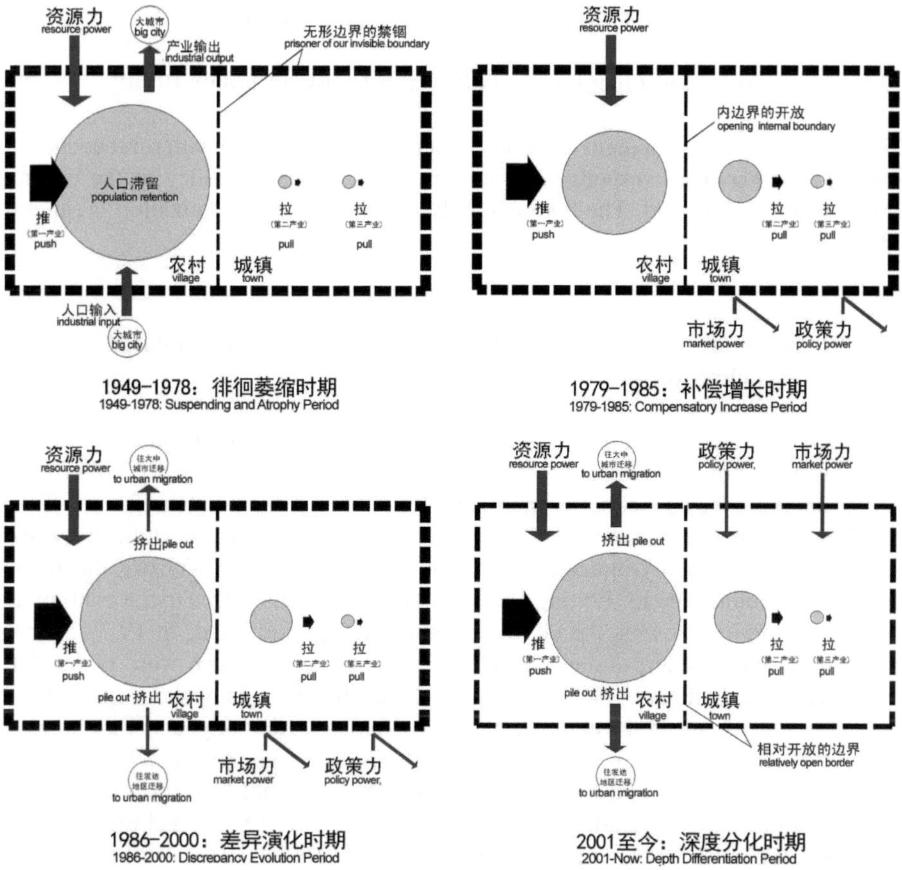

Fig. 9.3 Analysis on county self-organized mechanism of urbanization in Hubei Province

the state out of the county, which cannot form the development of related downstream industry; the outside effect of these enterprises cannot be shared by the county out of the local city but consume the surplus labor and product of Hubei province; furthermore, the price scissors of industrial and agricultural products forms the 'absorption-absorbed' relation of a big city such as Wuhan to county. Finally, the urban system is top-heavy under the system of valuing the city and ignoring the country; the big cities are developed quickly and the small towns shrink and degenerate (Fig. 9.3).

9.2.2 1979–1985: Compensatory Increase Period

9.2.2.1 Reform and Opening-up Policy and Dividend Policy Break

Since 1978, the development limit on small towns was reduced with rural economic reform, land contract responsibility system and broadening policy of the organic town forming standard. The automatically formed county urbanization is quickly developed.

9.2.2.2 Organic Towns Leap to Develop, County and City Updated Synchronously

In 1984, the state council readjusted and broadened the standard of small towns and approved *the report on adjusting organic towns* from the Ministry of Civil Affairs. After breaking the barrier between urban and rural areas, the number of small towns in Hubei quickly increased. In 1978, there were 105 small towns in Hubei province, which were all county-managed towns; in 1985, there were 858 organic towns (Fig. 9.4) (Committee for Editing Hubei Chorography 1997). Due to the continuous growth of organic towns, the city scale is accordingly adjusted. In 1979, Shashi, Xiangfan, and Yichang (county-level cities) in Hubei province were updated to prefecture-level cities. In the same year, Echeng, Jingmen, Laohekou, and Suixian were formed as county-level cities. In 1983, Xiaogan, Xianning, and Danjiangkou (county) and Enshi were updated to city status; Jingmen, Echeng, Suizhou and Laohekou were combined by county and city; Jingmen and Ezhou (Echeng) were updated to provincially administered municipalities. By 1985, Hubei province had 14 cities, including 1 megalopolis, 7 medium-sized cities and 6 small cities (county-level cities) (Gu Chaolin et al. 2012). Within a single county, take Hanchuan as

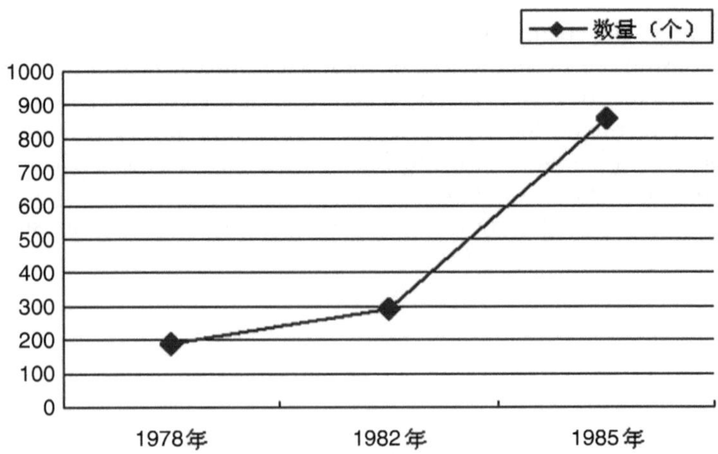

Fig. 9.4 Quantity changes of towns in Hubei Province from 1978 to 1985

example, there were 15 communes, 3 towns and 6 villages in 1981, and there were 8 districts, 8 towns and 6 towns by the end of 1985; the number of organic towns increased from 3 to 8, and the ratio of urban population increased from 7.98 % in 1978 to 28.56 % in 1985 (Geographic Name Leading Group in Hanchuan County 1981; Committee for Editing Hubei Hanchuan Chorography 1992).

9.2.2.3 Opening to Foreign Countries and Enlivening the Domestic Economy; Personal Competition and Self-Organization Creation

The open boundary is especially important to the self-organization process. After 1978, due to rural economic reform and the land contract responsibility system, the production activity of farmers was greatly stimulated and productivity was unprecedented. As a result, the land allowed per farmer was greatly reduced, thereby forming a surplus of labor. In order to maximize personal benefit, farmers started to join in the urbanization process and become its main driver. In 1984, the state council issued *a notice about settlement of farmers in town* that allowed farmers to work and go into business in and called on related departments to help the farmers do so, which broke the barrier between urban and rural areas to an extent. Towns were converted from closed form to opened form, in which the population could move and gather freely, and promoted industrial and commercial development. Small towns were gathered spatially for the first time and the self-organization process of urbanization started to form.

At the same time, the city forming standard was adjusted and the number of medium-sized and small cities increased, which shows the hierarchy and gradualness of the self-organization process of urbanization; this is a bottom-to-top process from low level to high level (Figs. 9.3 and 9.5).

9.2.3 1986–2000: Discrepancy Evolution Period

9.2.3.1 Eastward Movement of State Center, Initial Appearance of 'Center-edge' Pattern

Due to the reform and opening-up policy, market economy reform, and the globalization process, the southeastern coastal areas in China were quickly developed as overseas investment priorities due to their regional advantages The government also issued a series of opening policies for moving the development center to the east. The southeastern coastal areas became the national economic center, and the wide middle and western areas represented by Hubei province were marginalized in this competition.

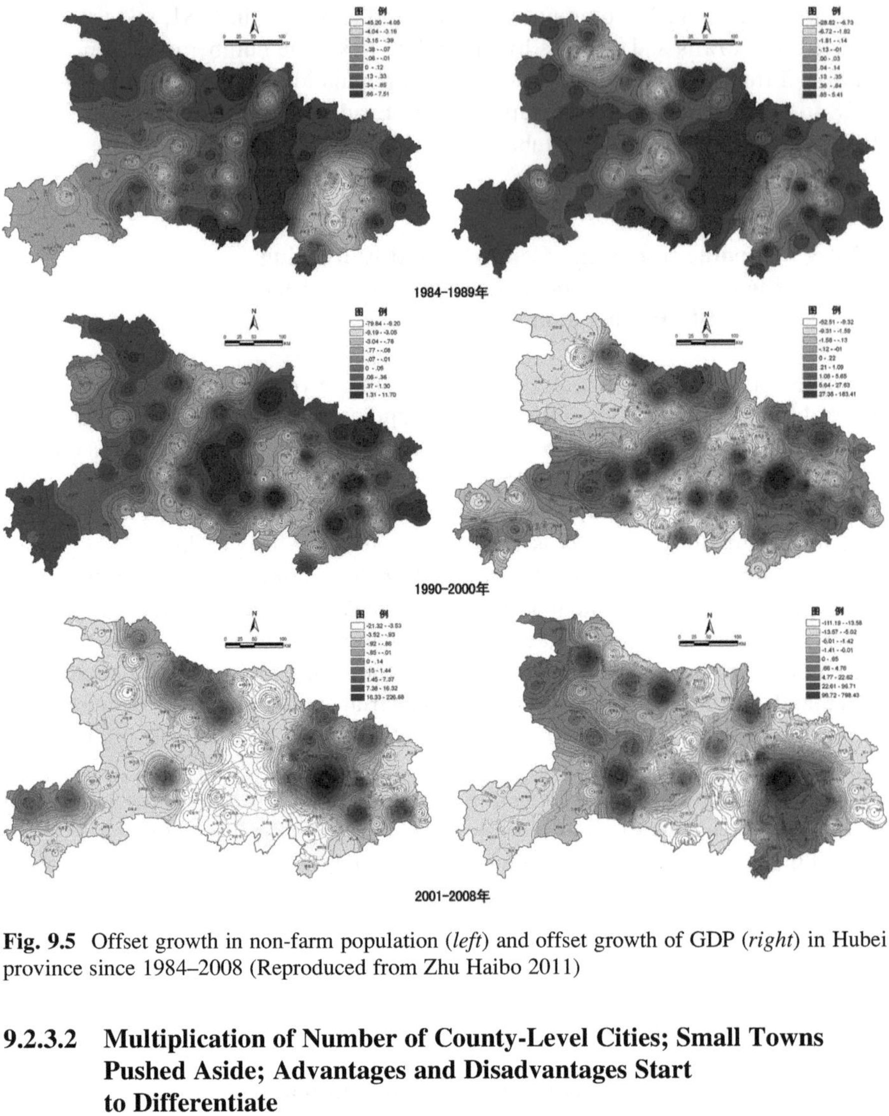

Fig. 9.5 Offset growth in non-farm population (*left*) and offset growth of GDP (*right*) in Hubei province since 1984–2008 (Reproduced from Zhu Haibo 2011)

9.2.3.2 Multiplication of Number of County-Level Cities; Small Towns Pushed Aside; Advantages and Disadvantages Start to Differentiate

At this period, the scaled development of township enterprises required concentration in towns with good regional conditions and great development potential, thereby promoting the spatial concentration of wealth and migration of the population to big towns and cities. In 1999, Hubei province had 36 cities, mainly of medium size (county-level cities) (Gu Chaolin et al. 2012). From 1986 to 1988, the number of county-level cities increased annually by 6, 5, and 4 (respectively). In 1993, the whole province had 22 small cities (Fig. 9.6). In the period 1979–1992, the annual increase rate of gross domestic product, financial income, industrial and

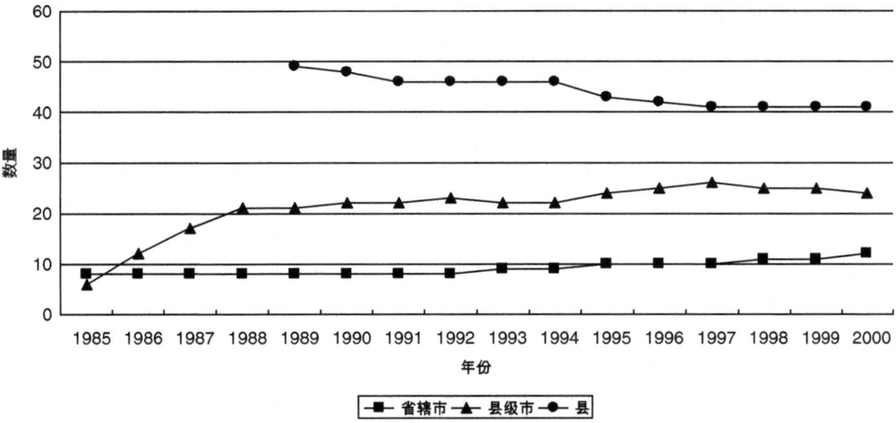

Fig. 9.6 Quantity changes of cities and counties in Hubei Province from 1985 to 2000

agricultural total output, and total volume of retail sales of county-level cities in the whole province were 9.58 %, 11.43 %, 12.14 % and 16.4 %, respectively. These were higher than the provincial index of 8.98 %, 10.96 %, 8.16 % and 13.73 % of said the same four items and higher than the growth in prefecture-level cities; county-level cities therefore caught up to the prefecture-level cities. In 2000, 6 counties in Hubei were listed in the top 100 national counties, including Xiantao city, Qianjiang city, Xiangyang city, Zaoyang city, Yichang city, and Tianmen city.

At the same time, small towns developed slowly, in comparison with strong counties and strong towns in coastal regions. While Hubei focused on medium-sized and small cities in this period, small towns had insufficient infrastructure, low household disposable income, slow industrial development, and large differences with coastal region. Different counties were compared: in 1985, the Xiantao (Mianyang at that time) was established with 12 communes, 2 towns and 8 farms; in 1986, the county was updated to city; in 1994, the Xiantao was listed as the provincially directly managed city; in 1999, the Xiantao was established with 19 towns, 6 villages, 4 streets and 2 farms (Mianyang Transportation Department 1985; Mianyang Geographic Name Leading Team Office 1982; Wu Decai and FU Xianrui 2000). The organic town was updated quickly, and urbanization speed obviously rapid. At the same time, the Yingshan county had no change, so the county under different conditions faced different development patterns, and the development paths started to differentiate.

9.2.3.3 Insufficient Power, Town Competition and Adaptability Expansion

The self-organization system has self-adaption for adjusting the power by an internal system when external power is insufficient; in this period, the Hubei county performed the adaptability expansion by differentiation.

At the beginning of the reform and opening-up policy, the manufacturing industry of the southeast coastal areas were quickly developed, based on global industrial division and industrialization, and quickly finished the local urbanization, which widened the difference between the middle part and eastern part. Hubei joined in the international division and was placed in an inferior geographical position. Many former industrial items were therefore tightly linked with the local comparative advantages, which increased production costs. The industrial output of the whole province mainly consisted of middle-stream and upstream products of the industrial chain. After losing the advantages of state controlled procurement and distribution, the benefit was hardly assured, and the Hubei product market continuously shrunk under the market economic environment. At the end of 1980s, the special pattern centered on coastal regions was formed. Hubei province formed a large labor flow group in the form of 'downwards going to south part for work.' If the Hubei county released prime power due to the open boundary at the former stage and benefited from rapid (leaping) development, at this stage, under the new national policy and market economy, Hubei county could not develop at the past speed. Because it lacked sufficient power and continuously consumed energy (a large amount of the population was urbanized in other towns), the Hubei county could only survive by adjusting its system and rebuilding its internal structure. In this stage, Hubei county mainly developed the medium-sized and small cities. The main urbanization place was converted from the small town to the medium-sized city and the small city. On one hand, this is the natural rule of self-organization development; on the other hand, under the strict situation of insufficient external power, this is an adaptive choice – quick development of a small number of medium-sized and small cities (Figs. 9.3 and 9.4).

9.2.4 2001-Present: Depth Differentiation Period

9.2.4.1 Continuously Increased Difference Between Coastal Area and Inland

After 2000, the difference between the coastal area and inland area has been increasing, with the middle and western areas becoming areas of high-speed economic increase and quick urbanization. At the same time that globalization drives international capital investment to expand towards the middle and western cities, the coastal industry spreads inland. In this new industrial gradient transition, some big and medium-sized cities in Hubei province are being quickly developed as investment 'hot cities.'

9.2.4.2 Polarization Development of County Surrounding (Megalopolis) Big City

In the new century, several (megalopolis) big city in Hubei province have featured quick development: Wuhan, Yichang, and Xiangyang. Along with industrial structure adjustment of the (megalopolis) big city, the comprehensive service function and radiation of the (megalopolis) big city improves the energy and generally enhances the economic development and population concentration of the surrounding county, thereby forming the central polarization development of county surrounding (megalopolis) big city (Fig. 9.5). The Hubei county urbanization is comprehensively sequenced; the top 20 counties are mainly concentrated in the following three areas: (1) eastern Hubei counties surrounding Wuhan, Ezhou, Huanggang, and Huangshi; (2) middle Hubei counties surrounding Yichang, Jingzhou, and Jingmen; (3) the northwest Hubei counties surrounding Xiangfang and Shiyan. These counties are close to big cities or located in the core city ring and are obviously affected by the big city. The counties located on plain and hill has obvious resource and geographic advantages. Take Huangpi and Daye as examples: in 2004, Huangpi contained 6 towns, 2 villages and 7 streets; up to the end of 2003, the administrative division contained 15 streets and 1 village (all organic towns are updated to streets) (Chen Qin and LI Zhongzhou 2009; Huangpi Geographic Name Leading Group 1981). In 2012, only Daye in Hubei was listed in the top 100 national counties in economic terms; Daye was ranked number 262 in 2005 and number 97 in 2011, an improvement of 165 places in 6 years.

Generally speaking, all factors are gathered to a few big and medium-sized cities and a small number of counties are radiated by a center city. Most of the counties urbanize slowly, which therefore broadens the difference. Since 2000, the population of small cities in the county of Hubei province was obviously reduced. In 2008, the ratio of population in small cities is 20.96 % of the population of the whole province and 6.97 % less than it was in 1999. In 2008, the population of 43 of the 51 small cities in Hubei was less than 0.15 million, the population of 22 small cities was less than 0.1 million, and there were 733 organic towns at a density of 39.4/km^2 (which is greater than the national density of 20.2/km^2). The average population in a town is less than 6000, and more than half of the county is in the form of 'dragging large cars by small horses' (Zhu Haibo 2011).

9.2.4.3 Population Loss Acceleration, Area Competition, Coexistence of Increase and Degeneration

With the development of the market economy, the competition between coastal areas and inland areas is getting fiercer; the economic power of coastal areas is stronger, so the population of the middle and western part flows outward continuously, the population difference is increased, and inland urbanization power is insufficient. The competition breaks the balance and accelerates the imbalance in higher levels, which is performed as coexistence of increase and degeneration, prosperity and depression. Within Hubei province, there is coexistence of polarization of counties surrounding (megalopolis) big city and the maintenance or depression of counties out of the way. The county seat is obviously faster than small towns inside counties. The development method of 'protecting the center by giving up surrounding parts' is a necessary self-protection measure under the loss acceleration condition (Figs. 9.3 and 9.4).

9.3 Analysis and Expectation of Urbanization System in Hubei County

9.3.1 External Function: Conversion from Protective Development to Extended Development

The external function of a self-organization system is a necessary condition, just like dinners for human to add energy. External function has different types, sizes, directions, and combinations. It makes different system internal structures and brings the system totally different results. Before the reform and opening-up policy, the government focused on urbanization and isolated cities from country. Therefore, the dual economy is obvious, the county was not positively supported but rather was exploited, and the self-organization was shrunk and degenerated. After the reform and opening-up policy, the county released the countryside 'push force,' which had been suppressed for a long time by the policy; in a short time, the small town quickly developed in a 'leaping' way. However, with the double loss of policy power and market power, the power became increasingly insufficient, and urbanization had to adjust the structure and continuously transferred the main battlefield from small town to small and medium-sized cities, and counties surrounding (megalopolis) big city by an adaptive expansion road. Currently, under the global financial crisis, the government has started to move the economic center inwards, and has changed development way from semi-market to full-market. Under the function of economic increase in middle China, industrial transition in coastal areas, and industrial expansion in big cities, the power resources are various and getting stronger. The Hubei county will be converted from the past 'protected' development to extended development and will gradually become stable and ordered.

9.3.2 Internal Function: Conversion from Low-Level Balance to Unbalanced Division

The self-organization system is structurally ordered by competition and cooperation of internal subsystems. Before the reform and opening-up policy, the government called for top-to-bottom urbanization according to the plan steps. The development was performed by integral benefit but not personal benefit, so self-organization was greatly limited. After the reform and opening-up policy, the rural economic reform and land contract responsibility system generated surplus labor which was gathered to small towns and formed the self-organization and personal competition class. However, the limited resources could not be distributed to every town equally – the towns with advantages attracted more population and industry; medium-sized and small cities developed well, small towns were suppressed, and the personal competition class was updated to town competition class. Then, the increased competition brought 'center-edge' where the increase and decline coexisted, the counties surrounding big city were developed quickly and integrated into the modern economy. The remote counties were urbanized slowly and even stayed in agricultural. There will be increasing polarization and difference between counties. It comes into area competition period. In the future, the area's spatial imbalance under the effect of macroscopical natural landscape, (megalopolis) big cities, and regional transportation will continuously work and form a new 'group-belt' spatial pattern. The counties in these areas will obviously develop first, and the level of county will be further divided. The counties surrounding the big city are divided into two layers at least: direct radiation and indirect radiation. The plain agriculture counties may be divided into three layers: middle plain, national channel, and southern flood discharge. The mountain area counties are divided into two layers at least: eastern near big city and western remote mountain. Generally speaking, the low-level balance in the past will be converted into imbalance, which increases the strength of the difference.

9.3.3 Boundary: Conversion from System Closing to Multi-channel Opening

The boundary is hardly apparent, especially when all system items are operated smoothly. When the system is operated un-smoothly, it is likely because the boundary is broken. The closed boundary has two results: (1) the external power cannot sufficiently input in the system; (2) factors in the system cannot be regrouped in sequence. Before the reform and opening-up policy, the Hubei county urbanization had two boundary problems: (1) the boundary between the county and the outside, and (2) the boundary between the village side and town within the county. At this period, the government focused on the development of big cities such as Wuhan, especially in terms of heavy industry, and limited the rural population's ability to move to a big city. Moreover, the government also changed towns into communes and encouraged young people to work in the rural areas, limiting the amount of people on the land and greatly obstructing free population migration. The urbanization process is full of twists and turns, mostly because of large effects from invisible policy boundaries. After the reform and opening-up policy, population mobility was greatly enhanced, but the policy dividing city from countryside represented by household registration systems still worked, so foreign urbanization and incomplete urbanization were frequent. The policy boundary between city and countryside was not completely eliminated and the visible boundary was to be broken. Hubei province is located in the middle part of China and surrounded by mountains so Hubei faces multiple boundary limits from transportation, position, and natural landforms. The high speed railway, international flights and ports along river must be constructed. Domestic and international connections should be made, and multi-channel and multi-level opening will improve the development of Hubei county in the future.

Acknowledgment Project aided by National Science Foundation of China (51178200; 51478199; 51408248).

References

Chen Qin, & LI Zhongzhou. (2009). *Huangpi general history*. Wuhan: Hubei People's Press.

Chen Yanguang. (2006). Self-organized characteristics and criteria of the development of Chinese cities. *City Planning Review, 6*, 24–30.

Committee for Editing Hubei Chorography. (1997). *Hubei chorography geography (I)*. Wuhan: Hubei People's Press.

Committee for Editing Hubei Hanchuan Chorography. (1992). *Hanchuan chorography*. Beijing: City Press.

Geographic Name Leading Group in Hanchuan County. (1981). *Geographic name Journal of Hubei Hanchuan*.

Gu Chaolin, Zhao Min, & Zhang Jingxiang. (2012). *Research on provincial territory urbanization strategic planning*. Nanjing: Southeast University Press.

Haken, H. (2005). *Erfolgsgeheimnisse der natur synergetik: die lehre vom zusammenwirken*. Shanghai: Shanghai Translation Publishing House.

Huangpi Geographic Name Leading Group. (1981). *Geographic name Journal of Hubei Huangpi*.

Mianyang Geographic Name Leading Team Office. (1982). *Geographic name Journal of Hubei Mianyang*.

Mianyang Transportation Department. (1985). *Mianyang Transportation Journal*.

Sun Zhihai. (2004). *Social evolution theory and model of self-organization*. Beijing: China Social Sciences Press.

Wu Decai, & FU Xianrui. (2000). *Xiantao education journal*. Wuhan: Wuhan University of Technology Press.

Wu Tong. (2001). *Research on self-organization method methodology*. Beijing: Tsinghua University Press.

Zhang Yongqiang. (2006). *Self-organization of city special development and city planning.* Nanjing: Southeast University Press.
Zhou Zhou. (2011). *Research on the development path of small towns in Hubei province.* Dissertation, Huazhong University of Science and Technology, Wuhan.
Zhu Haibo. (2011). *Research on the urbanization spatial evolution and development in Hubei province after the reform and opening policy.* Dissertation, Huazhong University of Science and Technology, Wuhan.

Chapter 10
From Three-Old Reconstruction to Expansion and Promotion: The Strategic Transformation of County-Level Cities in a Rapid Urbanization Area – A Case Study of Guangdong Province in Mainland China

Yikeng Luo

10.1 Introduction

Under the context of balancing new-type urbanization, industrialization, informatization and agricultural modernization in mainland China, it is important to improve the efficiency of urban development land use in county-level cities which are large in quantity and widely distributed. In 2011, the urbanization rate in Guangdong province[1] had reached 66.5 %, while mainland China had passed 50 % at the same time. This paper focuses on incremental land and stock land during new-type urbanization – incremental land refers to unused development land and stock land refers to the inefficiently used development land. With the continuous expansion of urban scale and the relative shortage of development land, it is becoming important to advance new-type urbanization to balance the development of incremental land and the redevelopment of stock land, on the basis of maintaining the steady rise of social economy and promoting the overall improvement of ecological environment, particularly in Guangdong as a typical rapid urbanization area.

Although county (xian cheng) does not belong to the statistical category of city in mainland China, county is frequently a basic unit of relatively complete socio-economic function which has the same function as city (Weng 2005). As the indispensable component of urban system, county not only bears the important responsibilities of urban-rural integration, but also shares the development pressures of megalopolis and metropolis. Hence, this paper considers that county is one

[1] Guangdong province, locating in south of mainland China, has 21 prefecture-level cities with the total resident population reaching 104.3 million in 2010. With its GDP reaching 6.22 trillion yuan in 2013, Guangdong province has been ranked number one for 25 consecutive years.

Y. Luo (✉)
School of Architecture, South China University of Technology, 510640 Guangzhou, China
e-mail: luoyikeng@qq.com

Q. Pan, J. Cao (eds.), *Recent Developments in Chinese Urban Planning*,
GeoJournal Library 114, DOI 10.1007/978-3-319-18470-8_10

part of county-level city (Xu 2001). According to the statistics of Ministry of Civil Affairs, there are 1999 county-level cities that include 1630 counties in mainland China by the end of 2011.

10.2 Basic Understanding of Urban Development Land Use in County-Level City

County-level cities face a lot of problems during new-type urbanization. For example, the developmental growth is relatively slow, the developmental scale is relatively small and the developmental quality is relatively low. Firstly, industrialization promotes urbanization but urbanization lags industrialization, leading to the weak attraction of county-level cities. Secondly, resources are overly centralized in megalopolises and metropolises, while county-level cities cannot get enough resources. The rapid urbanization area is the core region during new-type urbanization, which is pioneering, experimental and exemplary of its development (Wu et al. 2005). County-level cities in a rapid urbanization area present different characteristics and faces various challenges in urban development land use under the profound effect of regional growth poles such as Beijing-Tianjin-Hebei region, Yangtze River Delta and Pearl River Delta.

10.2.1 Basic Characteristics

The basic characteristics of urban development land use are abundant land resources, inefficient use and strong development potential in having further expansion in county-level cities. Firstly, land resources are relatively cheap and expanding space is relatively abundant. As a basic input factor to promote social and economic development, land resources, especially urban development land is still the developmental backbone in county-level cities. Meanwhile, county-level cities, which are near regional growth poles, keep their development potential with desireable location, cheap land and abundant space. Secondly, the spatial structure is in need of optimization and the development intensity is generally low. Due to the different levels of social and economic developments, the urban development land in megalopolises and metropolises is characteristically vertical and dense (Wang 2003), and emphasizes job-housing balance during spatial expansion; the urban development land in county-level city is more horizontal and less dense, and emphasizes functional zoning during spatial expansion. Finally, industrial land continues to expand and residential land continues to increase. The structure of GDP (gross domestic product) between primary industry, secondary industry and tertiary industry changes from “secondary, primary, tertiary” type in the beginning of reform to “secondary, tertiary, primary” type at present, especially in rapid urbanization area (Duan and Zhang 2007). At the same time, it is the opportunity for county-level city in rapid urbanization area to undertake the transfer of industries and labor force from regional growth pole, so that the urban development land especially industrial and residential land can keep sufficient growth momentum.

10.2.2 Main Challenges

The main challenges of urban development land use are balancing the development of incremental land and redeveloping stock land for county-level cities. These challenges are especially apparent in rapid urbanization areas under the background of fierce competition among cities and intense contradiction between people and land. In terms of the development of incremental land, deciding the scale of urban development land is always a difficult problem. At present, the scale of urban development land depends on the general plan for land use, and the spatial layout of urban development land is made according to the annual land use plan. However, breaking the urban development boundary and blindly expanding the scale of urban development land becomes the drawback of urban development land use. At the same time, the supply of land often exceeds the demand of land, because of the uncertainty of developmental direction and the inaccurate judgment of marketable demand. Redevelopment of stock land is considered to be an effective way to improve the efficiency of urban development land use in general. However, the confused property rights and complex benefits make it difficult to promote. In practice, only a few stock lands with excellent locations, high market valuations and good demonstration effects have had the opportunity for reconstruction, the remaining abundant and inefficient stock land cannot get improvement in the restrictions of manpower resources, financial resources and material resources.

10.3 Overview of "Three-Old Reconstruction" in Guangdong

Guangdong has become the pilot of economical and intensive land-use carried out by the Ministry of Natural Resources because of its developmental bottlenecks – shortage of land resources and inefficiency of land use – and the fundamental demand of optimizing urban space and promoting industrial upgrading. In 2009, People's Government of Guangdong made the policy of "Several Opinions on Promoting 'Three-old Reconstruction' to Facilitate Economical and Intensive Land-use", marking the official prelude of "three-old reconstruction". "Three-old reconstruction" refers to the redevelopment of old town, old factory and old village, aiming to redevelop stock land.

10.3.1 Purposes and Reasons

As the strategic action to promote industrial upgrading, institutional innovation and spatial adjustment (Chen 2010a), "three-old reconstruction" is a positive policy aimed at urban renewal (Liu et al. 2011), and the core is the innovation of land policy in Guangdong in the new period (Lai and Wu 2013). The contradiction between supply and demand of development land in incremental land is very prominent in Guangdong. The specified supply of development land is only 0.29 million MU according to "the Outline of the National Overall Planning on Land Use (2006–2020)," yet actual demand of development land is about 0.40 million MU average annually (Jiang 2008). The total "old town, old factory, old village" land area in stock land is more than 1.70 million MU, of which 0.34 million MU is old town, 0.81 million MU is old factory, and 0.64 million MU is old village in Guangdong according to the investigation (Zhao 2009). In this context, "the Work Program of Guangdong Facilitating Economical and Intensive Land-use as the Demonstration Province" made by People's Government of Guangdong and Ministry of Natural Resources, took the lead in putting forward the policy of "three-old reconstruction", aiming to promote the redevelopment of stock land, improve the efficiency of development land, accelerate the transformation of industrial structure, and enhance the environmental quality of urban human settlements.

10.3.2 Patterns and Effects

The patterns of "three-old reconstruction" are different from previous reconstructions in organization, planning and implementation. On the organization level, the offices of "three-old reconstruction", while integrating the functions of land department, development department, planning department, construction department, financial department and other relevant departments, are founded in various regions of Guangdong. On the planning level, the plans of "three-old reconstruction" are divided into the special plan and annual implementation plan in macro-level, and the regulatory plan and site plan in micro-level (Zhou and Fang 2011). On the implementation level, firstly, it simplifies the transformation procedure from "collective-owned development land" to "state-owned development land" in old village; secondly, it allows for completing the formalities of historical land according to the status quo and transferring land use rights by agreements; thirdly, "corner land (bian jiao di), sandwich land (jia xin di) and exclave land (cha hua di)" can be utilized flexibly, and scattered land can be applied to merged; finally, it establishes the return mechanism of land value increment (Wei and Wang 2011).

There are some good results of the economical and intensive land-use in "three-old reconstruction". According to "the 12th Five-year Plan of Land and Resources in Guangdong", the total scale of development land is within 1.934 million hectares and the total scale of new development land is no more than 0.107 million hectares, of which the total scale of "three-old reconstruction" is over 0.067 million hectares in 2015. By the end of June 2013, Guangdong had completed nearly 3000 projects, completing more than 0.15 million MU reconstruction area and economizing about 0.07 million MU development land in "three-old reconstruction" (Luo 2013).

10.3.3 Problems and Restrictions

The problems and restrictions of "three-old reconstruction" are limited space, complex benefits and its long duration. Firstly, the space is limited in "three-old reconstruction". The potential amount of stock land is theoretically large, but the spatial layout of stock land is always fragmented and dispersive, which seriously affects the development effectiveness. Hence, it is necessary to prevent extending the reconstructive range and to coordinate the comprehensive development benefit. Secondly, the benefits are complex in "three-old reconstruction". "Three-old reconstruction" implies the redistribution of urban-rural space and interest structure, which is bound to be a difficult process in space and interest coordination (Yang and Yuan 2010). Meanwhile, there are often unclear property rights boundaries, confused management authorities and serious illegal constructions, which increase the difficulties during "three-old reconstruction". Finally, the time period is long in "three-old reconstruction". Because it involves the demolition of house and the adjustments of land ownership (Wu and Wang 2013), the negotiation is relatively difficult, the investment is relatively large, and the period is relatively long. For example, Guangzhou has completed "three-old reconstruction" area of nearly 30 km^2 by the end of 2013, but it is only 5 % of the total scale of 554 km^2 (Zeng 2014).

It is worth further discussion as to whether or not "three-old reconstruction", which is the typical secondary urbanization pattern of limited space, complex benefit and long duration, is able to meet developmental demand opportunely and effectively in some small and medium cities, especially county-level cities in rapid urbanization areas. The megalopolises and metropolises such as Guangzhou and Shenzhen, which have a shortage of new urban development land, have to implement the policy to obtain incremental land, even though "three-old reconstruction" is relatively difficult. However, in order to enhance bearing capacity, competitive ability and radiation ability at present stage, most county-level cities with relatively abundant land resources focus their attention on expanding the scale of urban development land.

10.4 Necessity and Inevitability of "Expansion and Promotion" in Land

"Expansion and promotion" mainly includes three aspects: population, industry, and land. This paper discusses "expansion and promotion" from the perspective of land. Hence, this paper holds that "expansion and promotion" concretely includes developing incremental land moderately and redeveloping stock land opportunely to improve the efficiency of urban development land use. At present, Guangdong is actively approaching the policy of "expansion and promotion" in prefecture-level cities in east, west and north of Guangdong.[2] However, this paper maintains that there is also a certain inevitability and necessity to approach the policy of "expansion and promotion" in county-level cities. On one hand, county-level cities play a highlighting role in regional development, urban-rural integration, historical processes and land finance; on the other hand, "three-old reconstruction" which struggles with limited space, complex benefits and a long duration, is difficult to meet developmental demand opportunely and effectively in county-level city.

According to the statistics of "Guangdong Statistical Yearbook 2013", per capita GDP of Guangdong was 54,095 yuan in 2012. There are a total of 65 county-level cities (23 county cities, 39 counties, and 3 autonomous counties) in Guangdong, of which the GDP reaches 1.09 trillion yuan, accounting for 19.08 % of province's GDP. There are 50 county-level cities in east (9 county-level cities), west (12 - county-level cities) and north (29 county-level cities) of Guangdong, which become the basic units of social and economic development. Per capital GDP is about 15,000 yuan to 30,000 yuan (Fig. 10.1) and growth rate of per capita GDP is about 10–14 % (Fig. 10.2) in 50 county-level cities. Hence, they are the true portrayal of county-level cities in rapid urbanization areas, which are facing the opportunities and challenges in the development process.

10.4.1 Perspective of Regional Development

From the perspective of regional development, "expansion and promotion" will contribute to optimizing the existing urban system structure. At present, the problems of the urban system structure are mainly the following: the imbalance between regional growth pole and non-regional growth pole is widening and the uncoordinated development between cities is serious, because more and more

[2] Eastern Guangdong, with the total resident population reaching 16.89 million in 2010, covers 4 prefecture-level cities: Shantou, Jieyang, Chaozhou and Shanwei; Western Guangdong, with the total resident population reaching 15.26 million in 2010, covers 3 prefecture-level cities: Zhanjiang, Maoming and Yangjiang; Northern Guangdong, with the total resident population reaching 16.1 million in 2010, covers 5 prefecture-level cities: Yunfu, Qingyuan, Shaoguan, Heyuana and Meizhou.

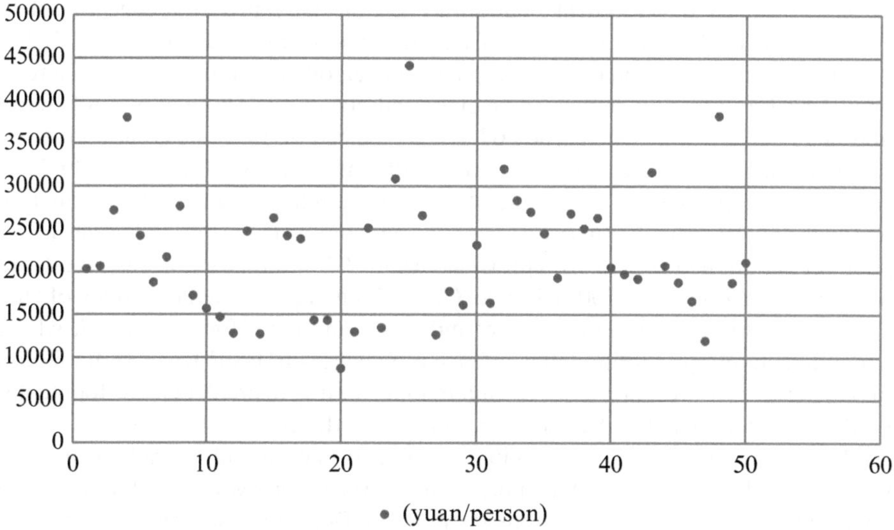

Fig. 10.1 Per capita GDP in 50 county-level cities in 2012

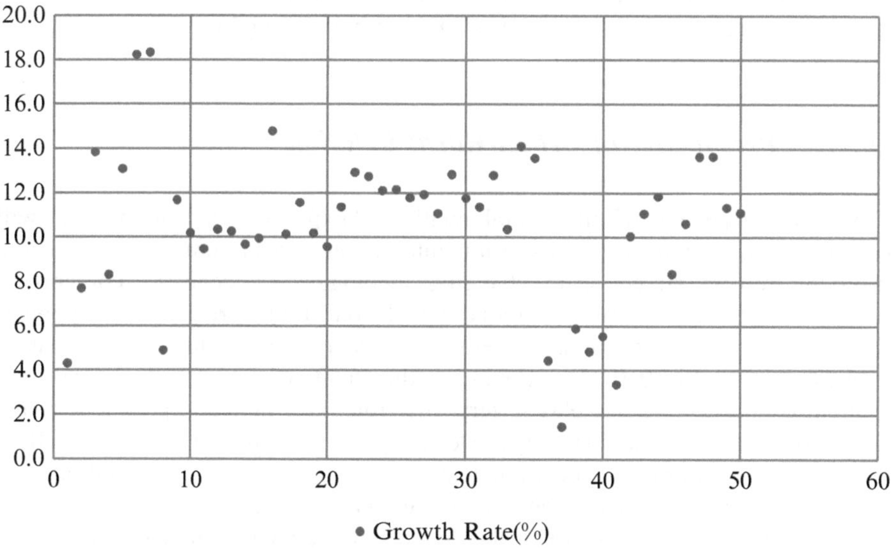

Fig. 10.2 Growth rates of per capita GDP in 50 county-level cities in 2012

development resources are over-centralized in the regional growth pole, especially in the megalopolis and metropolis.

With the transformation and upgrading of the Pearl River Delta, which is the regional growth pole, there will be more industry and labor force transferred to east, west and north of Guangdong. This requires county-level cities to be well-prepared to meet the new development opportunities, especially in the storage of urban development land. As the forefront of Reform and Opening, Guangdong is facing the problem of uncoordinated development in regions and cities, though there is great-leap-forward development in society and economy. Although the urbanization rate in Guangdong has reached 67.4 % by the end of 2012, areas east, west and north of Guangdong have a huge disparity, with rates of 83.84 %, 59.05 %, 39.72 % and 45.3 % respectively. As a whole, it is a long-standing problem in coordinating regional development because of the small economic scale, low industrial level, poor per capita income and difficult rural surplus labor migration in east, west and north of Guangdong (Liu 2009). Hence, Guangdong implemented the policy of "the Decision on Promoting the transfer of industries and labor force" in 2008, which aimed at actively promoting labor-intensive industries in Pearl River Delta region to transfer to east, west and north of Guangdong, aiming to break the developmental bottleneck of regional imbalance, accelerate the development in east, west and north of Guangdong while promoting the transformation and upgrading of Pearl River Delta. By the end of 2012, more than 6,000 enterprises with 5.59 million job opportunities have transferred from Pearl River Delta to east, west and north of Guangdong, which has created 36 industrial transfer parks and 2 economic special zones while generating over 1 trillion yuan industrial output (Xie 2012). According to the investigation of the spatial layout, almost a half of the 38 industrial parks located at county-level cities east, west and north of Guangdong.

10.4.2 Perspective of Urban-Rural Relation

From the perspective of urban-rural relations, "expansion and promotion" will contribute to strengthening urban-rural integration. Urban–rural integration is a profound reform aimed at to changing ideological conception, reorganizing resource allocation, optimizing productivity layout, transferring massive population, adjusting benefit relationship, reforming management system and innovating leadership style (Li 2010). Functioning as the strategic pivot of urban-rural integration, county-level cities play a very vital role. Li Xiaojiang, president of the China Academy of Urban Planning & Design, pointed out that in "2014 China Real Estate New Year Summit," according to a sample survey, 50 % of 260 million rural migrant workers do not leave their county and have stronger intentions of settling down in county-level cities, which should become the focus of new-type urbanization in mainland China. County-level cities are most likely to be the destination of rural migrant workers. Most rural migrant workers, since they have relatively low education, cannot meet industrial upgrading in megalopolises and metropolises. Instead, they prefer to settle down in county-level cities which can provide a salary income and in turn, local agricultural industrialization is promoted (Feng et al. 2010). However, it is very difficult to promote urban-rural integration in county-level cities, which display the prominent features of "small city, large country", because the large country is highly dependent on the small cities, while small cities can do little to help the large country. Hence, county-level cities that want to promote urban-rural integration should expand the scale of current development land and promote the efficiency of current development land use to meet a new round of population urbanization and industry urbanization.

Most county-level cities, which are small- and medium-sized cities, become the mechanisms that prevent heavy population flocking to metropolises and megalopolises, whose population is expanding and ecological environment is deteriorating. Hence, county-level cities are bound to become the fresh force in urban-rural integration to promote people-oriented urbanization, which is the main theme of next development stage. "Decision of the Central Committee of the Communist Party of China on Some Major Issues Concerning Comprehensively Deepening the Reform" adopted at the Third Plenary Session of the 18th Central Committee of the Communist Party of China, clearly points out that "completely relaxed restrictions on new residence registration in administrative townships and small cities, relaxed restrictions on new residence registration in medium-sized cities in an orderly way, lay down appropriate conditions for new residence registration in metropolises, and strictly control the population of megalopolises" – all of which are aiming to promote the urbanization of rural migrant workers. The Central Government Rural Work Conference in 2013 announced the plans to achieve the citizenization of about 100 million rural migrant workers, the reconstruction of shanty towns and urban villages, and the urbanization in the Midwest – the latter two affecting approximately 200 million people.

10.4.3 Perspective of Historical Process

From the perspective of historical process, "expansion and promotion" will contribute to promoting urban sustainable development. "The governance of county, the peace of country" is an old saying in China. As one of the basic pillars for national economy, county is the most stable organization in regime, representing the bridge linking municipal government to township government (Chen 2010a, b). With continuous promotion of the policies such as "expand the power of county" (kuo quan qiang xian) and "direct governance from province" (sheng zhi guan xian), that will be an opportunity to optimize the structure of urban development land in county-level cities. The historical evolution of urban development can be divided into three stages: the denotative expansion stage, referred to as the 1.0 era, which shows that urban population increases and urban development land expands; the denotative expansion and connotative promotion parallel stage, referred to as the 2.0 era, which shows that the incremental urban development land is limited by ecological environment effectively; and the connotative promotion stage, referred to as the 3.0 era, which shows that the scale of urban population tends to be stable and the demand of development land tends to be smaller. Although there are relatively large differences in social and economic development among county-level cities in east, west and north of Guangdong at present, as a whole, they are at the same stage – between the denotative expansion stage and the denotative expansion and connotative promotion parallel stage, or the 1.5 era, which is in huge demand of urban development land.

There are two measures of promoting the sustainable development of county-level cities. This paper suggests redeveloping stock land to improve the efficiency of urban development land opportunely, and developing incremental land to expand the scale of urban development land moderately. The report of "Urban China: Toward Efficient, Inclusive, and Sustainable Urbanization" released by the World Bank and Development Research Center of the State Council in 2014 points out that there is great space to promote urbanization in stock land due to the inefficiency of urban development land use. At the same time, "National New-type Urbanization Plan (2014–2020)" clearly puts forward the idea that the reconstruction of shanty towns and urban villages should be vigorously promoted, aiming to redevelop stock land. However, "three-old reconstruction" has proved that the redevelopment of stock land with limited space, complex benefit and long duration is difficult to meet developmental demand moderately and effectively.

10.4.4 Perspective of Land Finance

"Land finance" refers to the fact that local governments rely on transferring the land-use rights of state-owned development land to balance fiscal revenue and expenditure in mainland China – that is, the fiscal revenue of local governments greatly depends on the land-transferring fees which are derived from the development of incremental land. According to the statistics of Ministry of Finance, the total land-transferring fees were about 20 trillion yuan over the last 13 years. Meanwhile, the land-transferring fees increased from 129.6 billion yuan in 2001 to 4 trillion yuan in 2013. The developments of county-level cities greatly rely on the land-transferring fees in east, west and north of Guangdong.

However, "three-old reconstruction", as a typical model in the redevelopment of stock land, makes little contribution to "land finance", leading to the lack of motivation to promote it for local governments. "Expansion and promotion", which is a supplementary and a development to "three-old reconstruction", emphasizes both the development of incremental land and the redevelopment of stock land. Hence, compared to "three-old reconstruction", "expansion and promotion" is more appropriate for the current developmental status and better meets the developmental demand of county-level cities.

10.5 Three Aspects of "Expansion and Promotion" in Strategic Implementation

At present, county-level cities in rapid urbanization area not only need to develop incremental land moderately and reasonably, but also need to redevelop stock land opportunely and effectively on the basis of following the law of urbanization. This paper suggests analyzing the mutual interacting and restricting relationships between developing incremental land and redeveloping stock land – specifically from the perspectives of priority, subjectivity and hierarchy in strategic implementation, aiming to balance the development of incremental land and the redevelopment of stock land.

10.5.1 Priority of "Expansion and Promotion"

From the priority perspective, this paper suggests taking different measures to balance the urbanization of population, industry and land, depending of different circumstances. By focusing on the development of incremental land but ignoring the redevelopment of stock land, the urbanization of land is likely to be faster than the urbanization of population and industry. Redeveloping stock land but ignoring the development of incremental land, however, is prone to result in the shortage of land that would make it too difficult for county-level cities to accept the rural migrant workers and promote the industrial development opportunely and effectively. Hence, county-level cities in rapid urbanization areas that need to complete the function of city and enhance the capacity of city, should not only improve the efficiency of urban development land use through the redevelopment of stock land, but also expand the scale of city through the development of incremental land under the influence of the regional growth pole.

However, "easy to develop incremental land, but hard to redevelop stock land" has become the dilemma in strategic implementation. Hence, this paper argues that it is necessary to coordinate the priority of "expansion and promotion" properly. This will help avoid developing the incremental land excessively and redeveloping the stock land blindly. The priority of "expansion and promotion" in county-level city depends on the actual development of county-level city and the accurate judgment of regional development.

When the regional growth pole is in the phase of functional spillover, county-level cities which have undeveloped urban functions and have not formed scale industry clusters should promote the development of incremental land to ease the pressure of land supply; while county-level cities which have developed functions and have formed scale industry clusters should promote the redevelopment of stock land to improve the efficiency of urban development land use. When regional growth is in the phase of concentrating the surrounding resources, county-level cities which have undeveloped urban functions and have not formed scale industry clusters should control the scale of the development of incremental land to avoid becoming a "ghost town"; while county level city which have developed functions and formed scale industry clusters should conduct the type of the redevelopment of stock land to avoid becoming a "sleepers' town".

10.5.2 Subjectivity of "Expansion and Promotion"

From the perspective of subjectivity, this paper suggests coordinating the roles among government, market and society, and the developmental mechanisms of both non-profit and for-profit facilities to achieve the balance among social, economic and ecological benefits. In theory, the subjectivity of "expansion and promotion" in strategic implementation includes government (from top to bottom), market (from external to internal) and society (from bottom to top).

According to the objective property of "expansion and promotion", the development of incremental land and the redevelopment of stock land include both non-profit type and for-profit industries. With the limitation of financial resources, government is always ineffective in developing educational facilities, medical facilities, cultural facilities, and other non-profit facilities. With the difficulties in balancing all parties' interests, it is always difficult for market to develop residence, emporium, factory, and other for-profit facilities. Meanwhile, society is very limited in developing non-profit and for-profit facilities due to the imperfection of institutional guarantee.

This paper argues that the role of government should be transferred from developer to server during expansion and promotion. Government should become a guide in the development of incremental land. On one hand, government should induce market and society to develop for-profit facilities efficiently, aiming to promote the economic development; on the other hand, government should induce market and society to develop non-profit facilities by the measures of BT (Build-Transfer), BOT (Build-Operate-Transfer) and PPP (Public-Private-Partnership). Government should also become a coordinator in the redevelopment of stock land to protect the interests of the weak during the redevelopment of for-profit facilities or non-profit facilities, aiming to guarantee justice, equity and openness in redeveloping process.

10.5.3 Hierarchy of "Expansion and Promotion"

In the aspect of hierarchy, this paper suggests insisting on the comprehensive investigation, overall planning and ordered implementation, aiming to balance the development of incremental land and the redevelopment of stock land. It is important to follow the principle of "plan depends on investigation; implementation depends on plan" during "expansion and promotion". County-level cities not only become the final settlement for rural migrant workers, but also gather more labor-intensive industries during new-type urbanization.

Hence, it is a challenging job for county-level cities to coordinate the spatial layout of population and industry properly in the strategic implementation. Firstly, there is a need to distinguish the hierarchy of investigation, which includes the survey of external environments and internal environments. The policies and trends of regional development should be confirmed in the external environment survey, which is helpful to define the role and responsibility of the county-level city. The structure of population and industry, especially the scale of rural migrant workers, should be accurately predicted during the survey of internal environment. Secondly, there is a need to distinguish the hierarchy of planning, which includes the master plan, regulatory plan and other statutory plans. With the strict protection of prime cropland preservation area, this paper suggests optimizing the structure of urban development land and non-urban development land, analyzing the scale of incremental land and stock land, delimiting the border of urban development land, and formulating the intensity of urban development land during the planning process. Finally, there is a need to distinguish the hierarchy of implementation. "Expansion and promotion" should be carried out orderly with considering social cost and ecological cost. Meanwhile, it is necessary to provide financial support and make innovative policy during strategic implementation.

10.6 Conclusions

To sum up, this paper holds that the strategic transformation of county-level cities in rapid urbanization areas should be transferred from "three-old reconstruction" to "expansion and promotion", so that the county-level city can enhance bearing capacity, competitive ability and radiation ability on the basis of improving the efficiency of urban development land use. Under the background of strictly observing to the red line of 1.8 billion MU of arable land, the shortage of urban development land and the inefficient of urban development land use have become two problems that must be faced in mainland China. As the fresh force during new-type urbanization, county-level cities in rapid urbanization areas face the opportunities and challenges of "expansion and promotion" in population, industry and land, due to the highlighting role of regional development, urban-rural integration, historical process and land finance.

With respect to land, developing incremental land moderately and redeveloping stock land opportunely will optimize the structure of urban development land and improve the efficiency of urban development land use. Meanwhile, in the context of balancing population urbanization, industry urbanization and land urbanization, the priority, subjectivity and hierarchy of "expansion and promotion" in strategic implementation should be further clarified so as to promote the sustainable development of county-level city in rapid urbanization area.

Acknowledgement The author would like to express his sincere thanks to Meng Qi (Professor of South China University of Technology), Zhang Qiao, Chen Ke, Zheng Shujian and Huang Congjian for their contributions to this paper.

References

Chen, J. F. (2010a). *The study on reconstruction policies of old towns, old villages and old plants in Dongguan*. Guangdong: Sun Yat-sen University, Guangdong Province (in Chinese).

Chen, P. (2010b). Exploration of county level new countryside construction and planning based on urban-rural integration. *City Planning Review, 34*(2), 47–53 (in Chinese).

Duan, Z. G., & Zhang, W. (2007). Industrial development strategy and space layout of county-level city in rapid urbanization area: A case study of eight county-level cities in Jiangsu province. *Urban Problems, 2007*(4), 2–6 (in Chinese).

Feng, X. M., Wang, J., & Yue, K. (2010). Analysis on effects of county-level city development on farmers' income. *Journal of Anhui Agriculture Science, 38*(15), 8234–8236 (in Chinese).

Jiang, Y. F. (2008). Guangdong strictly controls the new development land. *Guangzhou Daily*. http://gzdaily.dayoo.com/html/2008-12/23/content_418740.htm

Lai, S. H., & Wu, J. (2013). Speed and benefit: Guangzhou "san-jiu" redevelopment strategies for new urbanization. *Planners, 29*(5), 36–41 (in Chinese).

Li, X. (2010). Coordinated urban-rural development with the guidance of urban-rural planning. *Shanghai Urban Management, 2010*(3), 2–3 (in Chinese).

Liu, L. (2009). Study of the regional linkage mechanism for industrial transfer and upgrade: Discussion on Guangdong's regional harmonious development mode. *International Economics and Trade Research, 25*(12), 9–13 (in Chinese).

Liu, Y. G., Huang, S. J., & Yuan, Y. (2011). Policy analysis of urban reconstruction: A case study of Dongguan. *Urban Insight, 2011*(2), 76–85 (in Chinese).

Luo, H. L. (2013). Guangdong has completed 3000 projects in "three-old reconstruction". *Guangzhou Daily*. http://gzdaily.dayoo.com/html/2013-06/24/content_2294643.htm

Wang, W. (2003). Planning management analysis of development land in county-level city. *Development of Small Cities & Towns, 2003*(11), 78–79 (in Chinese).

Wei, L., & Wang, S. F. (2011). The public intervention mechanism of "san-jiu" reconstruction from the urban planning perspective: Cases study of Guangdong. *South Architecture, 2011*(1), 26–29 (in Chinese).

Weng, L. (2005). *Spatio-temporal change of development level about city at county level in Zhejiang province*. Hangzhou: Zhejiang University, Zhejiang Province (in Chinese).

Wu, C. F., & Wang, Q. D. (2013). The principle and regulation of three-old lands reform in Guangdong. *Urban Problems, 2013*(10), 78–84 (in Chinese).

Wu, X. J., Zhang, W., Hu, H. B., & Chen, X. H. (2005). Method of county comprehensive planning in high-speed urbanization area. *City Planning Review, 29*(12), 58–63 (in Chinese).

Xie, M. (2012). "Double shift" open the 5 years transformation journey. *Nanfang Daily*. http://epaper.southcn.com/nfdaily/html/2012-12/08/content_7149669.htm

Xu, Q. (2001). Position and role of county-level city in developed area in the urbanization process. *Academia Bimestris, 2001*(6), 66–70 (in Chinese).

Yang, L., & Yuan, Q. F. (2010). Research on patterns of land integration in a context of "san-jiu" reconstruction in Pearl River Delta region: A case study of Lianjiao district in Nanhai, Foshan. *Urban Planning Forum, 2010*(2), 14–20 (in Chinese).

Zeng, N. (2014). "Three-old reconstruction" project completed only 5 %. *Nanfang Daily*. http://epaper.southcn.com/nfdaily/html/2014-01/23/content_7269556.htm

Zhao, Y. H. (2009). More than 1,700,000 MU land area of "san-jiu". *Yangcheng Evening News*. http://www.ycwb.com/ePaper/ycwb/html/2009-10/03/content_612717.htm

Zhou, X., & Fang, F. Y. (2011). Thinking about urban renewal from the 'three old reform' renewal in Guangdong. *Modern Urban Research, 2011*(8), 82–89 (in Chinese).

Chapter 11
Spatial Structure of Regional Economic Development in Henan Province, China

Li Hao and Qian Zhu

11.1 Introduction

While the most populous region in China, Henan Province has been at the middle or lower levels of the country's economic development for a long time. However, recently the regional economic structure has been changing significantly, as witnessed through rapid economic growth and regional development.

During the last decade, regional economic policy has become prevalent in China. Various regional economic policies, both on a national and provincial level, have been established in order to prompt regional economic development (Luo and Shen 2008); Henan Province falls within the jurisdiction of Central China Rising Strategy. Within the province, Central Plain Urban Cluster Planning (CPUC) (Fig. 11.1) has been launched and has significantly influenced regional economic development (Development and Reform Commission of Henan Province 2005). In the meantime, this provincial initiative is recognized as one of the key parts of the National Urban System Plan for 2005 through 2020.

Henan Province has been experiencing rapid economic growth and spatial transformations. However, the economic development is uneven within the region, as numerous spatial disparities exist. All this geographical inconsistency will have a long-lasting influence on the province's development. Thus, it becomes crucial for planners and policy makers to gain an in-depth understanding of this on-going

L. Hao (✉)
Research Center for Master Planning, Beijing Tsinghua Tongheng Urban Planning & Design Institute, 17th Floor East Tower, Building Jia1, East Qinghe Jiayuan, Haidian District, Beijing 100085, China
e-mail: interhow@hotmail.com

Q. Zhu
School of Planning, University of Waterloo, 200 University Avenue West, Waterloo, ON N2L 3G1, Canada
e-mail: z3qian@uwaterloo.ca

Q. Pan, J. Cao (eds.), *Recent Developments in Chinese Urban Planning*,
GeoJournal Library 114, DOI 10.1007/978-3-319-18470-8_11

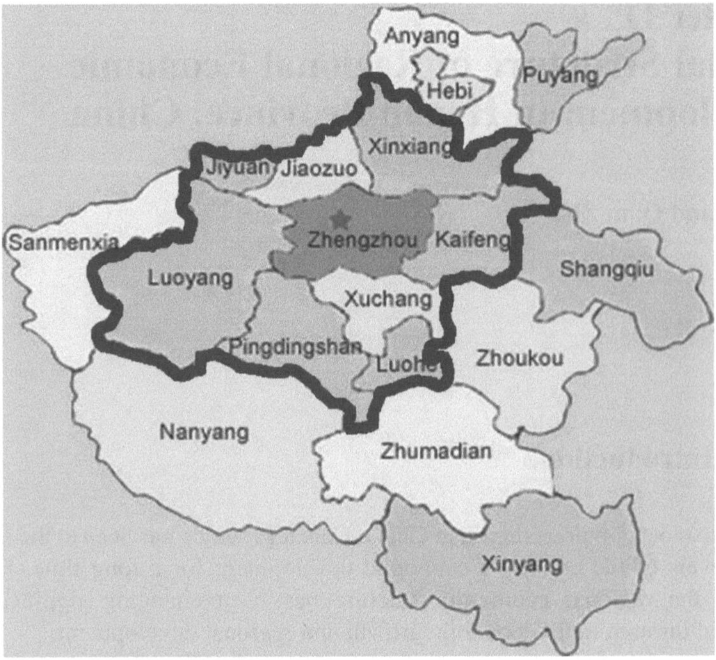

Fig. 11.1 Henan Province and Central Plain Urban Cluster (the scope within the *red* boundary) (Source: Original map from China Map (2012), modified by the author) (color figure online)

spatial trend. The study of Henan's regional economic structure will provide meaningful insights for revised regional planning policy.

This report will examine the changing spatial patterns of regional economic development in Henan Province since the late 1990s, and define the spatial trends. Through the study of the status quo, this report will identify the existing problems and attempt to provide solutions. Recommendations for regional development strategies will be provided later in this document.

11.2 Data Collection

All the economic and demographic data will be collected from an official data source: the Henan Statistics Year Book. The GIS data is from the National Fundamental Geographic Information System (http://nfgis.nsdi.gov.cn/).

11.3 Methodology

11.3.1 Spatial Units

China's administration jurisdiction system is very different from those of many western countries: there are many different kinds of cities on numerous levels. Henan Province has 18 prefecture-level cities, which have jurisdiction over 108 counties and county-level cities. In fact, both the prefecture-level city and county-level city are structured as cities within the planning system. In order to simplify these spatial units, only the municipal districts of the prefecture-level cities (*Shixiaqu* in Chinese) are used to examine the prefecture-level cities. Therefore, there were 126 basic geographical units in 2009. Previously, in 1999, there were 127 geographical units as Yancheng County was amalgamated into the urban area of the City of Luohe in 2004.

11.3.2 Indicator

The GDP (Gross National Product) per capita will be utilized as the main indicator in the analysis of the economic spatial structure, due to convenience and data availability. Although GDP per capita is not the only indicator, it succulently reflects the spatial economic structure, and it has significant implications for the understanding of the geographical dynamics.

Two kinds of statistical methods will be applied:

1. Mann-White Test

Mann-White testing is one of the most popular methods in investigating the significant difference between the values of two sets of data (Conover 1980). As the most important regional policy, CPUC's planning policy has a significant influence on the local economic development of the cities and counties within Henan. This report will divide all the spatial units into two groups: within CPUC and outside CPUC. The Mann-White test will be employed to examine whether the two groups of spatial units have significant differences in terms of GDP per capita or not.

2. Spatial Autocorrelation

Agglomeration or diffusion is a key issue in the study of regional economic structure. According to neoclassical economic growth theory, agglomeration means a geography unit is affected by its geographical neighbors. This phenomenon is a type of scale economy appearing in the spatial domain.

The spatial autocorrelation method advances the conventional statistical methods. Integrating spatial concerns into autocorrelation can accurately reflect the spatial pattern of the geographic units. This report will employ Anselin's (1995) methods and the software *Geoda* to investigated the spatial pattern of all the cities and counties' economic development. This software can identify regional homogeneity or heterogeneity and visualize the spatial patterns of the indicators.

11.4 Analysis

11.4.1 General Spatial Pattern

The histogram graphs of the geographical units' GDP per capita are presented in Fig. 11.2. The horizontal axis shows the GDP per capita while the vertical axis shows the frequency. The graph clearly illustrates that in both 1999 and 2009 the range of low GDP per capita has a high frequency, while the range of medium and high GDP per capita has low frequency. This is a typical pyramid structure, which means that the majority of cities and counties in Henan are not economically developed, and that moderately and significantly developed cities and counties are rare. This pattern implies significant disparity in regional economic development.

Based on the data from Henan Statistics Yearbook 2000 (Statistics Bureau of Henan Province 2000) and Henan Statistics Yearbook 2010 (Statistics Bureau of Henan Province 2010), Figs. 11.3 and 11.4 were generated by spatial analysis of GIS data. These maps, with graduated gray color, clearly demonstrate the general economic spatial structure.

In both of the maps in Fig. 11.3, the dark gray color implies a higher value. Thus, the maps illustrate serious social-spatial disparities within the region. The map on the right highlights greater populations in the east and south. Yet, the left map illustrates higher GDP per capita is concentrated in the north and west. Thus, on the spatial level, population concentrations and economic development are highly unbalanced. This is Henan's most critical contradiction, creating a serious restriction to regional development.

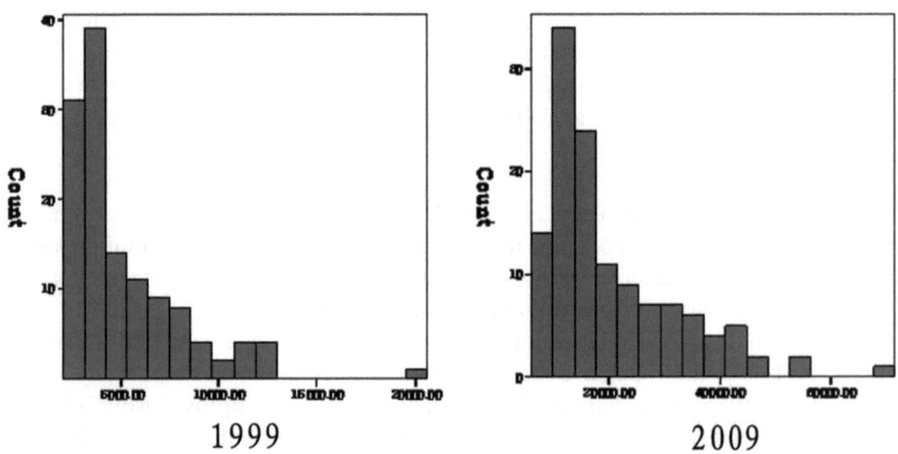

Fig. 11.2 Frequency histogram

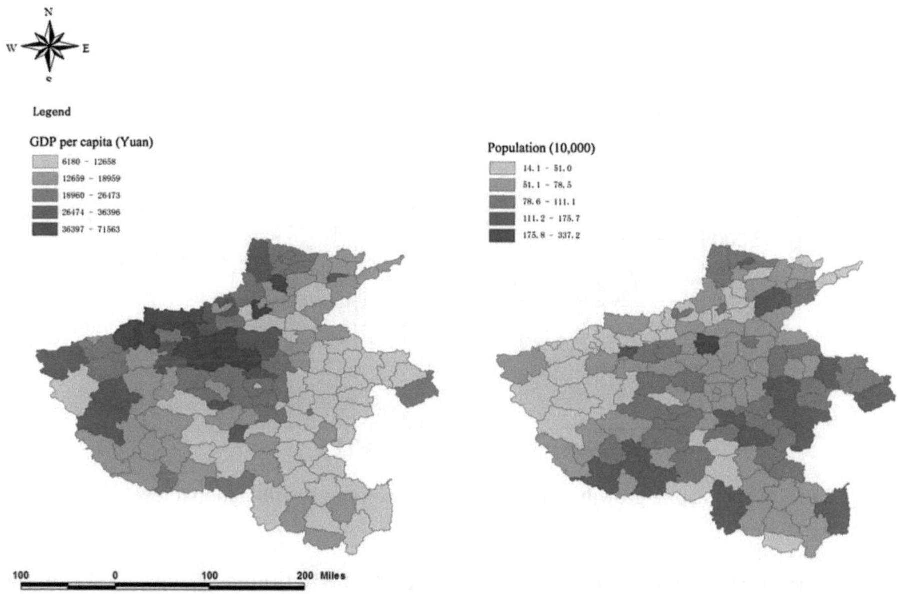

Fig. 11.3 GDP per capita distribution (*left*) and population distribution (*right*) in 2009

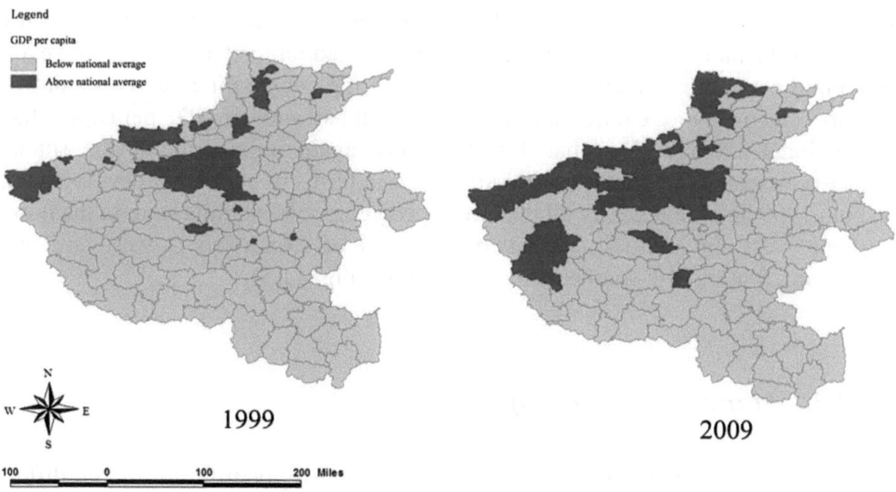

Fig. 11.4 GDP per capita compared with the national average (*red*: above the national average; *pink*: below the national average) (color figure online)

Figure 11.4 illustrates the regional economic development pattern from another perspective. The GDP per capita of each geographic unit is compared to the national average; the geographic units, which have a higher GDP per capita than the national average, are highlighted in red. In fact, the overall GDP per capita of

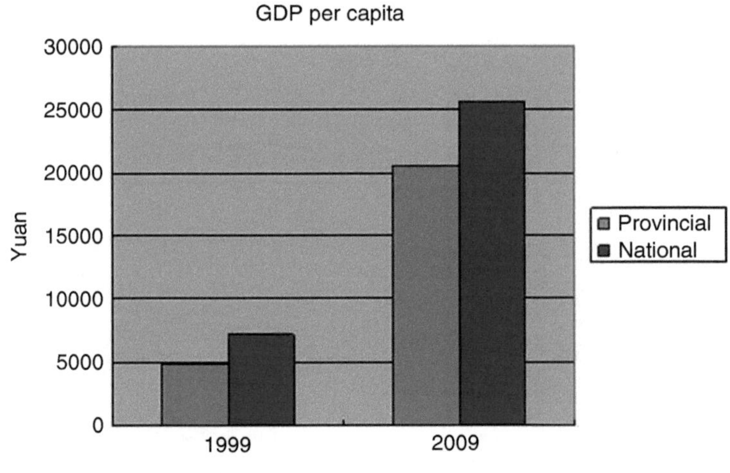

Fig. 11.5 GDP per capita growth of Henan Province compared with national level

Henan Province is at a medium level compared to the nationwide average. From the data in *Henan Statistics Yearbook 2000 and 2010*, and *China Statistics Yearbook 2010* (National Bureau of Statistics of China 2010), we can see that over the period of study (1999–2009), Henan has had a faster economic growth rate than the national average (Fig. 11.4). In 1999, the provincial GDP per capita was only 67.5 % of the national average; by 2009, this ratio had jumped to 80.5 %. However, this rapid economic growth is not shared by all the geographical units within the province, as most of the cities and/or counties still had a lower GDP per capita than the national average. The spatial pattern once again reflects the severe spatial disparity between the southeast and northwest in terms of economic development: in both 1999 and 2009 nearly all the developed geographical units were in the north and the west. In addition, the historical trend of expansion is westward and northward, leaving the southeast region lagging (Fig. 11.5).

11.4.2 Mann-White Testing

The Mann-White testing is done on 2 years' of data to identify the spatial differences. All the geographical units within CPUC will be compared with those outside CPUC. In this test, the working hypothesis is that the geographical units within CPUC have the same GDP per capita with those outside CPUC. Tables 11.1 and 11.2 provided the results from the statistical software SPSS.

The results in Table 11.1 vary significant at the 1 % confidence level, which means that at this confidence level the hypothesis must be rejected. This is to say; in 1999 all the geographical units within CPUC had significantly different GDP per capita than those outside CPUC. As the cities and/or counties within CPUC

Table 11.1 Mann-White test result for the year 1999

	VAR00008
Mann-Whitney U	1187.000
Wilcoxon W	3672.000
Z	−3.917
Asymp. Sig. (2-tailed)	.000

Table 11.2 Mann-White test result for the year 2009

	VAR00006
Mann-Whitney U	909.000
Wilcoxon W	3394.000
Z	−5.160
Asymp. Sig. (2-tailed)	.000

generally had high GDP per capita, their economic development levels were significantly higher than those of the cities outside CPUC.

The results in Table 11.2 show nearly the same situation: again, the hypothesis is rejected, showing that the cities and counties within CPUC still had significantly different economic status than their counterparts outside CPUC in 2009.

Thus there is a clear pattern: the CPUC regions and the non-CPUC regions differ in their economic development. This spatial disparity is very significant and there is no change during the period between 1999 and 2009.

11.4.3 *Spatial Autocorrelation*

11.4.3.1 Spatial Effect

The conventional regional research often overlooks the spatial effect in regional development. However, there are some rationales for taking spatial effects into account. The increasing intensity of trade between spatial neighbors; technology and knowledge spillover (Le Gallo and Ertur 2003); and industry cluster and agglomeration economies (Fujita and Thisse 2002) all contribute to the geographic dependency of spatial units. Thus, using geo-techniques which consider spatial effect can provide an in-depth understanding of the spatial pattern of regional development.

11.4.3.2 Global Spatial Autocorrelation

Global spatial autocorrelation refers to the spatial aggregation characteristics of all the spatial units, and it identifies the general degree of a spatial units' linear association with its neighbors (Yu and Wei 2008). This report employs the most popular indicator; Moran's I, to calculate the global spatial auto correlation. The function is as follows:

$I = \frac{\sum_{i}^{n}(xi - \bar{x})\sum_{j \neq i}^{n} W_{ij}(x_j - \bar{x})}{s^2 \sum_{i}^{n}\sum_{j \neq i}^{n} W_{ij}}$ In the function, $s^2 = \frac{1}{n}\sum_{i}^{n}(x_i - \bar{x})^2$ N is the number of observed samples; xi and xj are the observed value sat locations i and j; W_{ij} is the spatial weight set by the spatial location of i and j.

Given a certain confidence level, Moron's I has a value in the range of [−1, 1]. When it is positive, the sample shows positive spatial correlation. The negative value means negative spatial correlation, and the value of 0 means no significant spatial correlation.

From the software *Geoda*, Moran's I for Henan Province is 0.2272 in 1999 and 0.4951 in 2009 (Appendix B). This reflects a significant positive agglomeration effect for all the geographical units for those 2 years. In other words, it implies that cities and/or counties are often adjacent to ones with a similar economic development level. The increase of Moran's I between 1999 and 2009 indicates the spatial autocorrelation of the units has strengthened and that geographical units are increasingly dependent on their neighbors.

11.4.3.3 Local Autocorrelation

Since global autocorrelation only reflects the general pattern of the whole sample, and it cannot identify each specific unit's pattern, local autocorrelation is used to investigate single units' spatial characteristics. This report employs Anselin's (1995) Local Indicators of Spatial Association (LISA) method; the function is as follows:

$$LISA_i = Z_i \sum_{j \neq i}^{n} W_{ij} Z_j = Z_i Y_i$$

In this function $Z_i = \frac{x_i - \bar{x}}{s}$ and $Y_i = \sum_{j \neq i}^{n} W_{ij} Z_j$ are the weighted average of data deviation from other regions connected with the regional center si.

The LISA result is shown by Figs. 11.6 and 11.7. Figure 11.6 is the significance map and it shows the units which have significant spatial autocorrelation (the geographical units in green). The darker green means the spatial autocorrelation is more significant. Considering the two maps (1999 and 2009), an interesting change is illustrated: in 1999, the majority of the significant geographical units are at the 5 % confidence level (in light green), while in 2009, the majority of those units are at the 1% confidence level (in dark green). This change means the spatial autocorrelation in 2009 is more significant than 10 years previous.

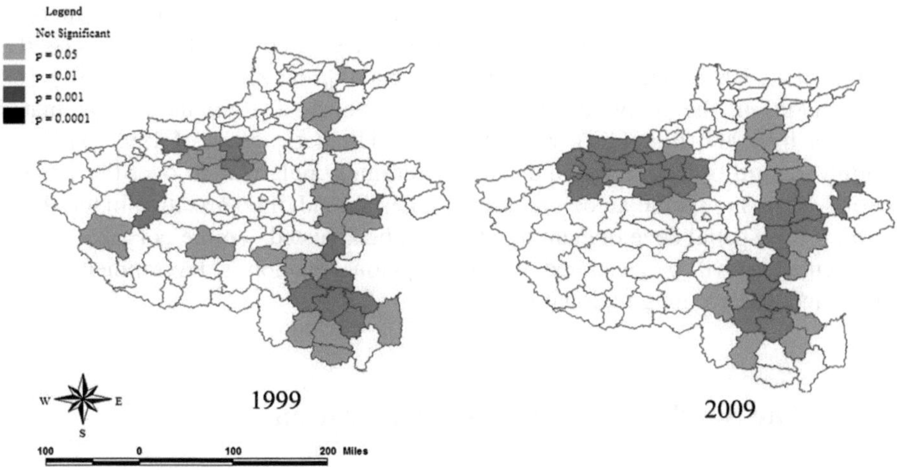

Fig. 11.6 LISA significance map

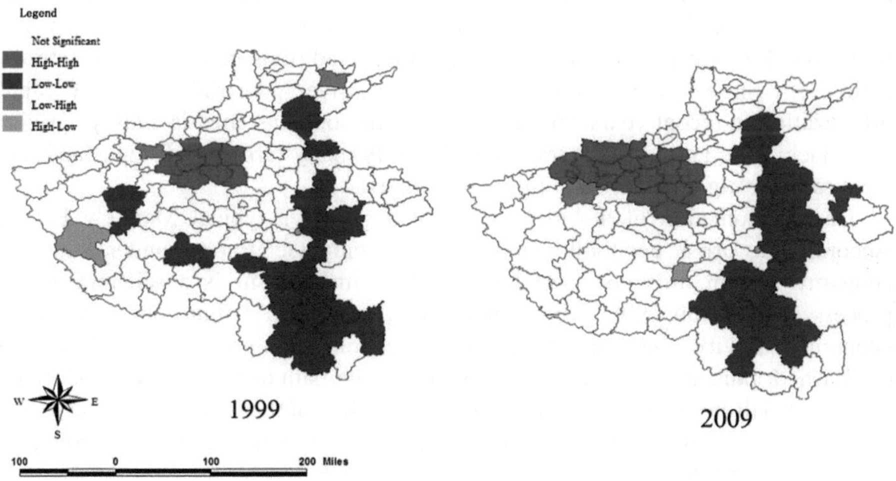

Fig. 11.7 LISA diagram

Figure 11.7 clearly defines the high-high status (hot spots) or low-low status (cold spots) geographical units in both 1999 and 2009, highlighting the two kinds of spatial agglomeration in the northwest and southeast, respectively. The areas in red (high-high status), constitute the core of CPUC, and are composed of the most economically developed cities and/or counties. Those spatial units show positive consistency and development trends. In contrast, the low-low areas (blue) are concentrated in the southeast region of Henan Province. These cities and/or counties have low economic development levels, in addition to experiencing the negative impacts from their less economically developed neighbors. Thus, there is a large depressed area where negative trends have formed. This spatial pattern is quite consistent with the core-periphery model in New Economy Geography (NEG) (Fujita et al. 1999), and it clearly illustrates spatial polarization.

In addition, the individual low-low status areas of 1999 had amalgamated together by 2009 (Fig. 11.7), while the high-high status areas of 1999 expanded northwestward by 2009. This indicates increasingly polarization over the 10 year period. Also, the spatial autocorrelation implies a trend of polarization in the short-term. If this spatial development trend cannot be adjusted, a more polarized binary differentiation will be generated, widening regional economic development differences and undermining regional equality.

11.5 The Southeast-Northwest Polarization

The northwestern and southeastern areas have distinct development patterns for several reasons. The northwestern area has significant natural resources, as there are abundant mineral deposits such as coal and iron, which are crucial for industrialization. There is also an established transportation network including railway and highways. Moreover, some cities have historical traditions in terms of industrialization. All these features contribute to the massive industrialization and rapid urbanization in recent years. On the contrary, the southeast area is mainly agricultural, lacking in an industrial tradition. Recently many farmers left their land and went to other big cities as rural migrant workers.

This phenomenon of a widening regional gap is a typical Matthew Effect. According to urban economics and New Economy Geography, under a market economy system, this core-periphery model's mechanism is a self-reinforcing process (O'Sullivan 1996; Krugman 1998). Due to the spatial concentration of economic activity, the disparity between locations will grow once it has been established; thus, the regional development is quite path dependent. The industrial northwest will be increasingly developed while the southeast will continue to lag behind. Thus, without policy intervention, the regional polarization will grow into the foreseeable future.

11.6 Discussion

The empirical study provides some significant findings on the spatial structure of Henan's economic development:

1. Fast economic growth and increasingly uneven geographic development

The statistical data illustrates the significant economic growth rate in Henan Province since the end of last century. However, the economic growth is based on an uneven economic distribution. Significant economic growth is only shared by certain geographic units, while the majority of units remain undeveloped. In the past, the uneven spatial development has been detected primarily on the national level; for instance, the developed east coast provinces versus the undeveloped western inland provinces. This paper identified the uneven regional development at the provincial level.

2. Significant spatial disparity

There are several kinds of disparities regarding the uneven spatial development. First, the disparity between demographic and economic distribution, as shown in Fig. 11.3, is the most significant social-spatial pattern of Henan. Second, the disparity between the southeast and the north west regions in terms of economic development is very significant and increasingly serious. Third, there is significant disparity between the area within CPUC and outside CPUC. This is related to the southeast-northwest disparity as CPUC is the major part of northwest region.

3. Current spatial strategy (CPUC) overlooks periphery areas and increases polarization

CPUC is the most important and influential regional development policy during the last decade in Henan Province. However, CPUC only contains the most developed cities and/or counties in the northwest. All those spatial units within CPUC receive favorable policy for economic development. For example, numerous new towns and industrial parks were planned and constructed within the geographical units within CPUC. According to *The development and planning outline of Central Plain Urban Cluster*, CPUC is expected to be the industrial, logistic, and financial centre of the whole region. In 2009, CPUC accounted for 57.1 % of the province's GDP. But, this share is to be increased to 70 % by 2020. Overall, the current regional planning is taking CPUC as the key area, thus overlooking the periphery of the province, especially the agricultural area in the southeast. In fact, in such a "developing" province of Henan, the public sector dominates the majority part of the provincial economy. Thus, Henan's regional development is very policy influenced. The current policies are not going to change the existing regional development trajectories, but will strengthen them. This urban centered and unbalanced regional strategy will inevitably lead an increasingly polarized regional development patterns.

Based on the empirical and detailed analysis, some reflections on policy can be addressed here. First, the regional policy should coordinate the core and peripheral areas. As the typical core-peripheral pattern forms, serious regional conflicts will arise if this ongoing trend continues to strengthen. Thus, regional coordination needs to be a focus of policy. Second, the provincial planning should prompt economic development in rural counties. In the polarization between northwest and southeast, the former area is composed of many cities, both at the prefecture and county levels; however, the southeast is mainly rural counties. Due to the dual regime of city and country, cities usually get more favorable policies while counties are overlooked. In order to ameliorate the worsening situation in the southeast, county development should be emphasized in future planning policy. Third, a more balanced regional policy is required to prompt regional justice. The regional polarization is actually harming the regional justice: the geographical units are not equally treated within a regional development framework. The current planning, which is overwhelmingly economic in orientation and urban centric, must be transformed to a more harmonious and people-oriented planning paradigm. The main objective for regional development should be changed from economic efficiency to regional equality.

Appendix

Appendix A. Cartogram map of GDP per capita distribution (left) and population distribution (right) in 2009

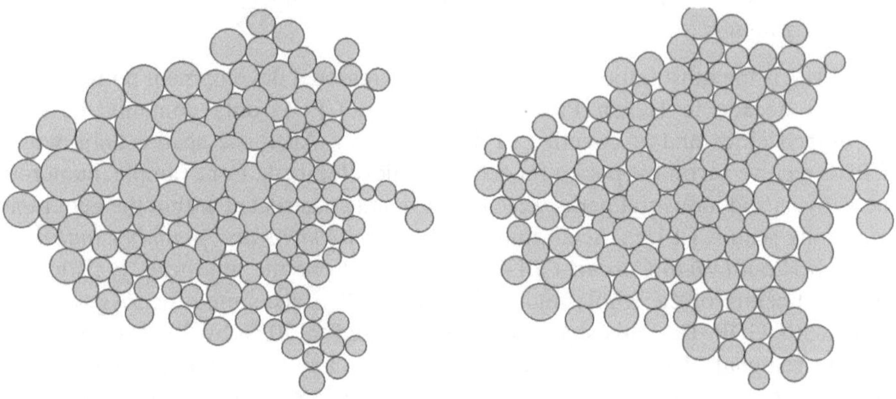

Appendix B. Moran's I scatter diagram (left: 1999; right: 2009)

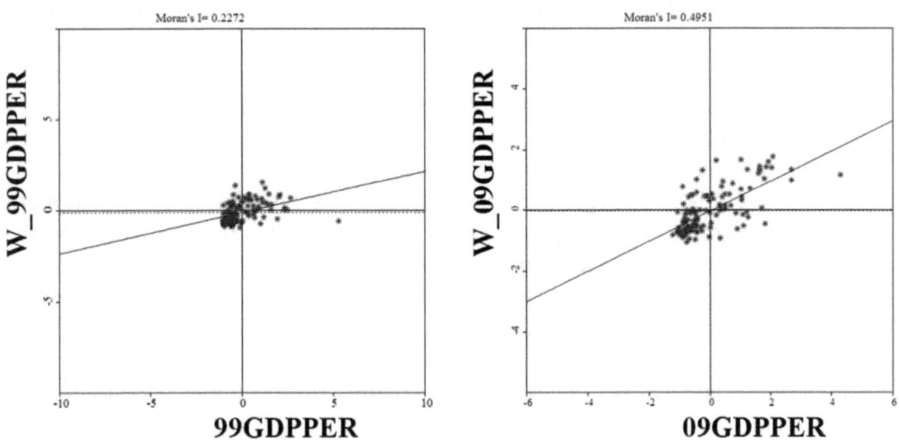

References

Anselin, L. (1995). Local indicators of spatial association—LISA. *Geographical Analysis, 27*(2), 93–115.

China Map. (2012). *Henan Province Map*. Retrieved on June 15, 2012, from http://www.china-emap.com/en/node/69/

Conover, W. J. (1980). *Practical nonparametric statistics* (2nd ed.). New York: Wiley.

Development and Reform Commission of Henan Province. (2005). *The development and planning outline of Central Plain Urban Cluster*. Retrieved on May 20, 2012, from http://www.hnrco.gov.cn/zycsq/city.asp

Fujita, M., & Thisse, J. (2002). *Economics of agglomeration: Cities, industrial location, and regional growth*. Cambridge: Cambridge University Press.

Fujita, M., Krugman, P., & Venables, A. (1999). *The spatial economy: Cities, regions, and international trade*. Cambridge, MA: MIT Press.

Krugman, P. (1998). What's new about the new economic geography? *Oxford Review of Economic Policy, 14*(2), 7–17.

Le Gallo, J., & Ertur, C. (2003). Exploratory spatial data analysis of the distribution of regional per capita GDP in Europe, 1980–1995. *Papers in Regional Science, 82*, 175–201.

Luo, & Shen. (2008). Why city-region planning does not work well in China: The case of Suzhou–Wuxi–Changzhou. *Cities, 25*(4), 207–210.

National Bureau of Statistics of China. (2010). *China statistical yearbook 2010*. Beijing: China Statistics Press.

O'Sullivan, A. (1996). *Urban economics* (3rd ed.). Chicago: Irwin.

Statistics Bureau of Henan Province. (2000). *Henan statistical yearbook 2000*. Beijing: China Statistics Press.

Statistics Bureau of Henan Province. (2010). *Henan statistical yearbook 2010*. Beijing: China Statistics Press.

Yu, D., & Wei, Y.-D. (2008). Spatial data analysis of regional development in Greater Beijing, China, in a GIS environment. *Papers in Regional Science, 87*(1), 97–117.

Chapter 12
Understanding Beijing's Urban Land Use Development from 2004–2013 Through Online Administrative Data Sources

Xiao Rong, Ying Jin, and Ying Long

12.1 Introduction

The decade of 2004–2013 saw one of the most rapid urban expansion periods in Beijing. According to official statistical yearbooks, Beijing's resident population grew from 14.9 million in 2004 to 21.1 million in 2013, i.e. at an annual rate of 5.1 % and adding 886,000 people each year.[1] Urban land use has expanded at a similar speed to surrounding rural areas, accompanied by significant urban redevelopment and densification within existing built-up areas. Understanding the patterns of population and urban land use growth during this decade has an immediate relevance to urban planning in the coming decades, and hence to the pressing policy objectives to enhance citizens' quality of life, social inclusiveness, environmental sustainability and economic vitality in Beijing and its wider hinterland.

Rapid urban development tends to create a significant gap between what takes place on the ground and the statistical data for strategic policy analysis. Typically, the data sources lag behind for one or more years, and for urban land use the data has been really patchy, in spite of great efforts by statisticians and scholars in the field. Studies on the geographical pattern of Beijing's development usually rely on satellite photos (e.g. Liu et al. 2000; Kuang et al. 2009) and confidential official data (e.g. Yan and Feng 2009), which are not generally open for public use. Satellite

[1] Anecdotal evidence suggests that this may have understated the residential population growth; in addition, there are several million of transient population staying in the municipality at any given time.

X. Rong • Y. Jin (✉)
Department of Architecture, University of Cambridge, Cambridge, UK
e-mail: Ying.Jin@aha.cam.ac.uk

Y. Long
Beijing Institute of City Planning, Beijing, China

Q. Pan, J. Cao (eds.), *Recent Developments in Chinese Urban Planning*,
GeoJournal Library 114, DOI 10.1007/978-3-319-18470-8_12

photos have limitations in categorising land use types. This hampers effective policy analysis in many areas of urban development where a rapid policy response is desirable.

However, the increasingly timely reporting of government administrative records for urban development permissions and urban land provisions, which arose from the Open Government initiative, has potentially opened up new opportunities to systematically monitor and analyse the emerging patterns of urban land use development in a timely fashion. Currently, these data sources are rarely used, in large part because the online data is crude and difficult to process. Yet, there has been research using similar data acquired from land bureaus mainly to examine city's land market through the land price and land intensity patterns (Ding 2004; Yan and Feng 2009; Wu et al. 2010; Gao et al. 2013; Dang et al. 2014).

This paper attempts to further fill the gap by examining the effectiveness and potential of using online administrative data sources to understand Beijing's urban expansion from 2004 when a number of such sources started to be released online. We focus on the pattern of land supply for housing and employment, which is one of the most essential factors affecting people's lives and the city's performance. This is also a weak spot in the field, as previous studies mainly approach this issue from the results of surveys (e.g. Chen and Meng 2011; Pan and Ge 2014) and census data (Feng and Zhou 2003), both of which are expensive to assemble and thus it is difficult to obtain frequent updates.

12.2 Data Assembly

This paper utilises urban land plot provision data collected from the Beijing Municipal Bureau of Land and Resources website, which provides land development and associated building floor space information. It covers the time after the reform of land provision.[2] Three datasets, respectively on land plot transaction,[3] land plot allocation[4] and land plot acquisition,[5] are published.

Land plot transaction and land plot allocation are the main types of urban land provision in transferring the right to use land. Land plot transaction deals with urban land plots which are to be used for profitable purposes, such as residential, industrial and commercial.[6] After the existing users or the primary land developers gain

[2] Most notably, the urban land development permission data published by the Beijing Municipal Commission of Urban Planning is not used because it does not include the land area and building floor space information in a manner required by this paper. We are separately studying the online data from that source for other purposes.

[3] http://www.bjgtj.gov.cn/publish/portal0/tab5422/

[4] http://www.bjgtj.gov.cn/publish/portal0/tab5163/

[5] http://www.bjgtj.gov.cn/publish/portal0/tab5164/

[6] Through this, the government grants 70-year leases for residential land use right, 50-year leases for industrial land use right and 40-year leases for retail land use right.

permission from the local government, the land plots go into open market for the potential recipients to bid on them. The bidding is conducted in three main ways: (1) invitation for tenders; (2) auction; and (3) notification (silent auction).[7] The recipients need to pay the government the market transaction price and the previous user/the primary land developer a specified fee (based on the construction cost of infrastructure on site).

Land plot allocation deals with urban land plots which are to be used for non-profitable, public purposes, such as infrastructure, government, affordable housing, etc.[8] The recipients do not need to pay a fee to the government, but in some cases they have to pay the previous user a small one-time fee for replacement cost and other expenses.

Land plot acquisition is a type of land provision specially adopted locally by Beijing municipality to allocate land through land acquisition. It deals with rural land which is collectively owned by villagers. Because the land use right of collectively-owned land is not transferable, the government usually has to first convert it into state-owned land through acquisition, and then provide it through transaction or allocation. However, in Beijing's case, the government directly allocates the rural land plots to the recipients, but the recipients have to pay compensation to the villages and to the primary developer of the land instead of the government which is usually responsible for these tasks in other Chinese cities.

In this paper, we choose to use the land plot transaction and allocation data because they provide a decade-long series since 2003 with reasonably consistent formats.[9] By contrast, the land plot acquisition data, which is separately collected and published by each city district, has little consistency in content and format, and

[7] The regulation 'Granting State-Owned Land Use Rights by Invitation for Tenders, Auction or Listing' was issued by the Chinese Ministry of Land and Resource on May 8, 2002, which became effective on July 2, 2003. Beijing issued a local regulation based on the national regulation on June 26, 2002, which became effective on Aug. 31, 2004. Prior to this, the land plot transaction for real estate was carried 'by way of agreement', in which the land was granted by the government when the developer and the previous land user reached an agreement on the payment. The payment included a nominal fee to the government, which was usually lower than the market value, and a one-time payment to the previous user, which was decided through negotiation.

[8] In this case, the recipients are given the allocated land-use right which has no time limit, but the land is not transferable to other uses or users.

[9] For land transaction data, we use the year the documents were signed for those transacted since 2008, and use the year they were published before 2008, to approximate the year of transaction. For land allocation data, the year is defined by the date of the document which approved the project; for the 173 projects without approval document dates, we use the year when they are published to approximate the year of allocation. There are three plots published before 2003 and without approval document number (one in 2001, and two in 2002). For land transaction data, the plots published in 2003 (1154 plots) are not included in the analysis because their location is only reported at the district level. The land data published in 2004 through 2013 is included in the analysis with one caveat: the data published online in 2004 is much larger than later years, which implies that the 2004 may have included transactions made in previous one or two years. For this reason, we have included all land allocation data available online (one record for 2001, two records for 2002, and all data from 2003 through 2014) in the analysis.

Table 12.1 Summary of the online administrative land plot data sources

	Transaction	Allocation	Acquisition
Year coverage	Published since 2003 to present	Published since 2003 to present (plus 1 site in 2001 and 2 sites in 2002)	Published in various years (earliest: Fangshan district, 2005; most recent: Miyun county, 2013)
Total number of plots until Feb 2014	7629	1324	891
Total number of plots that can be geo-coded below the district level since 2004–2013	6419	1313	/
Information	Recipient, location, land area, planned building floor space, planned land use, land transaction fee, boundary, signing time*, start time*, completion time*, FAR[a], number of page views (*: only available online since 2008)	Project name, user of land, location, land area, approval document number, approval authority, planned building floor space, planned land use, view times	Construction institution/company, project, planned land use, location, land area, land ownership, compensation, staff for resettlement, approval document
Included year	2004–2013 (Excluding 2003 data because of poor location information)	2003–2013 (including 1 in 2001 and 2 in 2002)	Not included in the analysis in this paper

[a]Floor-area ratio, i.e. the total building floor space area on site in square metre divided by the total land area in square metre

dates back to different years, typically 2009. The land plots in the acquisition data also include a substantial amount that will subsequently appear in the land transaction data (i.e. after primary land development) (Table 12.1). We will return later in the paper to the implications of leaving out this third source, and the prospects of using it in the future.

We first collected all the items published on the website.[10] After cleaning the addresses (mainly by confirming the integrity of the addresses and deleting unnecessary details), we used a VBA program[11] to geocode the address into longitude and latitude. By checking the longitude and latitude with the expected locations in the

[10] For this, we use a web-scraping software called LocoySpider (http://www.locoy.com/).

[11] http://www.masterable.com/1/post/2011/06/fusiontables-post.html. We tested geocoding using Esri Maps for Office (http://doc.arcgis.com/en/maps-for-office/) and the Google FusionTable VBA program, and based on the relative accuracy of the results preceded using the latter.

Table 12.2 Address misrecognition for geocoding before and after address cleaning

	Land transaction				Land allocation			
	Before cleaning		After cleaning		Before cleaning		After cleaning	
District level misrecognition	1339	22 %	0	0 %	205	15 %	0	0 %
Jiedao level and below misrecognition	1440	23 %	375	6 %	60	5 %	61	5 %
Total misrecognition	2779	45 %	375	6 %	265	20 %	61	5 %
Plots free from misrecognition	3433	55 %	5837	94 %	1059	80 %	1263	95 %
Total plots*	6212				1324			

*The comparison was done to the land transaction data up to August 2013 (excluding those from 2003 due to poor location information) and the land allocation data up to Feb 2014. That is why the totals are both different from the total number of plots we used for the final analysis as shown in Table 12.1

online maps, we further cleaned the addresses which were misrecognised by Google and controlled the level of accuracy in a semi-automatic way (Table 12.2).[12] The land use types were further categorised into 24 categories based on land use purposes,[13] and then combined into 7 main categories[14] (Table 12.3). Then the data were processed and analysed using Arc GIS.

Two main types of analysis were carried out – (1) the Rings of Growth which analyses the pattern of growth in different Ring Road areas to show any centralising/decentralising trends; (2) Growth Mapping which analyses the pattern of growth using a 3 km by 3 km grid overlaid on the Beijing map in addition to mapping all the projects by site to inspect growth trends with further spatial details.

[12] For one location (i.e. with same longitude and latitude) with more than 3 plots (exclusive of 3 for land transaction, and inclusive of 3 for land allocation), we check whether they have the same address, and if not whether they differ at the Jiedao level (sub-district) or district level. For those differ at the district level, we verify the district code and geocode again to make sure that they are geocoded into the correct address. For those differ at the Jiedao level, we carry out the same process for those with more than 10 (including 10) plots for one address. As the remaining misrecognitions are all at the Jiedao level and below and no more than 3 pots, and they are only 5–6 % of the total, we keep them as they are for the following analysis. Besides, we also correct the obviously wrong geocodes such as those suggesting locations outside Beijing. However, due to the large number of items, it is not possible to guarantee each and every address has been geocoded correctly.

[13] The most important land use purpose is chosen if there are more than one land use purposes.

[14] The main category of 'infrastructure' is not used in our analysis in the paper, as our focus is housing and employment land.

Table 12.3 Definition of land use categories

Main land use categories	Detailed land use categories
Housing	Housing
Offices and institutions	Office
	Institution
Manufacturing	Manufacturing
R&D	R&D (including higher education, and R&D)
Retail	Retail
Community services	Local education
	Hospital
	Public building
	Culture/Recreation
	Local commercial
	Hotel and restaurant
	Local facilities
	Mixed
Infrastructure	Facilities
	Parking
	Underground parking
	Transport
	Public transport
	Road/Railways
	Airport
	Storage
	Tourism
	Other

12.3 Trend Analysis

12.3.1 Overall Growth

From our data, we found that from 2004 to 2013, for housing and employment, 6719 projects acquired 21,342 ha of land through both land transaction and land allocation, which provide 337.6 million square metres of planned building floor space. Housing projects make up the largest proportion of land use at 32 %; the dominance of housing developments is further highlighted by its share in land area and planned building floor space, respectively 55 % and 68 % (Table 12.4). This is a trend shared by the majority of growing cities.

In order to show the trajectory of land transaction and allocation, we divided the decade into 4 periods, i.e. 2004 and before, 2005–2007, 2008–2010, and 2011–2013. In terms of the number of land plots, 2005–2007 is the highest (2108), whereas the other three periods are in the similar level of 1500 (Table 12.5). Land area has a similar pattern, with 2005–2007 ranking the highest (6129.0 ha). However, although the rest of the three periods have a similar number of projects,

Table 12.4 Summary of land area and planned building floor space by main type

	Number of projects		Land area (ha)		Planned building floor space (million m²)	
Housing	2111	32 %	11704.9	55 %	229.1	68 %
Offices and institutions	757	11 %	724.9	3 %	21.2	6 %
Manufacturing	1079	16 %	5049.3	24 %	25.1	8 %
R&D	344	5 %	1514.2	7 %	18.1	5 %
Retail	1412	21 %	1285.3	6 %	30.1	9 %
Community services	1016	15 %	1063.8	5 %	14.0	4 %
Total	6719	100 %	21342.4	100 %	337.6	100 %

Table 12.5 Number of plots of different land use types throughout the period

	2004 and before		2005–2007		2008–2010		2011–2013		Total	
Housing	810	38 %	414	20 %	401	19 %	486	23 %	2111	100 %
Offices and institutions	221	29 %	270	36 %	173	23 %	93	12 %	757	100 %
Manufacturing	116	11 %	436	40 %	284	26 %	243	23 %	1079	100 %
R&D	20	6 %	68	20 %	95	27 %	161	47 %	344	100 %
Retail	192	14 %	580	41 %	426	30 %	214	15 %	1412	100 %
Community services	173	17 %	340	34 %	195	19 %	308	30 %	1016	100 %
Total	1532	23 %	2108	31 %	1574	24 %	1505	22 %	6719	100 %

2004 and before has a much higher land area (5747.7 ha) than the other two because of some preceding developments that may have been included (4555.6 ha for 2008–2010, and 4910.0 ha for 2011–2013) (Table 12.6). Planned building floor space shows a very different picture. 2004 and before was the largest (108.4 million m^2), followed by 2011–2013 (87.9 million m^2), 2005–2007 (75.4 million m^2) and 2008–2010 (65.8 million m^2). Apart from the issue arising from the crude data, this is due to the composition of different types of projects, suburbanisation and the associated evolution of the FARs throughout the decade (Table 12.7).

As we are mostly interested in the size of the development for different categories, we will focus on planned building floor space for the analysis below. Overall, both housing and employment are developing at a steady pace over the decade; housing construction fluctuates more with lower ebbs in 2008–2010 as expected. As for employment land, the pace varies substantially between different categories. The development of offices and institutions decreased significantly from 60 % in 2004 and before to 6 % in 2011–2013. R&D increased significantly from 4 % in 2004 and before to 41 % in 2011–2013, whereas retail increased steadily although in a slower pace from 21 to 30 %. Manufacturing and community services varied from period to period, with the lowest in 2004 and before, and the highest in 2005–2007 and 2011–2013, with a drop in 2008–2010 (Table 12.7).

Table 12.6 Land area of different land use types throughout the period

(Unit: ha)	2004 and before		2005–2007		2008–2010		2011–2013		Total	
Housing	4251.3	36 %	2717.2	23 %	2057.8	18 %	2678.6	23 %	11704.9	100 %
Offices and institutions	329.5	45 %	204.6	28 %	113.2	16 %	77.6	11 %	724.9	100 %
Manufacturing	707.4	14 %	2027.1	40 %	1227.5	24 %	1087.3	22 %	5049.3	100 %
R&D	45.6	3 %	451.7	30 %	507.5	33 %	509.3	34 %	1514.2	100 %
Retail	309.5	24 %	344.7	27 %	359.6	28 %	271.5	21 %	1285.3	100 %
Community services	104.4	10 %	383.8	36 %	290.0	27 %	285.6	27 %	1063.8	100 %
Total	5747.7	27 %	6129.0	29 %	4555.6	21 %	4910.0	23 %	21342.4	100 %

Table 12.7 Planned building floor space of different land use types throughout the period

(Unit: million m^2)	2004 and before		2005–2007		2008–2010		2011–2013		Total	
Housing	83.6	37 %	46.9	20 %	41.8	18 %	56.9	25 %	229.1	100 %
Offices and institutions	12.8	60 %	4.6	22 %	2.4	12 %	1.4	6 %	21.2	100 %
Manufacturing	3.0	12 %	7.9	31 %	5.7	23 %	8.5	34 %	25.1	100 %
R&D	0.8	4 %	5.0	28 %	4.8	27 %	7.4	41 %	18.1	100 %
Retail	6.2	21 %	6.7	22 %	8.1	27 %	9.1	30 %	30.1	100 %
Community services	2.1	15 %	4.3	31 %	2.9	20 %	4.7	34 %	14.0	100 %
Total	108.4	32 %	75.4	22 %	65.8	20 %	87.9	26 %	337.6	100 %

Table 12.8 Rings of growth

Rings of growth	Area
City Centre	Area inside the 2nd Ring Road
Ring 2–3	Area between the 2nd and the 3rd Ring Road
Ring 3–4	Area between the 3rd and the 4th Ring Road
Ring 4–5	Area between the 4th and the 5th Ring Road
Ring 5–6	Area between the 5th and the 6th Ring Road
Hinterland	Area outside the 6th Ring Road but within Beijing municipality

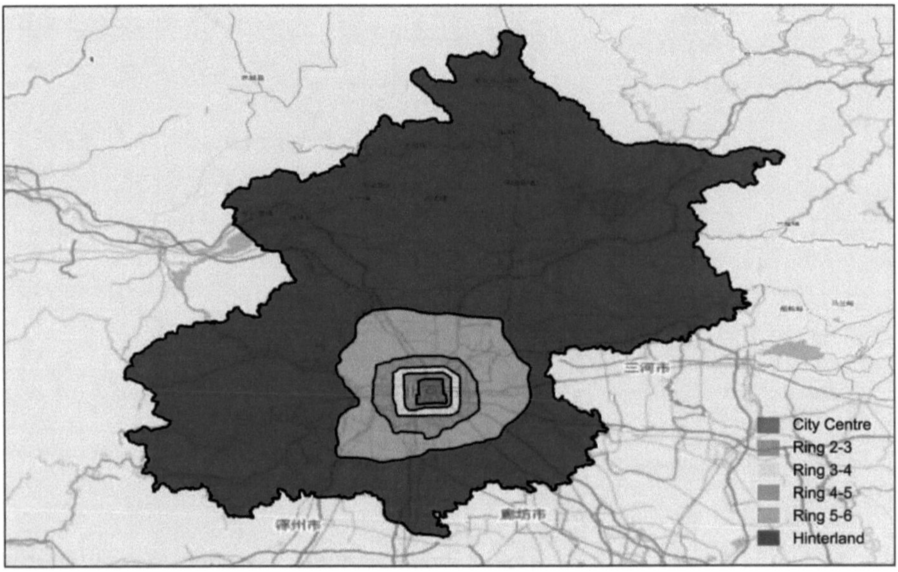

Fig. 12.1 Rings of growth

12.3.2 Rings of Growth

12.3.2.1 Ring Roads and Rings of Growth

In order to identify the trends of growth, we divide Beijing into 6 concentric rings in line with the wide ring roads which have been built out along with the outward urban expansion. The concentric rings for our analysis are: City Centre (i.e. inside Ring Road 2 which was built on the site of former city walls and moats), Ring 2–3, Ring 3–4, Ring 4–5, Ring 5–6, and the rest of Beijing municipality outside Ring Road 6 (Table 12.8, Fig. 12.1). The plan for the ring roads was first introduced by Soviet urban planning advisers to Beijing in the early 1950s following Moscow. The ring road pattern has remained in every version of Beijing's masterplan since then (Sit, 1996), and further extended. They not only serve as the physical boundaries of different patterns of land use development, but also embody psychological boundaries for business and residents, e.g. when discussing property prices.

12.3.2.2 Distribution of Planned Building Floor Space (BFS)

To analyse the geographical distribution as well as the rates of growth, we first compare the distribution of growth by land use type.

BFS of housing has shown a strong decentralising trend. There has been a steady build-up of the rates of growth starting from the centre, culminating to the area of Ring 5–6, which accounts for 38 % of the total growth. Beyond Ring Road 6 the share of BFS is 20 %, which is still higher than any area within Ring Road 5. In other words, 58 % of all planned new housing is located outside Ring Road 5.

BFS of employment, on the other hand, demonstrates a more mixed trend. Most types of employment are also decentralising. Manufacturing and R&D lead this decentralising trend, both with extremely low shares of total growth within Ring Road 4 (3 % and 5 % respectively). Most of the BFS for manufacturing is located in the hinterland (48 %), whereas R&D in Ring 5–6 (62 %). Retail and community services appear to decentralise at a slower rate, with relatively high growth in more central areas (8 % and 14 % respectively in the city centre), with the largest shares of growth concentrated in Ring 5–6 (31 % and 33 % respectively). The only exception are offices and institutions, which show a centralising trend. The city centre captures a share of 20 % of the BFS growth, and the areas within Ring Road 4 altogether have a 74 % share (Table 12.9, Fig. 12.2).

12.3.2.3 Changes in Growth Distribution Over Time

Building on the overall picture, we further investigate whether the decentralising or centralising trends are intensifying during the decade.

Figure 12.3, the stacked line diagram showing planned housing floor space development below, confirms a decentralising trend. It is clear that the rate of growth in Ring 5–6 and beyond started to accelerate around 2006.

For most employment land use types (Figs. 12.4, 12.5, 12.6, 12.7 and 12.8), the decentralising trend starts between 2005 and 2007, for example retail in 2005, community services in 2006 or 2007, R&D in 2007. Manufacturing stands out as the one land use type that has been decentralising throughout the decade. The land use types mentioned above have seen a build-out rate ramping up since 2008–2009, with the exception of community services for which the ramp-up started around 2010. It is worth noting that even for offices and institutions, areas outside Ring Road 4, especially Ring 5–6, have started to provide a larger proportion of BFS than areas within Ring Road 4 since 2009. However, the comparatively small amount of ramping up is not sufficient to change the overall centralising pattern.

Table 12.9 Spatial distribution of planned building floor space, 2004–2013

(Unit: million m^2)	City Centre		Ring 2–3		Ring 3–4		Ring 4–5		Ring 5–6		Hinterland		Total	
Housing	10.3	4.5 %	17.4	7.6 %	26.8	11.7 %	40.5	17.7 %	87.8	38.3 %	46.4	20.3 %	229.1	100.0 %
Offices and institutions	4.3	20.3 %	5.6	26.2 %	5.8	27.1 %	2.1	9.8 %	2.8	13.0 %	0.7	3.4 %	21.2	100.0 %
Manufacturing	0.2	0.6 %	0.2	1.0 %	0.4	1.7 %	5.0	20.1 %	7.1	28.2 %	12.2	48.4 %	25.1	100.0 %
R&D	0.1	0.5 %	0.1	0.6 %	0.7	3.9 %	2.7	14.8 %	11.1	61.5 %	3.4	18.7 %	18.1	100.0 %
Retail	2.5	8.3 %	2.3	7.7 %	5.5	18.3 %	7.0	23.1 %	9.2	30.6 %	3.6	12.0 %	30.1	100.0 %
Community services	2.0	14.2 %	1.1	7.7 %	1.7	12.4 %	2.7	19.2 %	4.5	32.5 %	2.0	14.0 %	14.0	100.0 %
Total	19.3	5.7 %	26.7	7.9 %	40.9	12.1 %	59.9	17.8 %	122.5	36.3 %	68.3	20.2 %	337.6	100.0 %

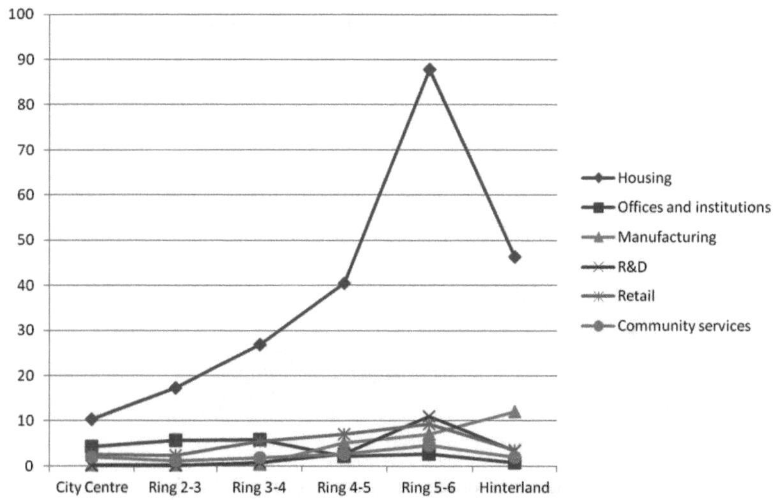

Fig. 12.2 Distribution of planned BFS for each land use type, 2004–2013 (Unit: million m^2)

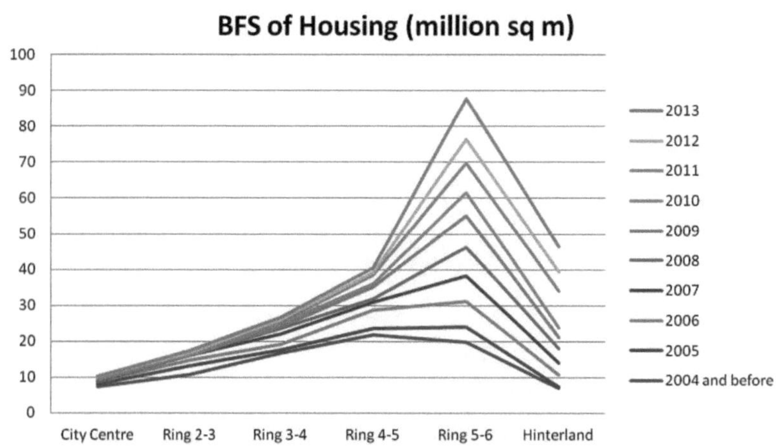

Fig. 12.3 Cumulative distribution of housing building floor space by year

12.3.3 Growth Patterns by Neighbourhood and Site

As the pattern of growth varies from land use type to land use type, we further map the growth for each land use type in progressively more detailed locations – first by neighbourhood and then by site (where feasible). For mapping the neighbourhood, we overlaid a 3 km by 3 km grid onto the municipal area and summed all the building floor space of each land use category in each grid cell. In addition, we also mapped the developments in their places using 5 dot sizes to demonstrate the

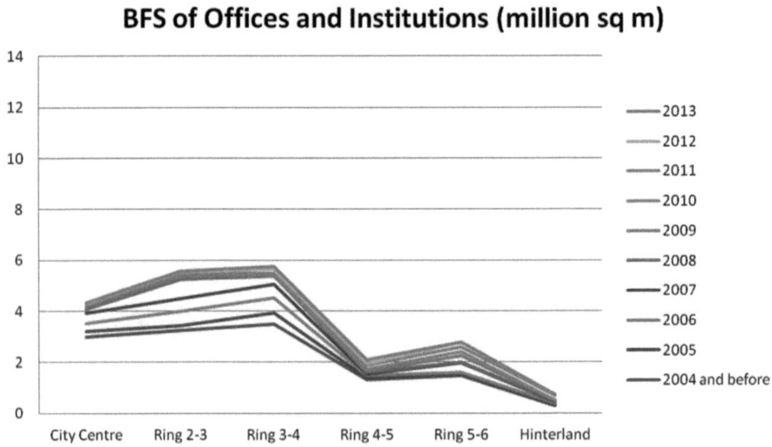

Fig. 12.4 Cumulative distribution of offices and institutions building floor space by year

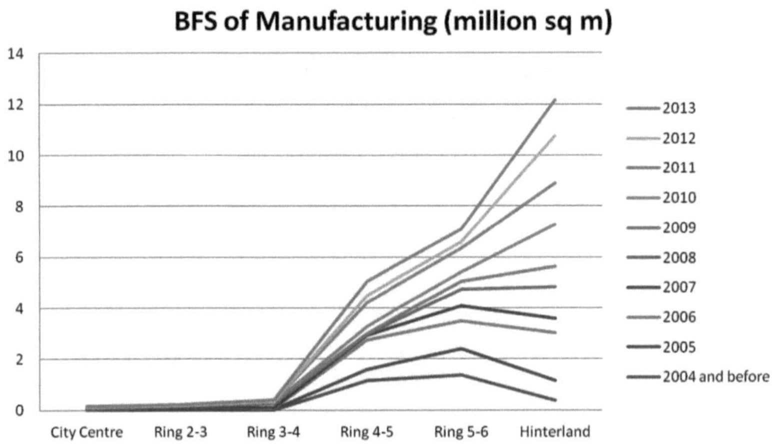

Fig. 12.5 Cumulative distribution of manufacturing building floor space by year

5 levels of planned project size where the sites are coloured from green if the planned project was before or in 2004 to red if in 2013.

For housing, the growth covers almost all the grids within Ring Road 5 and the majority of the grids between Ring Road 5 and 6 (Fig. 12.9). The locations of the projects suggest that the major housing project sites follow the main road arteries, such as the ring roads, the East-West axis extending from the Chang'an Boulevard, and the radial expressways outside Ring Road 5 and especially Ring Road 6 (Fig. 12.10). The majority of the recent projects are of increasingly larger scale and located further away from the city centre. Also worth noting are the numerous small central city rehabilitation projects in historic residential areas inside Ring Road 2.

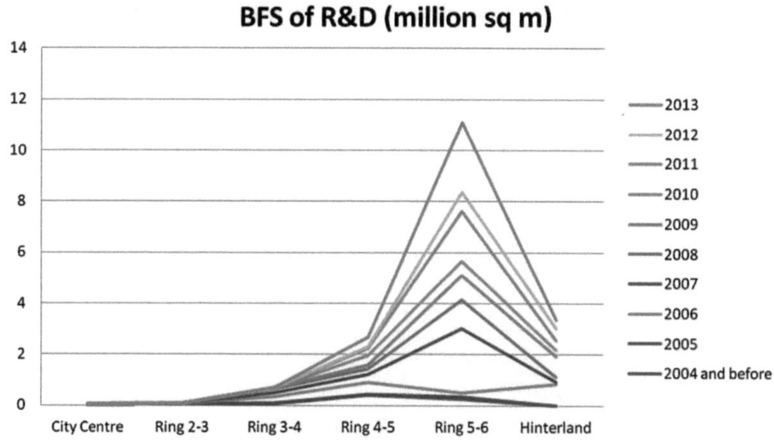

Fig. 12.6 Cumulative distribution of R&D building floor space by year

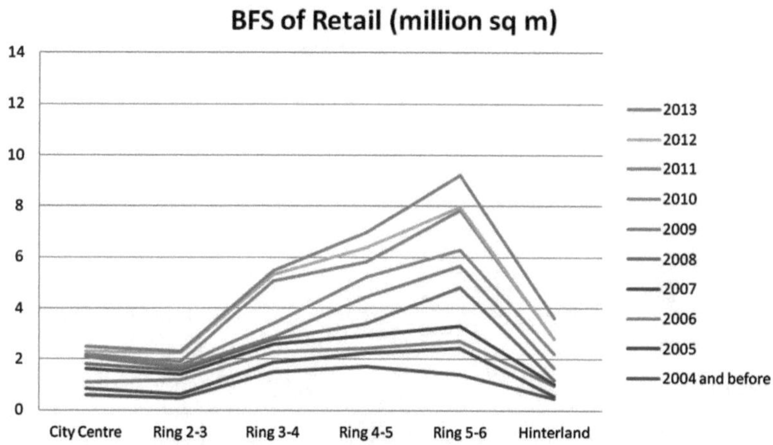

Fig. 12.7 Cumulative distribution of retail building floor space by year

Offices and institutions are mainly growing through densification in the northern part of the area encircled by Ring Road 5, particularly within Ring Road 4 (Fig. 12.11). This occurs in clusters along the Chang'an Boulevard and primarily to the north of it. Outside Ring Road 5, there are a few development clusters along the main transport corridors, although the sizes of the projects are smaller. The development cycle appears to have been affected by the world financial crisis of 2008: the majority of developments took place before 2008, and after 2008 the projects tend to be smaller and mainly located within the existing clusters (Fig. 12.12).

Manufacturing is expanding out in a dispersed way. Most of the growth is along and outside Ring Road 5 towards the south (Fig. 12.13). Within Ring Road 4, small

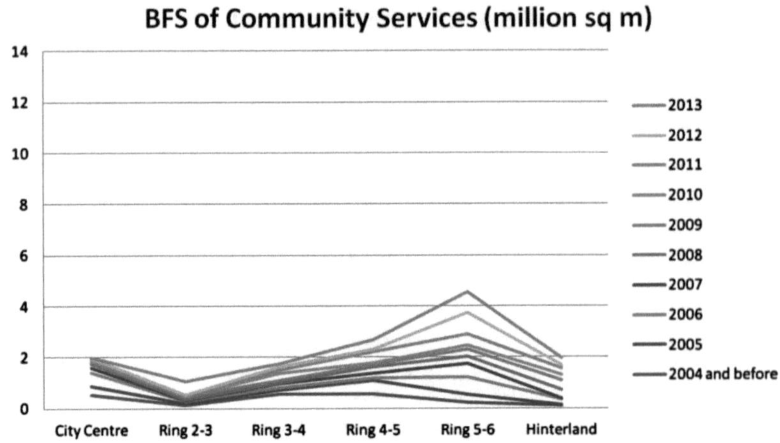

Fig. 12.8 Cumulative distribution of community services building floor space by year

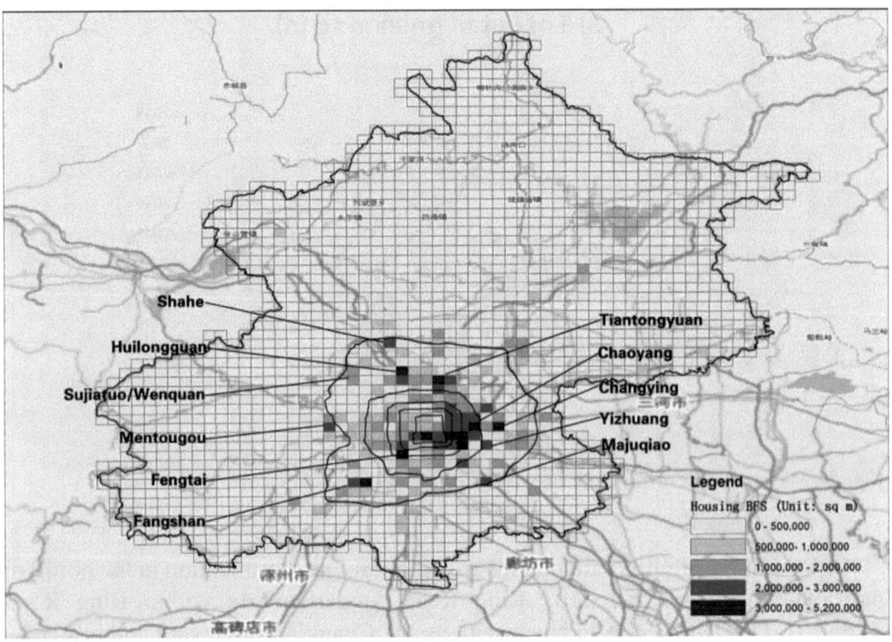

Fig. 12.9 Grid mapping of housing growth (BFS)

projects are located round Ring Road 2 and 3. In the Ring 4–5 area, large projects have started to emerge and grow along all the major road corridors. Among those, recent projects seem to have the potential to form the nuclei of future clusters (Fig. 12.14).

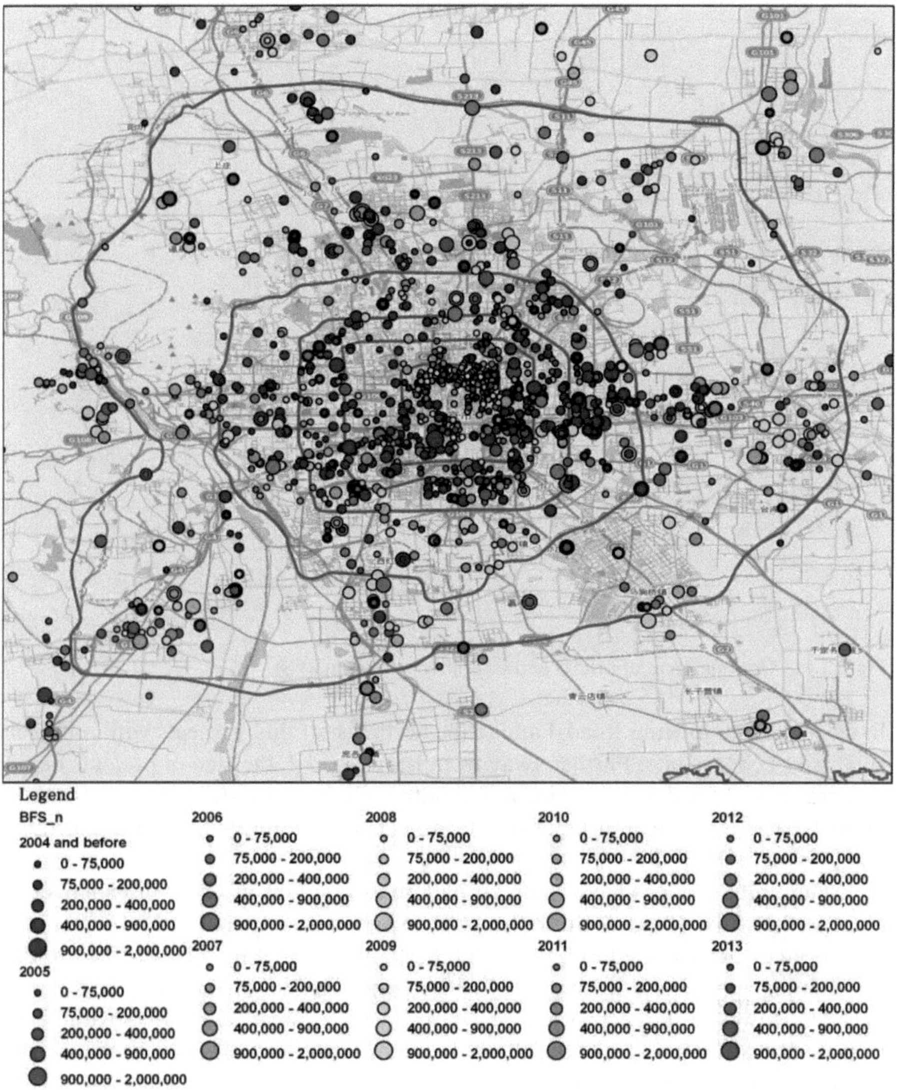

Fig. 12.10 Planned housing floor space by site

R&D, although large in the total volume of building floor space, actually consists of a very small number of projects, which are very dispersed (Fig. 12.15). Most large projects are outside Ring Road 4, occurring in clusters along the transport corridors. There are also the obvious cluster along the northwest axis of G6 Beijing-Lhasa Expressway, and the newly emerging clusters in Xiaotangshan, Wangjing, Wenquan, Yizhuang and Fangshan (Fig. 12.16).

Retail developments cover the areas within Ring Road 4 and extend further towards northeast as well as towards the west in Ring 4–5 and beyond (Fig. 12.17).

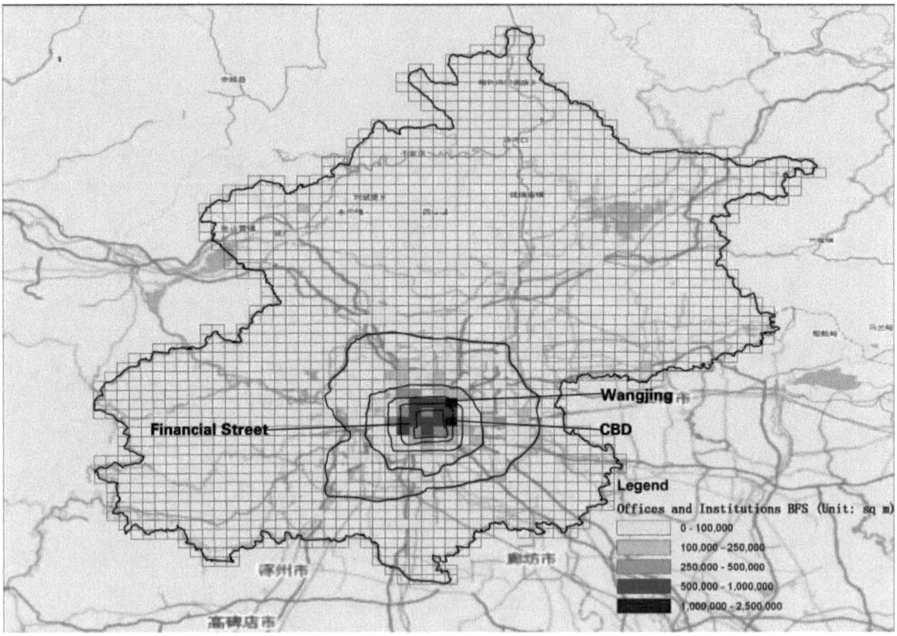

Fig. 12.11 Grid mapping of offices and institutions growth (BFS)

In the areas within Ring Road 4 and to the northeast of this, projects with relatively small sizes congregate. Further away from Ring Road 4, the size of projects begins to increase and the density begins to fall, although there are still clusters of smaller projects. There are several large recent projects expanding out along major transport corridors in all directions (Fig. 12.18).

Community services have a similar pattern as retail, as most developments take place within Ring Road 4. But rather than expanding northeast and west, community services expand mainly towards the north and west (Fig. 12.19). The projects are smaller and denser in the areas within Ring Road 4 than those in the areas outside. However, large projects are more evenly spread compared with retail. Outside Ring Road 4, projects follow the main transport corridors, more towards the west than east (Fig. 12.20).

12.3.4 Growth Clusters

For most land use types, specific clusters of growth can be observed. This can be compared with what is known about the main locations of housing and employment land growth in the municipality.

For housing, there is a large cluster in Chaoyang district between Ring Road 2 and 5, and another in the south of the Chang'an Boulevard axis towards the west

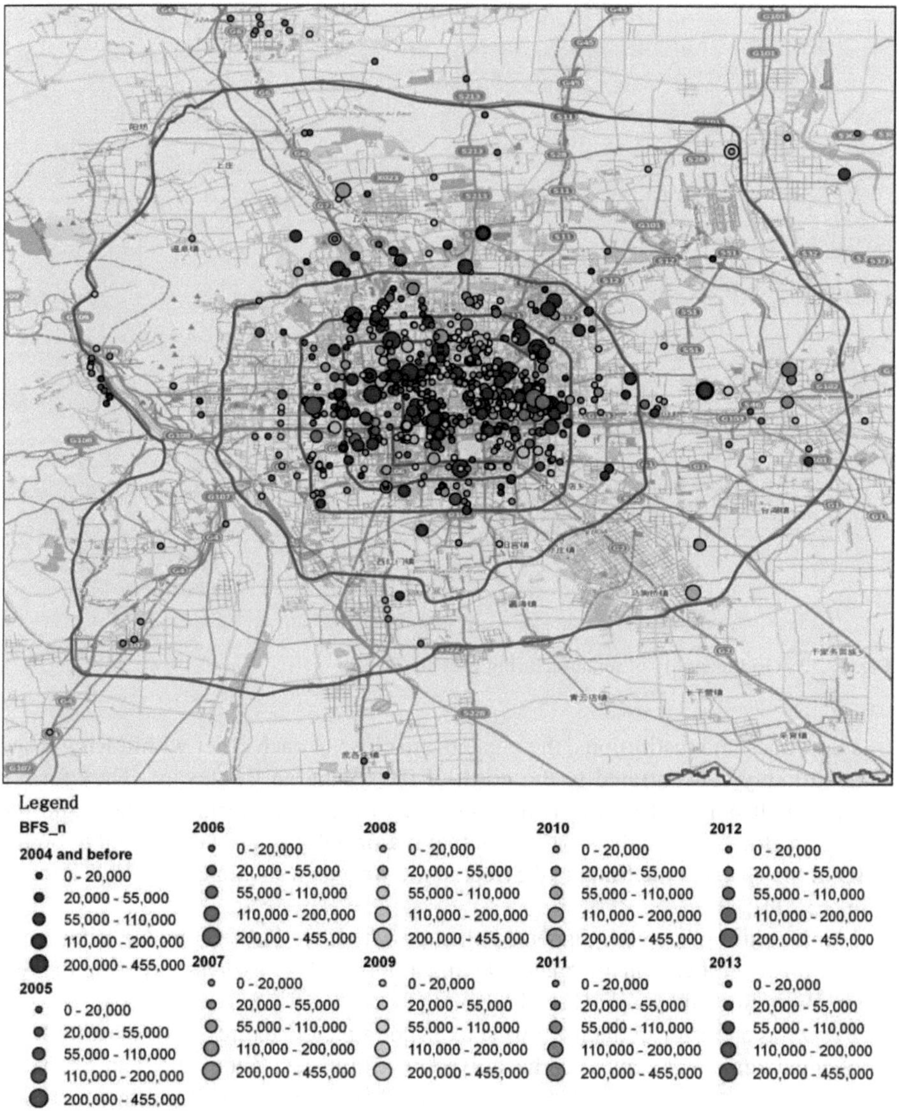

Fig. 12.12 Planned offices and institutions floor space by site

until the 2nd Ring Road. Further out, there are two further rings of clusters: the first ring is formed by comparatively smaller clusters which belong to the larger residential districts planned and developed in the early part of the decade, such as Changying, Tiantongyuan, Huilongguan and Fengtai; the second ring consists of newer clusters of even larger planned residential districts in the latter half of the decade – the developments are further away from the city centre, such as Yizhuang, Majuqiao, Shahe, Sujiatuo/Wenquan, Mentougou and Fangshan (Fig. 12.9).

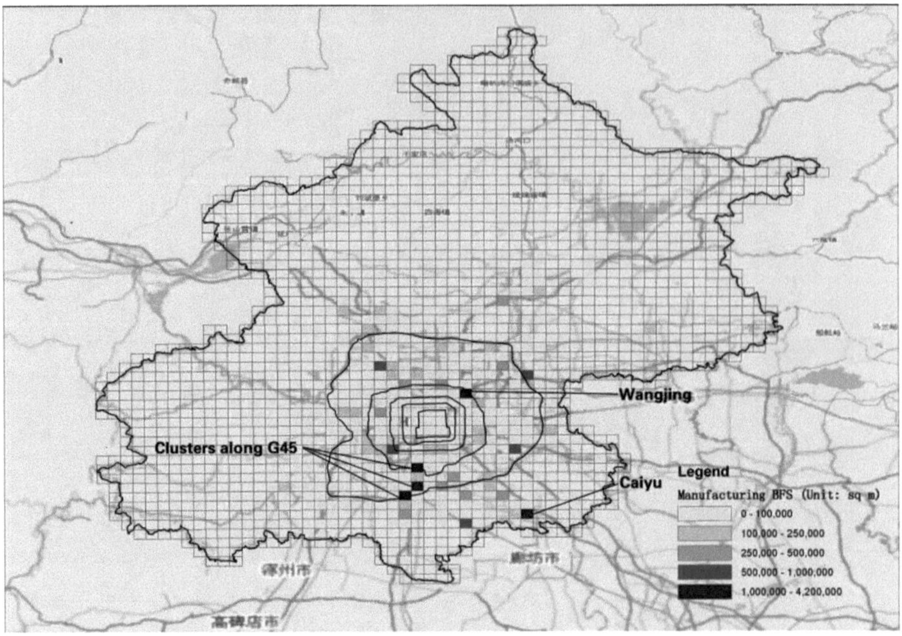

Fig. 12.13 Grid mapping of manufacturing growth (BFS)

For offices and institutions, the clusters are close to each other within Ring Road 4, remarkably concentrated to the north of the Chang'an Boulevard. They form a box-like belt bounded by the Ring 2–3 area in the west, north and east, and the Chang'an Boulevard to the south, with two major poles of growth centred upon the CBD in the east and the Financial Street in the west as well as a cluster in Wangjing (Fig. 12.11).

Manufacturing has very dispersed clusters which are located outside Ring Road 4. The most significant is the one in Daxing along the G45 expressway. And the rest are in Wangjing and Caiyu (Fig. 12.13).

R&D has a cluster in the northwest axis along the G6 expressway, starting from Zhongguancun town centre to the university campuses, Zhongguancun Software Park, Zhongguancun Life Science Park, and Changping. Apart from this, there are clusters emerging in Xiaotangshan (Weilai Science and Technology City), Wangjing, Wenquan, and Yizhuang, which are all part of the Zhongguancun Technology Park, as well as Fangshan, which is a planned new centre for higher education (Fig. 12.15).

For retail, the city centre and the northeast quarter of the Ring 2–3 area and the Ring 3–4 area are filled with small projects with the particular centres in CBD and the Financial Street. Outside this, there are a ring of clusters including Tiantongyuan, Huilongguan, Weigongcun, Shijingshan, Fengtai, Xihongmen, Wangjing and Pinggu. As expected, the larger clusters coincide with the clusters of housing or offices and institutions (Fig. 12.17).

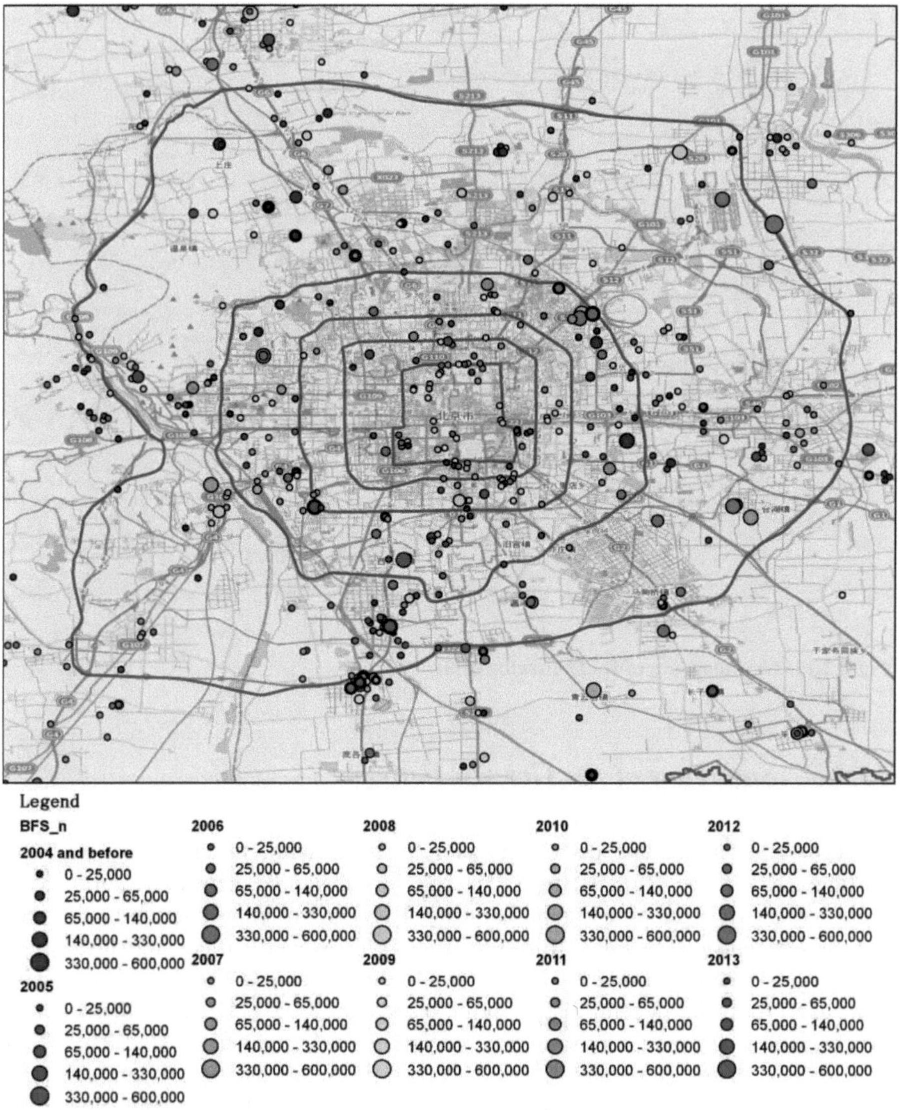

Fig. 12.14 Planned manufacturing floor space by site

The clusters of community services follow housing and retail to a certain extent. The city centre, the northeast half of the Ring 2–3 area and the Ring 3–4 area are filled with projects. Immediately outside this core, there is a ring of small clusters in Olympic Park, Changchunqiao, Fengtai and Xiaohongmen/Shibalidian. Further out in Ring 5–6, there is another ring of clusters, such as Huilongguan, Tiantongyuan, Houshayu, Tongzhou, Taihu and Fangshan (Fig. 12.19).

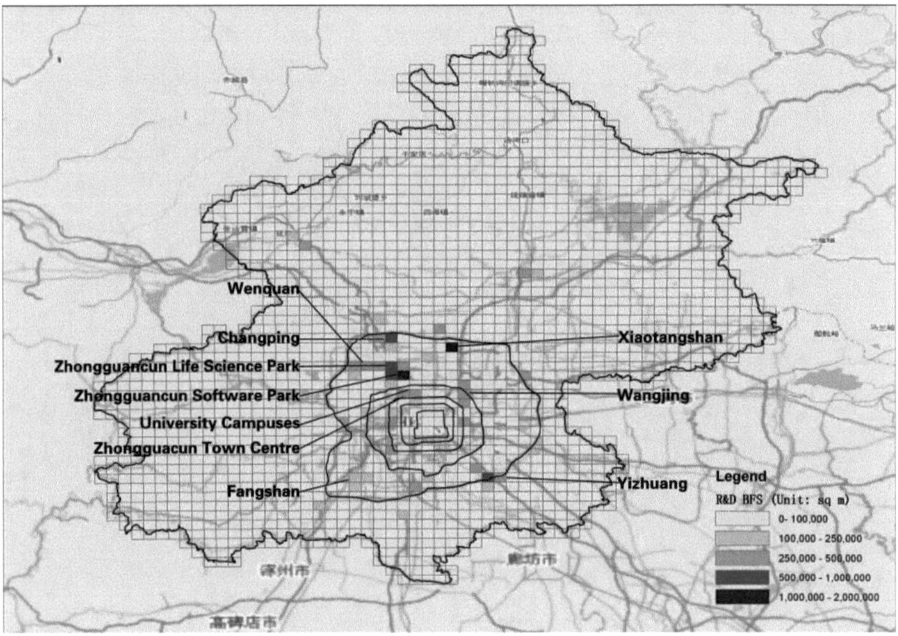

Fig. 12.15 Grid mapping of R&D growth (BFS)

12.3.5 Summary of the Growth Patterns

In summary, in the last decade, housing was growing faster in the outskirts, particularly outside Ring Road 5, leading to substantial decentralisation. Most employment types, such as manufacturing, R&D, retail and community services, are also decentralising but at different speeds. However, offices and institutions tend to grow more prominently in and around the city centre. For the decentralising land use types, the decentralising trend started in 2005–2007, and intensified in 2008–2009. Even manufacturing has seen more BFS planned outside Ring Road 4. The detailed spatial pattern of growth varies from land use type to land use type, in terms of overall locational distribution, project size, density, and clusters of growth. In particular, there appears to be a general spatial mismatch between the trend of housing development in the outskirts and that of offices and institutions, which tend to focus around the northern part of the Ring 2–3 area as well as the city centre.

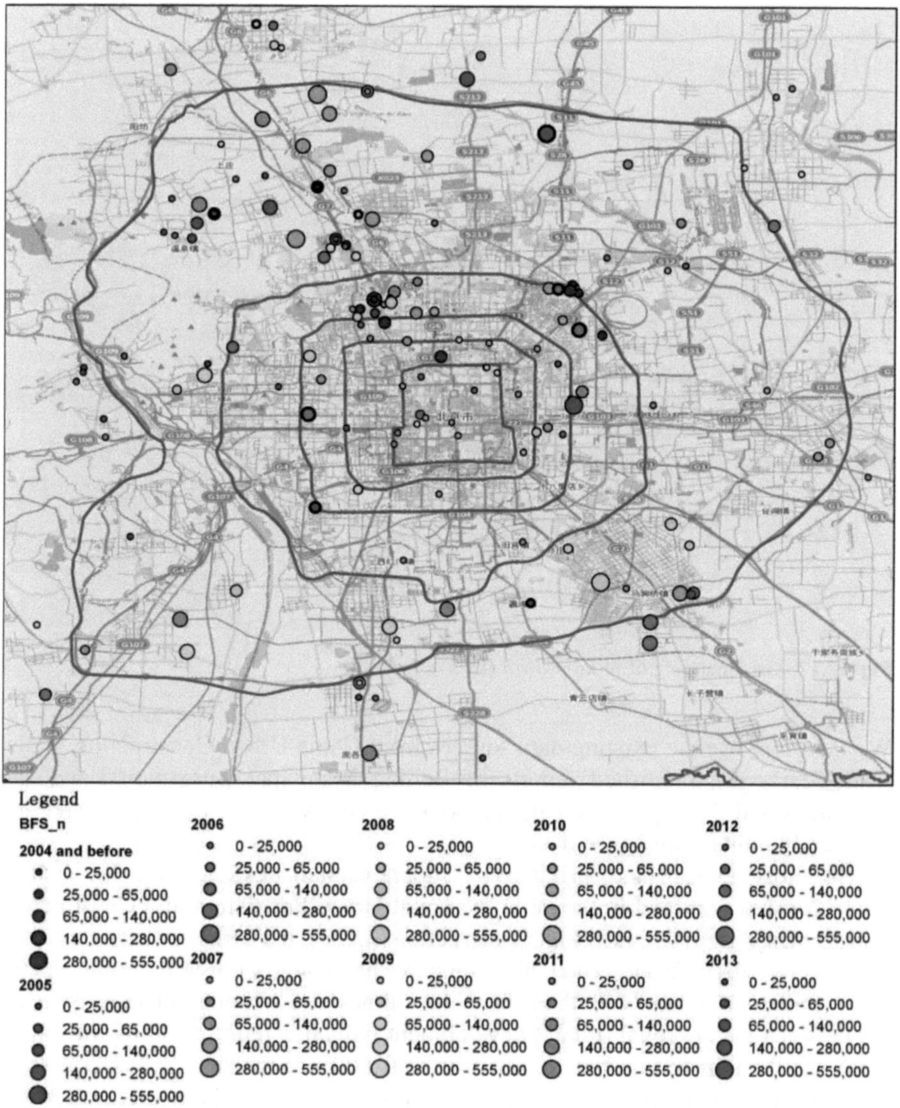

Fig. 12.16 Planned R&D floor space by site

12.4 Comparison With Published Data on Land Development

Although the data is corroborated by our knowledge of the implementation of the urban development plans as shown in Sect. 3.4, it would be necessary to examine how well the online data compares with the sources published in official statistics.

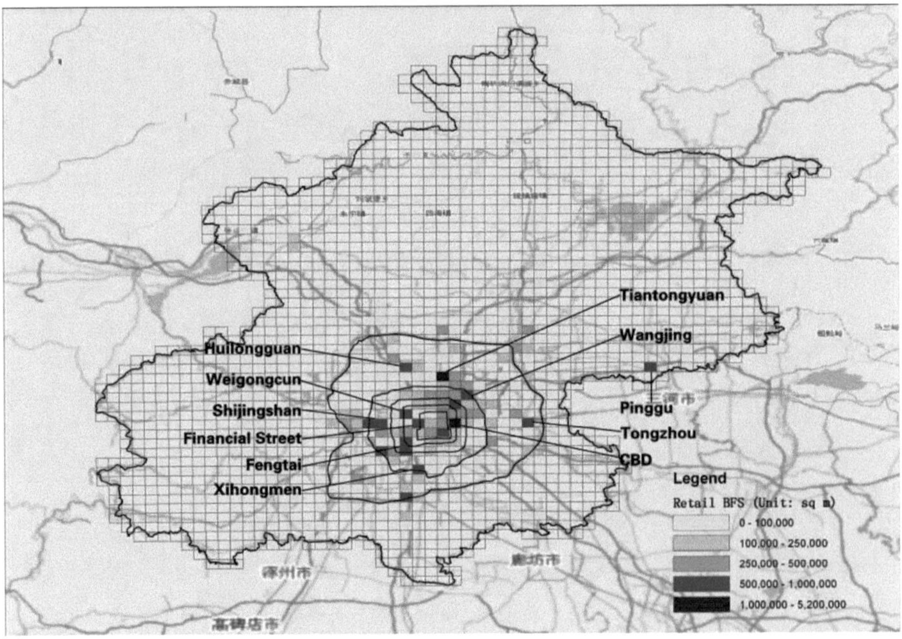

Fig. 12.17 Grid mapping of retail growth (BFS)

We therefore use the existing data sources from China Urban Construction Statistical Yearbooks, published reports of Beijing Municipal Bureau of Land and Resources and Beijing Statistical Yearbooks to compare with the online land and planned building floor space data.

The most relevant and, in theory, comparable statistics data is in the area of urban construction land in the table 'National Urban Population and Construction Land by City' from China Urban Construction Statistical Yearbook (CUCSY 2007, 2008, 2009, 2010, 2011, 2012). We compare the change of area for each type of urban construction land with those from the online data. Although there is usually a lag between the approval of land development (which is represented by the online data) and the registration for the actual use of the land (which is then recorded in CUCSY), the regulations ensure the approved land is used fairly swiftly in Beijing. Therefore we choose to compare the land change between 2005 and 2010 of online data with CUCSY data from 2006 to 2011, i.e. assuming there is a 1-year lag. In general, the online data reports lower total change than CUCSY, mainly because the former has more limited categories; but the changes in existing types of land use match well, e.g. for residential, public facilities and manufacturing (Table 12.10). More specifically, online data reports higher figures in residential, public facility, storage and municipal utility land development, similar figures in manufacturing, and lower figures in categories such as transport infrastructures.

There are several possible reasons for the mismatches for the existing land use types. The underestimation of the change for transport infrastructures is mainly

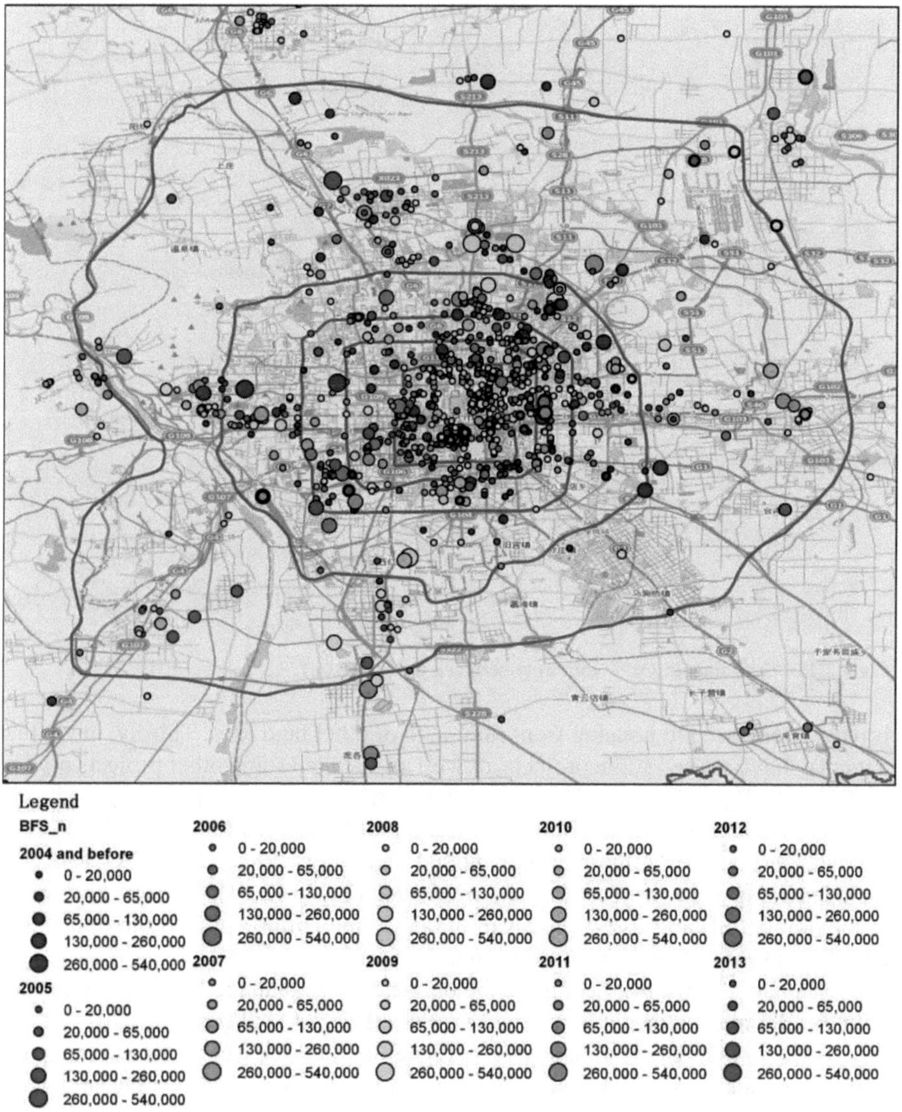

Fig. 12.18 Planned retail floor space by site

caused by the exclusion of land acquisition data, which tends to be used for non-profitable projects such as transport infrastructure taking place in rural areas. This also contributes to the underestimation of the total land area change. For the overestimation of housing land and public facility land, there are more complicated reasons behind this. First, this suggests that the online data may provide more comprehensive coverage than CUCSY. Secondly, the online data contains redeveloped housing land, whereas the CUCSY data only includes the land which

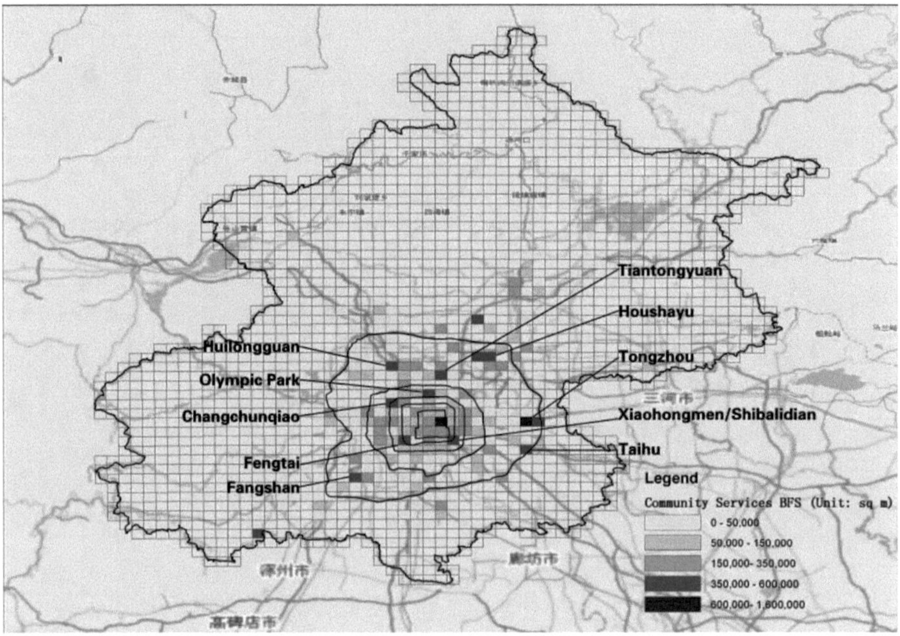

Fig. 12.19 Grid mapping of community services growth (BFS)

is newly converted to housing from rural land or other land uses. Thirdly, the online data may have missed some projects, as well as included some other projects missed by CUCSY. By comparing the total land area from the online data with the report from the Beijing Municipal Bureau of Land and Resources (2009, 2011, 2012),[15] we can see that the online data has missed some projects for land allocated, and included some extra projects for land transacted (Table 12.11). Moreover, changes in the amount of land use do occur in the development process, so the online and CUCSY sources are not expected to match exactly.

For housing, we further compare the online data on planned housing floor space with the Beijing Statistical Yearbooks (2006, 2007, 2008, 2009, 2010, 2011, 2012, 2013). The online data is almost always higher than the Yearbook data in land area developed for market housing, although lower for land purchasing cost (Table 12.12), even if we consider the time lag. This confirms the overestimation of housing land observed in the comparison between online data and CUCSY data, which suggests that online data may be more comprehensively, particularly in the outskirts where the published data may be weak. In addition, the Yearbook data is restricted to reported figures from the real estate companies, whereas the online data

[15] There are only reports which contain information of land provision through land transaction, land allocation and land allocation through land acquisition in 2009, 2011 and 2012. The report for 2013 only includes the first three quarters which is not comparable with our data.

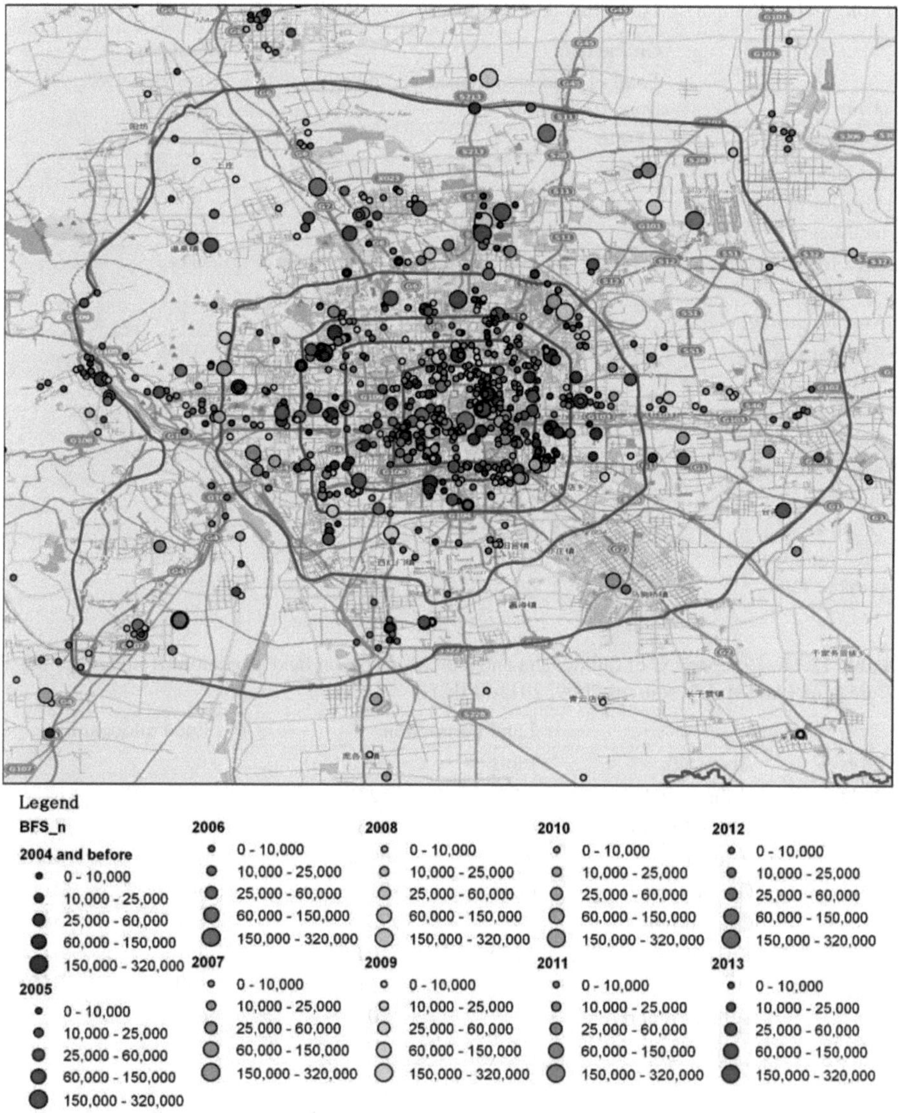

Fig. 12.20 Planned community services floor space by site

account for other types of developers such as non-real estate companies (especially state-owned enterprises controlled by central government) and institutions.

However, the online data is always lower in building floor space for housing (both market and non-market housing) than the Yearbook data (Table 12.13). This is contrary to the overestimation of land area for housing observed above. Our conjecture is that the online data on land has a better coverage, and the land recipients increase the building floor space (or FAR) after they have obtained the

Table 12.10 Comparison of online data and statistics on land area change (with 1 year lag)

(Unit: ha)	Online data Total of land area transacted and allocated from 2005 to 2010	Statistics Change between 2006 and 2011
Subtotal	12,990	17,164
Residential area	5288	4170
Area for public facilities	2868	2356
Area for industrial operation	3110	3142
Area for storage	186	−104
Area for traffic system	468	1459
Area for roads and plazas	0	3289
Area for municipal utilities	459	216
Green area	0	2662
Area for specific-purpose land	0	−26

Table 12.11 Comparison of online data and Bureau report on land provision (Beijing Municipal Bureau of Land and Resources 2009, 2011, 2012)

		Land transaction		Land allocation	
Plots	Area (ha)	Plots	Area (ha)		
2009	Online total	541	1271	87	475
	Official report	677	1854	133	722
	Percentage	80 %	69 %	65 %	66 %
2011	Online total	604	2156	68	215
	Official report	351	2086	121	586
	Percentage	172 %	103 %	56 %	37 %
2012	Online total	402	1118	139	345
	Official report	/	1077	/	450
	Percentage	/	104 %	/	77 %

land, as the planned building floor space suggested in land provision may not be legally binding and can be revised upwards in further planning permission applications.

In general, the verification results show that the online data does not fully represent what is taking place on the ground. This implies several issues with the data. Firstly, whilst the building floor space in the online data may be different from the actual practice on the ground, the land area in the online data for housing, public facilities, industrial operations and storage may be closer to reality than published statistics. Secondly, the exclusion of land acquisition data may underestimate the land provided, especially for transport infrastructure, and also for affordable housing and municipal utilities. Thirdly, for land allocation and land transaction, the

Table 12.12 Comparison of online data and statistics on real estate land development and land purchasing cost (2005–2012)

	Unit: ha	Unit: 100,000,000 Yuan			
	Land allocated and transacted	Land developed	Land purchasing cost		Land purchasing cost
	Online data (market housing)	Statistics (market housing)[a]	Online data (housing only)	Online data (including office, retail, community facilities and etc.)	Statistics
2005	565.5	314.2	43.9	75.3	239.8
2006	1181.2	840.5	116.3	165.7	477.9
2007	970.5	248.7	252.1	313.4	644.7
2008	779.7	351.5	466.1	539.0	639.0
2009	772.1	364.0	517.7	695.0	587.7
2010	506.1	/	640.5	840.7	1292.7
2011	1078.1	/	799.7	1267.5	1301.2
2012	697.8	/	424.7	656.9	1102.7

[a]For both statistics and online data, the land is for market housing projects, which also include a small number of office and commercial buildings as facilities

Table 12.13 Comparison of online data and statistics on housing building floor space (2008–2011)

	Total		Market housing		Non-market housing	
Unit: ha	Online data	Statistics	Online data	Statistics	Online data	Statistics
2008	1453.7	2177.4	1148.5	1565.3	305.2	612.1
2009	1599.9	2212.6	1174.0	1380.3	425.9	832.3
2010	1128.7	2999.3	851.3	2063.4	277.4	935.9
2011	2330.3	4047.4	1479.4	2596.4	850.9	1451.0

data has a small difference with the actual land allocation and land transaction. However, albeit that the difference is small, there is currently no way to identify which projects have been left out or need to be excluded. Fourthly, the land provision data by nature cannot fully reflect what is actually constructed on the ground in the particular year, because the developers and other recipients may stock the land or change the intensity of land use development.

This will particularly affect our understanding of the urban developments in the urban fringe. On the one hand, the exclusion of land acquisition data may underestimate the decentralising trend of housing, community services, etc. On the other hand, although the online data may better reflect the land development in land areas in outskirts compared to the official data, it may underestimate the land development in building floor space in the outskirts as the regulation of building floor space may be less strict in urban fringes than more central areas.

However, the land statistics has always been notoriously difficult. Despite the weaknesses, we are confident that we have captured the main land uses for housing and employment land, given the comparison above and the nature of the online data. In other words, the online data is starting to give us an insight into the patterns of growth for Beijing in the last decade.

It would seem that the pattern it captured is sensible, as some of it has been seen in many other historical precedents in developed countries. The finance and business centres in global cities such as New York, London, and Tokyo have been strengthening its concentration, followed by high-end hotels, retailers and cultural buildings. High technology sectors (i.e. R&D in our categories) tend to agglomerate in outlying clusters and corridors near major international airports, leading universities, etc., for example Silicon Valley in California and Heathrow-M4 Corridor (Cervero 1998). Other types of employment, relatively low-skilled services and human capital-intensive industry such as manufacturing, have been decentralising (Cervero 1998; Glaeser and Kahn 2001). On the other hand, in most European and US cities, housing has gone through almost a century of suburbanisation. Large-scale public housing projects were mainly built in the suburbs, for example new towns and council housing estates in London and grand-ensembles in Paris.

The trend also conforms to the dynamics of different industries. There are two counter-acting factors, which make decentralisation more common in manufacturing and less common in offices and institutions. One factor is the lower cost and higher abundance of land in the suburbs. The other factor is the nature of industries. Skill intensive industries, such as finance and business, prefer higher speed of the flow of ideas for instance through face-to-face contact, and easy access to specialised skills (Glaeser and Kahn 2001). In the mean time, decline in environmental quality of the densely built city centre, lifestyle changes, and the relative ease of developing suburban projects contribute to the decentralisation of housing (Camagni et al. 2002). On top of the market mechanisms, government policies also play an important role in Beijing. The manufacturing relocation policy, the new town policy, and the relocation policy for city centre residents, strengthen the decentralisation.

In summary, the land provision data has captured the main projects, and is able to demonstrate the growth pattern of Beijing's urban land use development. However, it has several weaknesses, including missing land acquisition data, and differences with the actual land development because of the nature of land provision data. More importantly, the online data reports more market housing land and less housing floor space. This would suggest that – on the one hand, more market housing land suggests that the online data may cover more comprehensively than reported in published statistics; on the other hand, less housing floor space suggests that the online data may have underestimated the actual size of housing on the ground. Given the data available, we are not able to solve this puzzle, because the planning permission data published online does not include land area or building floor space information. Therefore, it remains a critical issue to look at in the next step.

For future analysis and application, the data needs to be further compared with construction data/statistics, and adjusted and used accordingly. First of all, the puzzle for housing mentioned above needs to be solved, which then will allow us to estimate more precisely about building floor space. Secondly, the data needs to be compared in more detail with construction statistics for other land use type than housing, which still needs to be looked for. Thirdly, the development in the urban fringe, especially of affordable housing and community services, needs to be better estimated from the limited land acquisition data, either from the years which have complete data for each district, or from the districts which have the complete data for all the years. However, both the exclusion of land acquisition data and the underestimation of building floor space for housing suggest the contrast between centralisation and decentralisation trends may be even stronger.

12.5 Implications for Understanding the Growth Patterns

The increasing separation of housing and offices/institutions, which is demonstrated in Beijing's urban land provision, implies potential problems. The different pace and degree of decentralisation of housing and employment apart from offices and institutions, as well as the continuous development of offices and institutions in more central areas, are likely to add pressure on the city's transport, as the commuting travel demand will rise in the form of both more people travelling in cars and public transport and longer journeys. Given the city has already been suffering from traffic jams on the road and at-capacity flows on the main commuter metro lines, the continuing increasing travel demand would be a huge challenge. On the other hand, residents have already been suffering from long commuting time. According to the China Academy of Science (NIU 2012), Beijing's average travel time to work was 52 min in 2012, ranking it the first among all Chinese cities. In addition, Beijing is also considering increasing the metro fares to mitigate public spending in metro subsidy. The increase in either travel time or travel cost could result in significant impacts on the well-being of residents.

Yet, this is not yet a clear picture. Beijing is learning the lesson from historical precedents trying to solve the present and potential problems. The land plots allocated for affordable housing show a less decentralising trend compared with those for the market housing (Fig. 12.21). Also the city has been constructing public transport at rapid speed. Moreover, lower-end employment, especially manufacturing, is decentralising, which may provide jobs for people who live nearby.

Most importantly, people will change their location choice for working and living as well as adjust their daily activity pattern according to the price, travel time, comfort level, etc. For example, the estimated two million people who live in antiquated underground bomb shelters, and the expanding group-renting population, have found their own way to solve the contradiction of high housing costs, long commuting time and high commuting costs. Obviously, neither is a desirable choice in terms of living conditions which may also undermine the residents' well-being, and may cause potential problems which the city might not be able to estimate and solve because of its informality. Yet, there might be other adjustments which may mitigate the potential problem.

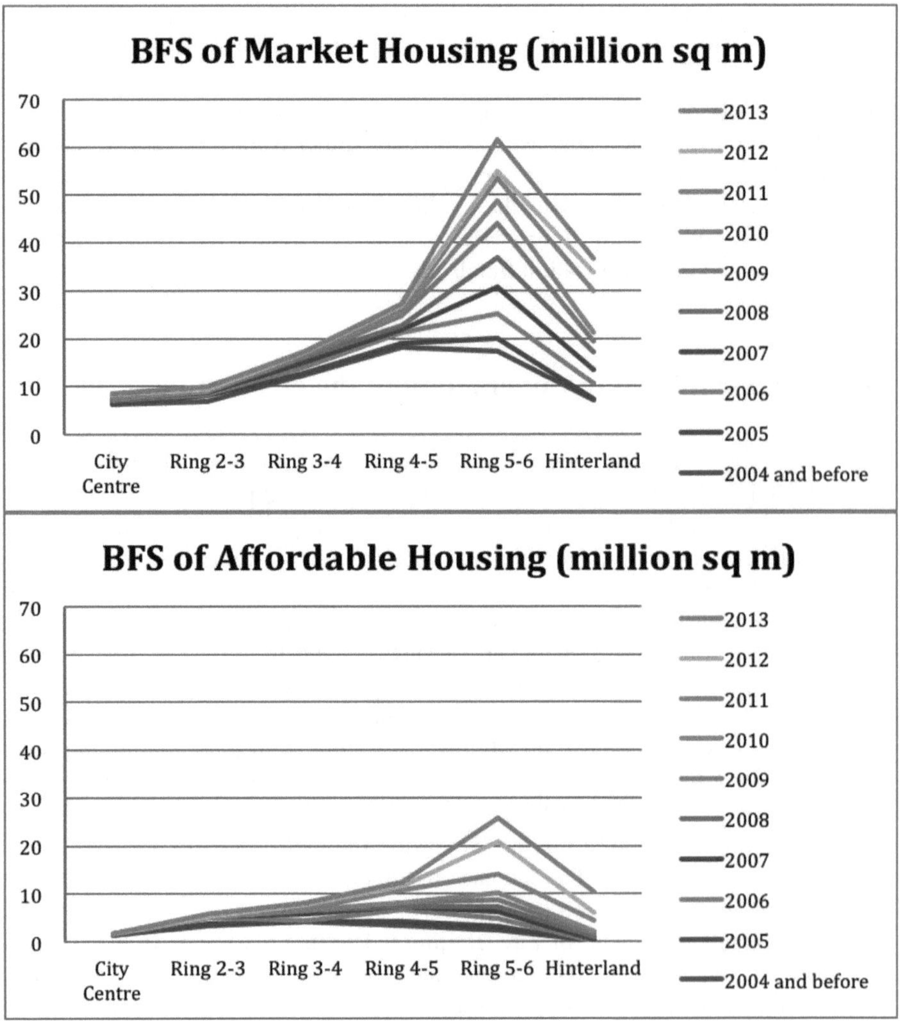

Fig. 12.21 Comparison of planned BFS distribution of market housing and affordable housing

Therefore, in order to better examine the impact of the current development trend, we propose to use a land use and transport simulation model that is based on robust economic and behavioural theories regarding individual, corporate and institutional behaviours, i.e. a recursive spatial equilibrium model (Jin et al 2013). Through this, we are able to simulate residents' adjustments, as well as to overcome the significant gaps and weaknesses of the online data through a new filtering and infilling method, and analyse the impact of the current development trend in comparison with other alternative strategies. With the aid of a theoretically robust simulation model, the online administrative data can make a significant and timely contribution to the understanding of growth patterns, and to core policy analysis.

12.6 Conclusions

This paper is a first attempt to use the novel online land plot provision data (land plot transaction and land plot allocation) to analyse the growth patterns of housing and employment land in Beijing in the last decade. In general, the online data demonstrate that housing is developing and decentralising at a faster rate than employment when measured in terms of both land area provided and square metres of floor space to be constructed. Land for manufacturing, retail and community, are also decentralising at a faster rate than average for employment land as a whole. For higher-end services, R&D is decentralising, but concentrated in very few projects in the outskirts of the city. The land for offices and institutions is clearly developing in a centralising way.

This analysis demonstrates that the online administrative data for land cannot precisely capture the on the ground construction taking place each year, and may underestimate the actual constructed building floor space as well as developments of housing and community services at the urban fringes. These weaknesses may be overcome by more detailed field work. However, when the data is fully corrected and validated we expect that the contrasts between the centralizing and decentralizing trends would be even stronger.

This trend together with the intensification of decentralisation since 2008–2009 means increasing separation of housing and office/institution sector employment, which may exert an even higher demand on the city's transport systems, both on the road and the metro. Next steps include use of a land use and transport simulation model to assess the impacts of alternative growth patterns and policies that will promote economic efficiency, social fairness and environmental sustainability of the city.

Acknowledgments Xiao Rong wishes to acknowledge the funding support of the Raymond and Helen Kwok Research Scholarship which is administered through Jesus College, Cambridge and the Cambridge Overseas Trust. Ying Jin wishes to acknowledge partial funding support of the Tranche 1 research grant funded by EPSRC Centre for Smart Infrastructure and Construction at Cambridge University (EPSRC reference EP/K000314/1), and of the research grant from the Tsinghua-Cambridge-MIT Low Carbon Energy Alliance (Tsinghua University reference: TCMA2010). Ying Long wishes to acknowledge the support of the China Scholarship Council in funding his academic visit to the Martin Centre for Architectural and Urban Studies, Cambridge University. We also wish to acknowledge Beijing City Lab who started the initiative of collecting the land transaction data up to 21 Oct 2013. The usual disclaimers apply and the authors alone are responsible for any views expressed and any errors remaining.

References

Beijing Municipal Bureau of Land and Resources. (2009). *Land Provision Report*. http://www.bjgtj.gov.cn/tabid/5356/InfoID/61877/settingmoduleid/10750/frtid/5238/Default.aspx

Beijing Municipal Bureau of Land and Resources. (2011). *Land Provision Report*. http://www.bjgtj.gov.cn/tabid/5356/InfoID/152327/settingmoduleid/10750/frtid/5238/Default.aspx

Beijing Municipal Bureau of Land and Resources. (2012). *Land Provision Report*. http://www.bjgtj.gov.cn/tabid/5356/InfoID/151824/settingmoduleid/10750/frtid/5238/Default.aspx

Beijing Municipal Bureau of Statistics. (2006). *Beijing statistical yearbook 2006*. Beijing: China Statistics Press.

Beijing Municipal Bureau of Statistics. (2007). *Beijing statistical yearbook 2007*. Beijing: China Statistics Press.

Beijing Municipal Bureau of Statistics. (2008). *Beijing statistical yearbook 2008*. Beijing: China Statistics Press.

Beijing Municipal Bureau of Statistics. (2009). *Beijing statistical yearbook 2009*. Beijing: China Statistics Press.

Beijing Municipal Bureau of Statistics. (2010). *Beijing statistical yearbook 2010*. Beijing: China Statistics Press.

Beijing Municipal Bureau of Statistics. (2011). *Beijing statistical yearbook 2011*. Beijing: China Statistics Press.

Beijing Municipal Bureau of Statistics. (2012). *Beijing statistical yearbook 2012*. Beijing: China Statistics Press.

Beijing Municipal Bureau of Statistics. (2013). *Beijing statistical yearbook 2013*. Beijing: China Statistics Press.

Camagni, R., Gibelli, M. C., & Rigamonti, P. (2002). Urban mobility and urban form: the social and environmental costs of different patterns of urban expansion. *Ecological economics, 40*(2), 199–216.

Cervero, R. (1998). *The planned city: Coping with decentralization: An American perspective*. Berkeley: University of California Transportation Center.

Chen, L., & Meng, X. C. (2011). A study on the job-housing spatial balance of Beijing. *Progress in Geography, 30*(10), 1210–1217.

Dang, Y., Liu, Z., & Zhang, W. (2014). Land-based interests and the spatial distribution of affordable housing development: The case of Beijing, China. *Habitat International, 44*, 137–145.

Ding, C. (2004). Urban spatial development in the land policy reform era: evidence from Beijing. *Urban studies, 41*(10), 1889–1907.

Feng, J., & Zhou, Y. (2003). The latest development in demographic spatial distribution in Beijing in the 1990s [J]. *City Planning Review, 5*, 16.

Gao, J. L., Chen, J. L., & YANG, D. H. (2013). Spatial distribution of urban land prices in Nanjing. *Progress in Geography, 32*(3), 361–371.

Glaeser, E. L., & Kahn, M. E. (2001). *Decentralized employment and the transformation of the American city* (No. w8117). Cambridge, MA: National Bureau of Economic Research.

Jin, Y., Echenique, M., & Hargreaves, A. (2013). A recursive spatial equilibrium model for planning large-scale urban change. *Environment and Planning B: Planning and Design, 40* (6), 1027–1050.

Kuang, W., Liu, J., Shao, Q., & Sun, C. (2009). Spatio-temporal patterns and driving forces of urban expansion in Beijing Central City since 1932. *Journal of Geo-Information Science, 4*, 005.

Liu, S. H., Wu, C. J., & Shen, H. Q. (2000). A GIS based model of urban land use growth in Beijing. *Acta Geographica Sinica-Chinese Edition, 55*(4), 416–426.

Ministry of Housing and Urban–rural Development, P.R. China. (2007). *China urban construction statistical yearbook 2007*. Beijing: China Planning Press.
Ministry of Housing and Urban–rural Development, P.R. China. (2008). *China urban construction statistical yearbook 2008*. Beijing: China Planning Press.
Ministry of Housing and Urban–rural Development, P.R. China. (2009). *China urban construction statistical yearbook 2009*. Beijing: China Planning Press.
Ministry of Housing and Urban–rural Development, P.R. China. (2010). *China urban construction statistical yearbook 2010*. Beijing: China Planning Press.
Ministry of Housing and Urban–rural Development, P.R. China. (2011). *China urban construction statistical yearbook 2011*. Beijing: China Planning Press.
Ministry of Housing and Urban–rural Development, P.R. China. (2012). *China urban construction statistical yearbook 2012*. Beijing: China Planning Press.
NIU, W. Y. (2012). *China new urbanisation report*. Beijing: Science Press.
Pan, H., & Ge, Y. (2014). Jobs-housing balance and job accessibility in Beijing. In *Transportation Research Board 93rd annual meeting* (No. 14-5416). Washington, DC: Transportation Research Board.
Record of land acquisition in Beijing. http://www.bjgtj.gov.cn/publish/portal0/tab5164/
Record of land allocation in Beijing. http://www.bjgtj.gov.cn/publish/portal0/tab5163/
Record of land transaction in Beijing. http://www.bjgtj.gov.cn/publish/portal0/tab5422/
Wu, W. J., Zhang, W. Z., Liu, Z. L., & Huang, X. W. (2010). Tempo-spatial analysis of the residential land's spatial pattern in Beijing. *Geographical Research, 29*(4), 683–692.
Yan, Y. T., & Feng, C. C. (2009). Research on spatial structure of urban land use intensity of Beijing. *China Land Science, 3*, 008.

Chapter 13
A Configurational Accessibility Study of Road and Metro Network in Shanghai, China

Lingzhu Zhang, Alain Chiaradia, and Yu Zhuang

13.1 Introduction

Accessibility remains a key challenge when looking at the relationship between transportation and land use, especially in metropolitan areas. Although the use of urban design and associated accessibility influence travel makes intuitive sense, researchers have found it extremely difficult to provide clear evidence of the influence of urban form. At the same time, it is generally recognized that land use patterns and transportation patterns are closely related through accessibility change. The spatial organization of human activities creates a pattern of personal travel and goods transport, thus influences the mobility behavior of actors such as households and firms. Conversely the availability of infrastructure makes certain locations more accessible.

Major cities in China, most notably Beijing and Shanghai, are building and expanding rail transit systems as a strategy to reduce the negative environmental and social consequences of fast-paced motorization. Transit systems play an important role in urban public policy. They are increasingly seen as an essential element in policy packages that aim to reduce congestion, make more efficient use of road space, reduce pollution, and keep the lid on increasing energy consumption in the transportation sector and as a counter-measure to automobile-driven and automobile-dependent suburbanization.

Sponsored by Natural Science Foundation of China, No. 51178318.

L. Zhang (✉) • Y. Zhuang
College of Architecture and Urban Planning, Tongji University, Shanghai, China
e-mail: abc30303@163.com; arch-urban@163.com

A. Chiaradia
School of Planning and Geography, Cardiff University, Cardiff, UK
e-mail: ChiaradiaAJ@cardiff.ac.uk

Q. Pan, J. Cao (eds.), *Recent Developments in Chinese Urban Planning*,
GeoJournal Library 114, DOI 10.1007/978-3-319-18470-8_13

13.1.1 Objectives of This Paper

This paper identifies spatially disaggregated micro-macro relative accessibility relationships between urban block size, road and metro-line network design, metro stations and bus stop locations, commercial land use locations distribution and station usage in Shanghai.

13.1.2 Structure of the Paper

The paper is organized as follows, in the next section, we set out the literature context of the present paper. In the subsequent two sections, we introduce the methodology, the data and software. In Sect. 13.4, urban block size, the road network accessibility is computed. The relationships between road network accessibility and commercial land use, between road network accessibility and metro stations, bus stops are then presented in Sects. 13.5 and 13.6. Regression analysis is applied in Sect. 13.6 to discuss the interaction between the above systems. In the last section, we discuss and contextualize the analysis results.

13.2 Context and literature

13.2.1 The Relation Between Built Environment and Transportation

A recent meta-analysis of more than 200 studies of the built environment–travel environment literature (Ewing and Cervero 2010) found that of all of the environmental variables considered, none on their own has a significant impact. Still, the combined effect of several such variables on travel could be quite large. Consistent with prior work, it was found that "vehicle miles travelled is most strongly related to the measures of accessibility to destinations and secondarily to the street network design variables. Walking is most related to the measures of land use diversity, intersection density, and the number of destinations within walking distance. Travel by bus and train use are equally related to proximity to transit access points and street network design variables, with land use diversity a secondary factor." Surprisingly, population composition and job densities were found to be only weakly associated with travel behavior once these other variables were controlled for. We can identify a set of key indicators: road network design variable, accessibility indices, urban block size, non-residential land use location diversity and clustering, and public transport access point accessibility. In Shanghai, studies of travel behavior and built environment (Pan et al. 2009; Zacharias 2005) and suburbanization (Cervero and Day 2008) corroborate the meta-analysis. These indicators seem to be usable in Shanghai. However, the studies did not make an attempt to assess the accessibility evaluation methods or the details of urban layout.

13.2.2 The Relationships Between Transport Networks, Land Use and Accessibility

The impact of transport on land use is well recognized (Hansen 1959; Banister 1995; Giuliano 2004; Wegener and Furst 1999; Geurs and van Wee 2004; Borzachiello et al. 2010). While accessibility change due to new infrastructure is instantaneous, it is obvious that land use has much longer investment cycle lag in taking advantage of these accessibility changes. Exactly how developments in the transport system influence the locational behavior of landowners, investors, firms, and households is less clearly understood as other influences such as planning, take a longer time to have an impact.

The idea of the **"land use transport feedback cycle"** (Giuliano 2004; Meyer and Miller 2001; Wegener and Fuerst 2004) is often used to illustrate the complex relationship between land use, transport and change in accessibility. In this cycle, land use and accessibility patterns both influence each other. Land use patterns are partly conditional on accessibility advantage, which causes the locational sorting of human activities such as living, working, shopping, education, and leisure; in other words, the distribution of human activities reflect the different requirements of and competition for accessibility advantage. The transport system changes the pattern of accessibility by overcoming, with different level of ease, the distance between the locations where these activities take place. The increase and clustering of activities create new travel demand and, consequently, a need for transportation services, whether in the form of new infrastructure or more efficient operation of existing facilities which in turn change accessibility. The resulting increase in accessibility co-determines the location decisions of landowners, investors, households and firms and so results in changes of the land use, starting the cycle again. This process continues until a (provisional) equilibrium is reached or until some external factor intervenes (Meyer and Miller 2001). A key to understanding this cycle is understanding change in accessibility and the co-variation of land use density and diversity.

13.2.3 Beyond TOD, Exploring Node-Place Effect of Metro Stations in Shanghai

In Shanghai, the first metro line opened in 1993. In 2014, there were 14 metro lines and 329 stations with an operating route length of 538 km, making it the longest total system length in the world. The ridership was 2.5 billion in 2013. On a normal weekday over 8 million people use the Shanghai metro. At the end of 1993, the population of Shanghai was 13 million (Yeung 1996); in 2013, it was estimated to be 23.9 million.[1] The metro system has radically changed the accessibility pattern of the Shanghai metropolitan area.

As noted by a number of researchers, transit oriented development (TOD) has become widely adopted in China (Cervero and Day 2008; Pan and Ren 2005; Thomas and Deakin 2008). Pan et al. (2011) reviewed the origination and development of TOD in the USA, studying the case of Songjiang District in Shanghai. In a recent publication, Zhang (2007) explored a specifically "Chinese edition of transit-oriented development". He argued that most of the TOD performance standards developed in the United States were not applicable directly to China. In their place, he presents a modified TOD model derived from the experience of urban development around transit in Hong Kong and Taipei, Taiwan. This modified model was characterized by such key features as differentiated density, dock-sized district, 'delicate' design, diverse destination and distributed dividends. Parallel to the TOD tradition in the USA and its adaptation in China, a node and place model was developed in Europe (Bertolini 1996; Bertolini and Spit 1998; Trip 2007). The Node and Place model is elaborated within an economic framing. This is relevant to Shanghai in the transitioning from industrial to service economies and the rise of creative industries and city competiveness (Florida 2000; Storper and Venables 2004). It enables a multi-scale spatial approach in relating station and station surroundings, land use mix and intensity, accessibility and the urban buzz of place.

In this context the station emerge as a new central place in metropolitan cities. It becomes both a hub of networks due to their high accessibility by different modes of transport mirrored by a broad range of users and also one of the very few places in the contemporary city where the participants in its increasingly heterogeneous communities still physically meet (Bertolini 1996). The fact that all these people pass through public transportation nodes does not necessarily imply, of course, that people are also interacting with each other there. However, these intense and diverse flows of people do have the potential of translating into equally intense and diverse patterns of human interaction. If the right conditions are met, social, cultural and economic activities still requiring physical proximity can thrive in these areas. Moreover, this potential can be realized in a relatively sustainable way, as it can be coupled with environmentally more effective transport, land-use patterns and urban structure and design. More importantly these locations appear to be the locus where the micro-foundations of urban agglomerations conceptualized as "matching, learning and sharing" (Duranton and Puga 2004), have a great potential to unfold.

The station, which is the transport service's access point, together with its surroundings, are the interface which produces and attracts movements of significant magnitude, and therefore constitutes opportunities for mixed use and commercial development. In Shanghai, this has been documented as a dynamic coupling

[1] http://worldpopulationreview.com/world-cities/shanghai-population/

process (Pan and Ren 2005). The node-place model follows the reasoning of the transport land use feedback cycle outlined above. This feedback pattern is mediated by planning regulation and development conditions fuelled by increase in property value around transit station that are well documented in the US, China (Pan et al. 2014) and in the UK (Network Rail 2011). In Shanghai, exploratory studies are showing a similar pattern (Pan and Zhang 2008; Pan et al. 2014). All of this conform to spatial and urban economics theory.

13.2.4 Toward Operationalizing Node and Place in Shanghai

Florida (2000) and Storper and Venables (2004) used a range of indicators of 'place' at metropolitan scale to monitor policy effectiveness, but these are not usable by planners and urban designers (Trip 2007). Zhang (2007) while giving guidance, provides a set of indicators to operationalize an adapted TOD model in China. In the most recent node and place model study (Chorus and Bertolini 2011), the 'node' value of a station is evaluated with transport mode frequency and range, proximity to CBD by metro, bus stops and lines from the station; that is, the transport networks accessibility position of the station. The 'place' value is evaluated by population and job levels around the station, and mixed use clustering – the functional position of the station. The functional diversity and intensity dimensions – also found in Zhang (2007) as differentiated density and diverse destination – while part of place-making and useful to planner, remain much less operational for urban designer. An important point to note is that if population composition and density are dynamically coupled with station accessibility levels, this would clarify why these variables may become less significant in the meta-literature review (Ewing and Cervero 2010) as population composition and density will be highly co-linear to level of accessibility and connectivity of the station itself.

Beyond accessibility, what will differentiate the stations will be their place value. A review of place literature and the related aspects of development form while relevant to this discussion is beyond the scope of this paper. Aspect of development forms that embed place ontologies can be categorized as follow: 'place as visual attributes', 'place as product', 'place as process', 'place as values' (Arefi and Triantafillou 2005). For an introduction to a discussion of the analysis of quality of place around station see Trip (2007, pp. 67–83) which echo Zhang's "delicate design" (2007). Zhang's last dimension, the value capture has been widely researched in the UK. In the UK, a growing body of seminal research investigated the economic value of urban design. The research ranges from the value of street public realm improvement (CABE 2007; Transport for London 2011), to the social and environmental value of park and public space (CABE Space 2003), the value of green space (GLA Economics 2003, 2010; Dunse et al. 2007; Rogers et al. 2012; CABE Space 2009; Jim and Chen 2010), the value of blue space (Garrod and Willis 1994; Fisher 1999; Rouwendal et al. 2014; Goetgeluk et al. 2005), the value of station investment (Network Rail 2011), the value of housing and urban layout

(CABE et al. 2003; The Prince's Foundation for the Built Environment 2007; Chiaradia et al. 2013), the value of mixed use street (Jones et al. 2007; Chiaradia et al. 2012), the value of urban design (CABE et al. 2001; British Council for Offices 2006), and more recently resilient urban form, governance and the creation of long term value (Grosvenor 2013). All of these studies link "delicate design" characteristics of the built environment to economic value. They investigate the relationship between physical configuration or condition (e.g. layout, perceived street quality, etc.) to economic value and in some studies, the social and environmental economic value. They also employ various methodologies in answering these questions, drawing on different data sources in different ways. However, they all link "urban design" with "value" through inferring relationships from a small sample to large sample. A variety of methods are used and there is an explicit recognition of design value. Overall they are more robust and detailed than earlier research examining the "Value of Urban Design" (CABE et al. 2001). Within these studies and along several literature review on urban design value have been published (CABE 2003; Ministry for the Environment, NZ 2005; McIntyre 2006). While the results are fragmented, some of these conceptualization and assessment techniques of good design, correlates against economic value findings and economic underpinning. While yet to be consolidated by further research, they have been integrated and operationalized by consultancies (CBuchanan 2008; Tribal Urban Studio and CBuchanan 2008; Amion Consulting et al. 2007) advising local authorities on capturing the value of public investment in good urban design (Chiaradia et al. 2015). Following Zhang (2007) we doubt that these studies are directly transferable to Shanghai; more research is needed. While there are similarities between TOD, Zhang's adapted TOD model (2007) and the node and place model (Bertolini 1999), it is the node and place model that here provides the basis of a systematized approach.

In this paper which is the first part of a detailed study on the synergy between node and places around metro station in Shanghai, our analytical focus and contribution is on place characteristics and aspect of development form that are structural to metro station as central place. Metro station and their surrounding in Shanghai have very different urban block size and form, street network layout and resulting multi-scale accessibility. The importance of urban block size in relationship to central place i.e. central business district has been studied in the American and Australian cities (Siskna 1990, 1997). Our research questions in relation to metro stations and their surroundings are: do the different urban block and form, street layout, including multi-level layout work equally well? Do different block size and form and street layout affect the functioning of circulation patterns? Do particular block forms and sizes create favorable or optimum conditions for one or more of these aspects? Are there any desirable or undesirable consequences which result from the choice of block sizes and form, street layout, multi-level layout?

13.2.5 Urban Block Size and Form Dynamics and Accessibility Measures

Siskna's (1997) study revealed that block size and form in American and Australian city center have crucial and predictable effects on evolutionary urban patterns. Extensive alterations to the original layout can occur through successive modifications often by un-coordinated actions that might lead to optimal collective patterns over time, yet this is not always the case and there is a need for better coordination. Two main dynamic processes are observed. In cities with small and medium initial block sizes, the street and block layout has remained intact whereas in cities with large initial blocks the layout has been considerably modified by addition of street and alleys, creating smaller block and sub-blocks. Smaller medium blocks are more suitable than larger blocks for the general functioning of the center, which are areas of intense pedestrian activity as they produce and permit fine meshed circulation, affording better change in travel direction which is a good indicator of ease of movement and thus accessibility. Small blocks increase dispersion, reduce congestion and enable better level of services. This principle can also be seen applied within department stores and shopping malls, where small display islands minimize distance while maximizing display surface by providing a greater length of island perimeter. Contemporary development has included the seen the commercial centre, multi-level shopping and entertainment mall directly connected to station. While these developments have often high internal permeability they often encapsulate this increased permeability within a building which itself acts as impermeable superblock. The "delicate design" and node and place model can be improved by including an assessment of the urban block for accessibility of the station and its surroundings, whether of its size and form, or as multi-level complex environment.

In previous decades, various definitions of the accessibility, as well as indicators, have been developed and used to describe accessibility (Reggiani 1998; Geurs and Ritsema 2001; Geurs and van Wee 2004). Most accessibility definitions derived from the seminal work of Hansen (1959), which first defines accessibility as "the potential of opportunities for interaction". More precisely, Hansen defined that "the accessibility at area A to a particular type of activity at area 1 (say employment) is directly proportional to the size of the activity at area 1 (number of jobs) and inversely proportional to some function of the distance separating area A from area 1. The total accessibility to employment at Area A is the summation of the accessibility to each of the individual areas (1 to n) around area A." The accessibility at location A varies directly with the sizes of the other locations (1 to n), and inversely with the spatial separation between A and (1 to n).

Size is measured with respect to quantities such as employment, retail floor area, population, retail sales, etc., while spatial separation is measured with respect to distance, travel cost, travel time and other similar spatial metrics variables. Recognize the co-determination of accessibility and land use, it is necessary to use indices of accessibility that are not weighted by land use to identify the role of urban block and size impact on street network layout configuration in shaping accessibility. The alternative to Hansen's weighted version of accessibility (type 1) is an unweighted accessibility measure (type 2) which omits the size variable (Ingram 1971). This focuses on the spatial separation variable (Pooler 1995). Spatial separation is easy to understand and calculate. This is of particular interest in intra-urban situations to disentangle the role of accessibility in the potential for interaction between land use diversity and intensity which are thickly and continuously intertwined with transport service access points.

On network, the first unweighted definition of accessibility can be found in Shimbel (1953). Within the context of network analysis using graph theory, Shimbel defines the unweighted accessibility of a network vertex with respect to the sum of the distances at that vertex:

$$A_i = \sum_{\substack{j=1 \\ j \neq i}}^{n} d_{ij} \quad i = 1,\ 2,\ \ldots, n \tag{13.1}$$

where d_{ij} is the shortest path from vertex i to vertex j.

In order to measure the overall network dispersion, Shimbel defines another elementary measure:

$$A_i = \sum_{i=1}^{n} \sum_{\substack{j=1 \\ j \neq i}}^{n} d_{ij} \quad i = 1,\ 2,\ \ldots, n \tag{13.2}$$

Equation (13.1), representing the simple sum of the distances, is described usually as a measure of the compactness of network relative to each vertex.

Equation (13.2) measure the overall "network dispersion". This is the mean shortest path length on network. Christaller (1933–1966) and Reilly (1931) were also early pioneers of "mean shortest path length".

Shimbel defines "stress" on a vertex k as the "count of all the minimum paths which pass through site k", then we have a measure of the "stress" which site k must undergo during the activity of the network. In transport this is also called 'path overlap'. This is a definition of structural potential flow in route assignment through the network sampling each link origin as all or nothing to link destination (Pooler 1995). Shimbel's (1953) work on network was applied later in transport network analysis (Kansky 1963), and by geographers for the analysis of networks geography. For an extensive review see Haggett and Chorley (1969).

More recently, analysis on network has been called differently, such as closeness and betweenness centrality analysis (Cutini 2001; Porta et al. 2006; Newman et al. 2006; Borzachiello et al. 2010; Xiao et al. 2013). These analysis are often referenced in Social Network Analysis studies (Bavelas 1950; Freeman 1977). The foundation of such analysis is by the mathematician Euler who, in 1736, solved analytically the "first travelling salesman problem" for Konigsberg, inventing at once network codification, graph theory and transport network analysis (Coupy 1851). Euler also showed that network layout can make certain travel route patterns impossible.

13.3 Methodology

In Shanghai, we propose to follow a three-step methodology described below:
We calculate:

- Calculate urban block size and show a thematic block size of Shanghai within the outer ring.
- Street network accessibility unweighted by land use as betweenness centrality at micro, meso, and macro spatial scale for the networks (radius at 600 m, 2000 m, 5000 m, n). These correspond to walking, short cycling, e-bike and bus ride, short to long car trip.
- Metro lines accessibility unweighted by land use as closeness and betweenness centrality.

13.3.1 What Spatial Unit?

The spatial unit of analyses the standard transportation node/link central path between two junctions. A street link has length, angularity along its path, connectivity with other links, incidence with other links at junction. The node/link principle can be extended to pedestrian and cycling path network and to complex multi-level environment.

13.3.2 What Analytical Catchment Should We Use?

Zhang (2007) in the section on "Docksized district" discuss the issue on choosing an analytical catchment area (e.g. 400 m) from the station. He recommend that the underlying behavioral principle should be defined by people's willingness to walk.

Calthorpe (1993, p. 53) used 600 m as comfortable walking distance in his TOD theory. Studies in Shanghai showed that 500–600 m is also the comfortable walking distance (Pan et al. 2007; Bian 2006; Liu 2012). Moreover 600 m is half of the median distance between Shanghai's metro stations. In Shanghai, when trip distance increases to 2000–2500 m, people switch from walking to cycling (Pan et al. 2003; Zacharias 2005). The majority of within-city trips are less than 5 km in Shanghai, and this is also the upper limit of non-motorized trips (Zacharias 2005).

The micro, meso and macro radii are associated with different uses of the road network, 800 m, 2000 m, 9000–18,000 m have been used as walking, cycling, car trip distance in Wuhan (Xiao et al. 2013). Thus we propose to use the radius (600 m, 2000 m, 5000 m) which related to people travel behavior in Shanghai.

However, we agree with Zhang (2007) that the willingness to walk or cycle is strongly affected by other "Delicate Design" factors affecting level of services

(LOS) of walking, for example, perceived safety, security, architectural interest, pedestrian-scale lighting and amenities, and presence of other pedestrians (Krambeck 2006). For this level detail a whole level of very detailed descriptors will be required that can be accommodated easily with the link codification (Lin and Moudon 2010; Parks and Schofer 2006). See for example, Link and Place model (Boujenko 2007) developed in the UK which has become part of transport planning policy (DfT 2007; CIHT 2010).

13.3.3 What Metrics?

In using accessibility analyses i.e. mean shortest path and resulting structural flow, the underlying metric should relate to route choice behaviors (walking, cycling, driving, and riding a bus). In the recent literature we found three types of metric: topological, Euclidean, and Angular. Euclidean is the standard shortest distance often criticized because it does not account for the potential value of speed (which does not apply to walking) and is rather blind to geometry, while topological metric is capturing directness, it has been criticized because it is blind to Euclidean change and least angular metric is associated with capturing both geometric directness and the geometry of speed. Ideally it would be best to combine Euclidean and Angular. Angular metric associated with and Euclidean radius is appropriate to take account of Euclidean distance changes. An analysis of shortest paths in Shanghai according to these three metrics provides the results shown in Table 13.1. The results demonstrate that Angular metric is also a very good proxy for Euclidean metric. This should be the urban design metric of choice as it is simple and does account best for change in design.

For the 600 m radius, 88 % of the top 30 % links with the shortest Angular paths are also the top 30 % of the Euclidean shortest path. It is 81 % for 2000 m and 77 % for 5000 m. This is particular to Shanghai and may change from city to city making comparison unlikely.

Repeating the analysis for the top 50 %, Angular metric shares of Euclidean metric is higher TOP 50 % (Table 13.2).

These auto-correlation between metrics might explain why the literature is so uncertain about which metric to use. Comparison between metrics and cities is impossible before identifying these auto-correlation levels.

We explore the relationship between accessibility indices and Metro station, bus stops, Commercial land use location distribution in order to understand the mutual relationships between network (configuration), metro station usage, and land use using frequency distribution analyses. This is because circular causation imply generalized spatial auto-correlations. Hence the point here is not to eliminate auto-correlation because it does not suit normative statistical analyses techniques but on the contrary we want to systematically understand the profile of the auto-correlations across the different analysis.

Table 13.1 TOP 30 %

	R600 m	R2000 m	R5000 m
ANG vs EUC	88 %	81 %	77 %
ANG vs TOPO	73 %	66 %	63 %
EUC vs TOPO	76 %	71 %	65 %

Table 13.2 TOP 50 %

	R600 m	R2000 m	R5000 m
ANG vs EUC	92 %	87 %	85 %
ANG vs TOPO	79 %	76 %	78 %
EUC vs TOPO	81 %	80 %	79 %

13.3.4 Data and Software

All data are brought into a geo-database using ArcGIS software. ArcGIS is used to store, compute centrality indices and used to visualize the results.

Spatial accessibility variables computed using Spatial Design Network Analysis (sDNA)[2] are used in this study. It is a set of multi-level spatial analysis techniques for urban networks. SDNA calculate centrality closeness and betweenness centrality on network with user defined radius with different metrics: Euclidean, Angular, and Topological distance as travel budget.

13.3.5 Mean Angular Distance (MAD)

MAD is defined as the mean (averaged per link) of the angular distance from each origin link to each possible destination falling within the network radius of the origin. It is an accessibility measure, in that lower values of MAD indicate straighter paths to destinations within the radius. Thus,

$$\boldsymbol{SAD}(\boldsymbol{x}) = \sum_{y \in \mathbf{Rx}} \mathrm{d}_\theta(\mathrm{x}, \mathrm{y})\mathrm{P}(y) \tag{13.3}$$

Where SAD(x) is the SAD for link x, $\boldsymbol{y} \in \mathbf{Rx}$ is each other link y in Rx the radius surrounding x, $d_\theta(x, y)$ is the shortest possible angular distance along a route from x to y, and P(y) is the proportion of y falling within the radius.

[2] sDNA is a plugin for ArcGIS, Autocad, and open source GIS (QGIS) it uses the Shapefile (.shp) or .gdb files and link/node standard to analyze the spatial networks design characteristics using centrality measures and other measures such as severance. It provides many control variables. The software is freely available after registration at www.cardiff.ac.uk/sdna/with full specifications.

13.3.6 Angular Betweenness (BtA)

Angular betweenness measures the frequency with which each link x falls on the shortest angular path between each pair of other links y and z, provided the Euclidean distance from y to z is within the network radius. For BtA, the network radius can be regarded as a kind of maximum trip length. Thus,

$$\boldsymbol{BtA}(\boldsymbol{x}) = \sum_{y \in \mathbf{N}} \sum_{z \in \mathbf{Ry}} \mathrm{P}(z) OD(y, z, x) \tag{13.4}$$

Where BtA(x) is the angular betweenness of link x, **N** is the set of all links in the network, **Ry** is the set of all links within the defined radius of link y, P(z) is the proportion of y falling within the radius from y, and OD(y,z,x) is defined as

$$\boldsymbol{OD}(\boldsymbol{y}, \boldsymbol{z}, \boldsymbol{x}) = \begin{cases} 1, \; \textit{if x is on the shortest angular path from y to z} \\ 1/2, \textit{if } x \equiv y \not\equiv z \\ 1/2, \textit{if } x \equiv z \not\equiv y \\ 1/3, \textit{if } x \equiv y \equiv z \\ 0, \; \textit{otherwise} \end{cases} \tag{13.5}$$

The 1/2 and 1/3 contributions to OD(y,z,x) handle the cases of routes which terminate on the link of interest, and routes from a link to itself. (1/3 represents the average traffic for each point on a link assuming traffic is generated by the product of origin and destination link proportion). All measures were computed with sDNA software (Chiaradia et al. 2014; Cooper et al. 2014).

13.4 Block Size and Road Network Accessibility Analysis

Using road center lines node/link to represent the road network of Shanghai, Fig. 13.1a, b show the road network and the urban blocks shaped by the road network. The color of urban blocks from light gray to dark gray represents the size of blocks. According to urban economics theory, it is unsurprising to find that the road density in the center of Shanghai is much higher than in the suburban areas. As agglomeration increases, the increase in network density acts as dispersion to mitigate congestion and increase level of proximity of a larger number of people. As population density increases faster than road density, there is an economy of scale in the network increased density (Chiaradia et al. 2013).

As identified by Siskna (1997) in the US and Australia, in Shanghai smaller block size are mostly located in the center, where people density is higher. High small block density is also related to high junction density and high network density. In Shanghai within the outer ring the block size average is 65,900 m^2, the median is 22,718 m^2.

In the Inner ring the block size average is 36,300 m^2 (−45 %), the median is 15,966 m^2 (−30 %).

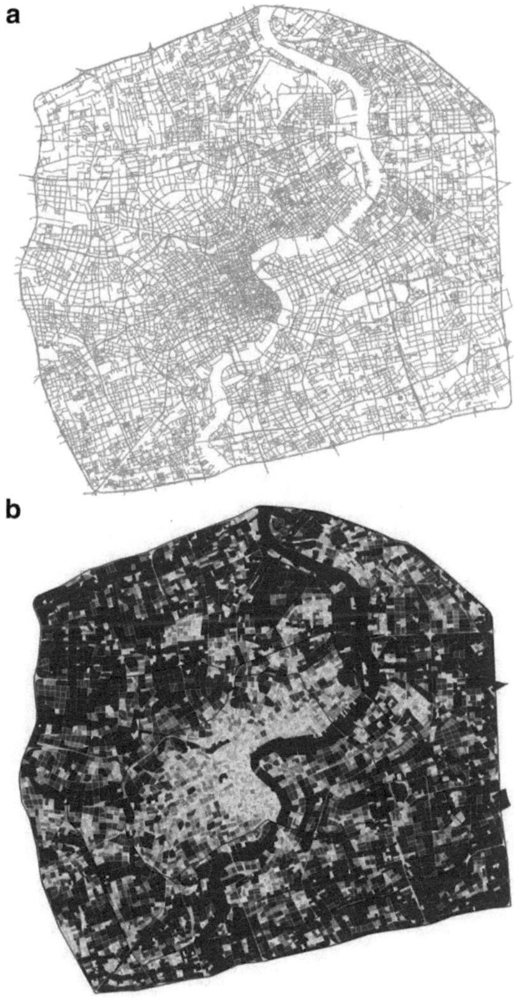

Fig. 13.1 Road Network (**a**) and Urban Block Size (**b**) of Shanghai in 2010

13.4.1 Potential Flow as Centrality Betweenness

Figure 13.2 shows from cold color (light gray) to warm color (dark gray), the level of betweenness centrality. **Betweenness (BtAW)** measures how often each link is used on Angular shortest paths from all links to the other links in the radius and then all links are ranked from high to low levels. BtAW is a measure of shortest path overlap on each link which indicates potential flow level of each link.

Links with high betweenness at a micro level (radius: 600 m) identifies links that are closely clustered (i.e. short street, small block and high junction density). As the radius increases to the meso level (2000 m), links with high level of betweenness are identified and, the inter-neighborhood street and roads are still very dense in central Shanghai. We can clearly see how the pattern of block size in Fig. 13.1

Fig. 13.2 Betweenness of Shanghai road network in 2010 (BtAW, radius: 600 m, 2000 m, 5000 m)

a

b

c

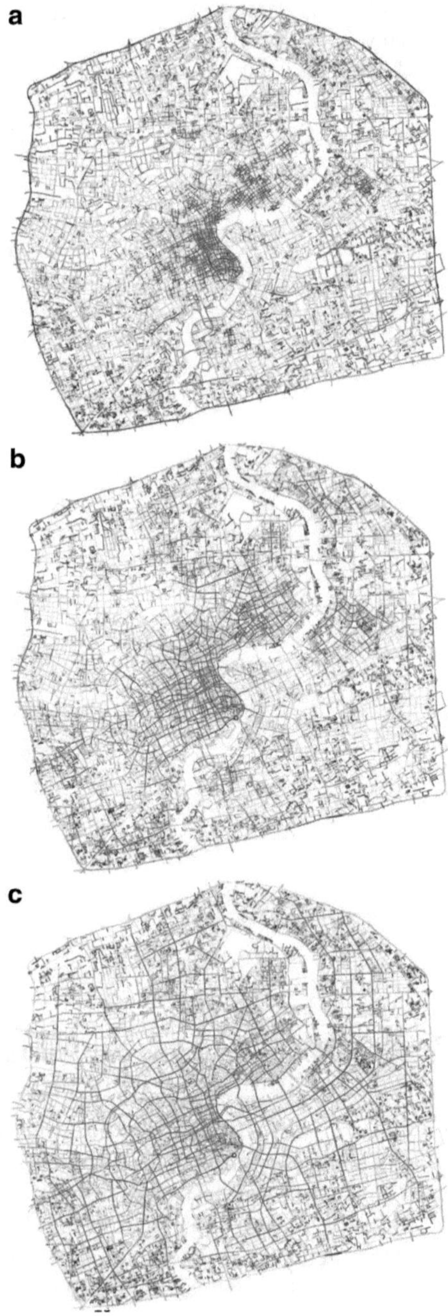

relate to the pattern of local accessibility in Fig. 13.2. These relationships are somehow overlooked.

At macro level (5000 m), the main arterial network which allows one to move from one part of Shanghai to the others is identified. The arterials are evenly distributed in the whole city as they follow the national design standard used by transport planners with still a strong focus on central Shanghai.

13.5 Commercial Land Use & Flow Potential

Figure 13.3 shows the land-use map and commercial land use location (illustrated in red) of Shanghai in 2010. The commercial designation includes both commercial and office. The urban land has been planned and transformed from a "single-core" pattern around the city center into a "multi-core" pattern – (i.e., urban land expansion around the city center and sub-centers). The four sub-centers are known as Xujiahui, Zhenru, Wujiaochang, and Huamu. Some large commercial clusters are also located in proximity of the ring roads.

To understand the relationship between commercial land use and flow potential, we divided the roads into 2 categories according to their flow potential: the 50 % highest flow potential roads (50 %) and the 50 % lower flow potential roads (50 %), then calculated the frequency distribution of commercial block. From Fig. 13.4a, we can see that across radii 72 % of the commercial land use locate at the roads with highest flow potential. We repeat the frequency distribution analysis with deciles, Fig. 13.4b shows the density distribution of commercial land use location by flow potential (level 1 highest, level 10 lowest). Commercial land use maximize locational advantage from micro to macro flow potential level.

There are 17 %, 19 %, 19 % commercial land use locate on the link with highest 10 % betweenness at different scales (radius 600, 2000 m, 5000 m). Cumulatively, the top 30 % of the links with highest flow potential have almost 60 % of commercial land use locate on them.

13.6 Metro Stations, Bus Stops & Flow Potential

In Shanghai metro system there are 329 metro stations in operation[3] and more than 200 of the stations inside the outer-ring of shanghai (Fig. 13.5a), 85–43 % of which locate in the center of Shanghai (inside the inner-ring). The average distance between metro stations is 1.2 km with the standard deviation of 400 m.

[3] http://www.shmetro.com/

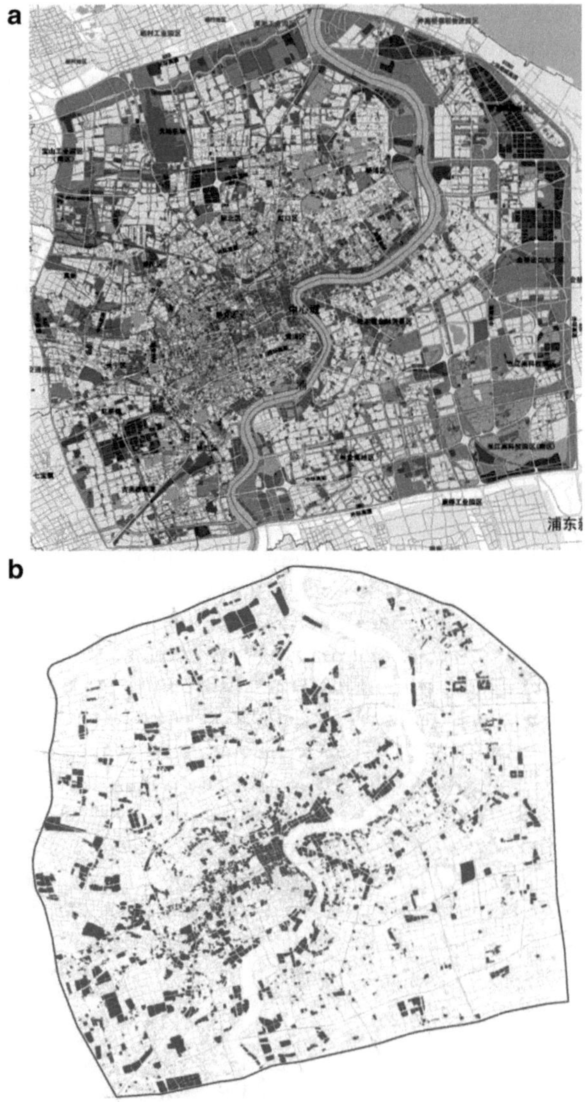

Fig. 13.3 Urban Land use (**a**) and Commercial land use location (**b**) of Shanghai in 2010

There are more than 4000 bus stops inside the outer-ring of Shanghai (Fig. 13.5b), more than 1000 are located in the center of Shanghai (25 % of the bus stops on 18 % of the land – inner ring is 115 km^2 and outer Ring area is 630 km^2).

Figure 13.6 provides the density distribution of metro stations and bus stops in relation to link multi-level flow potential levels (divided into 10 deciles, with level 1 denoted as "highest" and level 10 denoted as "lowest").

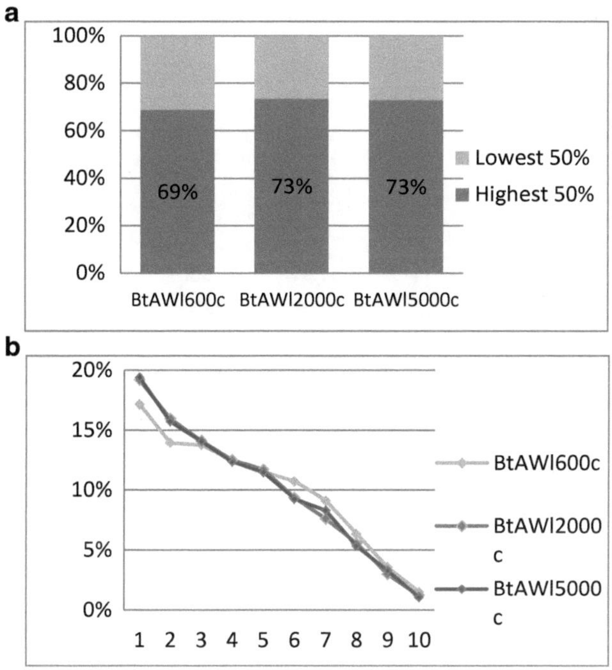

Fig. 13.4 The density distribution of commercial land use in relation to link multi-level Betweenness centrality ((**a**) divide road accessibility into 2 categories; (**b**) divide road accessibility into 10 categories)

It can be easily seen from the Fig. 13.6 that the relationship between link flow potential and metro stations and bus stops location profile are very different according to radius. Only 13 % of metro stations and 8 % bus stops are located at links with the highest 10 % flow potential at radius of 600 m, which would correspond best to a very walkable neighborhood. This is understandable as most of the construction of the metro lines have used a cut and cover construction technique, and thus the metro stations mainly have been following the arterial roads. This is reflected in their name which most of the time contain road and not and rarely a place name. An aspect that may bias this result is that the metro station are often coupled with large shopping and entertainment mall. The details of the permeability network is missing in the general network codification.

For bus stops location, this is because bus lines have to mitigate proximity to patronage and service speed and thus cannot be located on very local streets. The links with 30 % highest level of flow potential, at radius 5000 m, are where 76 % of the station are located. Respectively for 20 % and 10 % it is 64 % and 38 %.

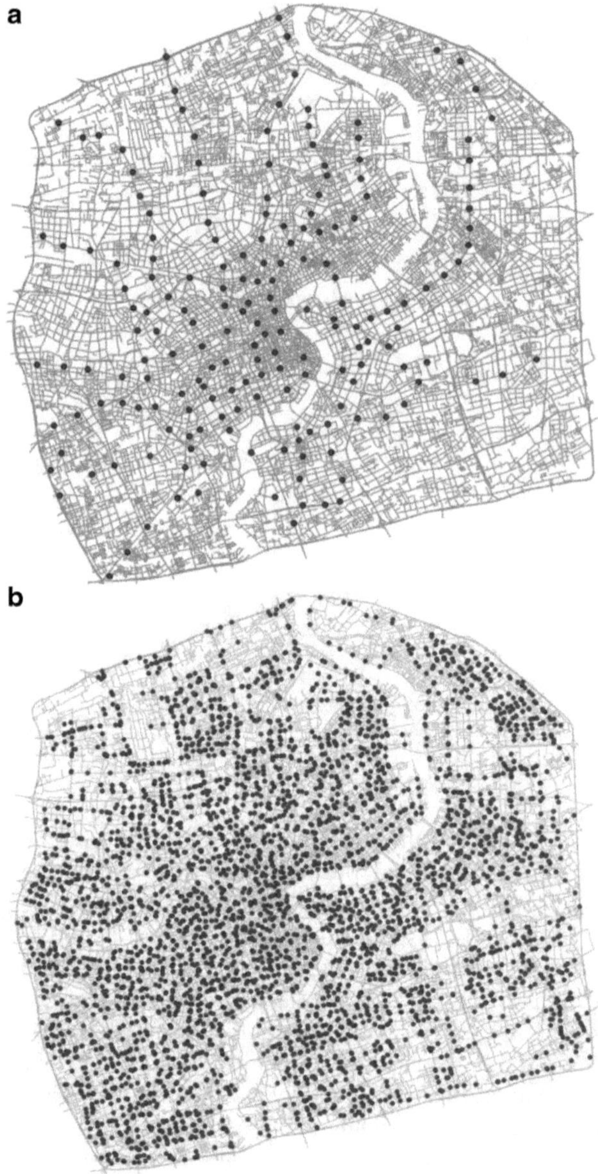

Fig. 13.5 Metro Stations (**a**) and Bus Stops (**b**) in Shanghai (2013)

These are very remarkable spatial distribution because of the asymmetry of the probability distribution. For bus stops the links with 30 % highest level of flow potential (5000 m) are where 65 % of the bus stops are located.

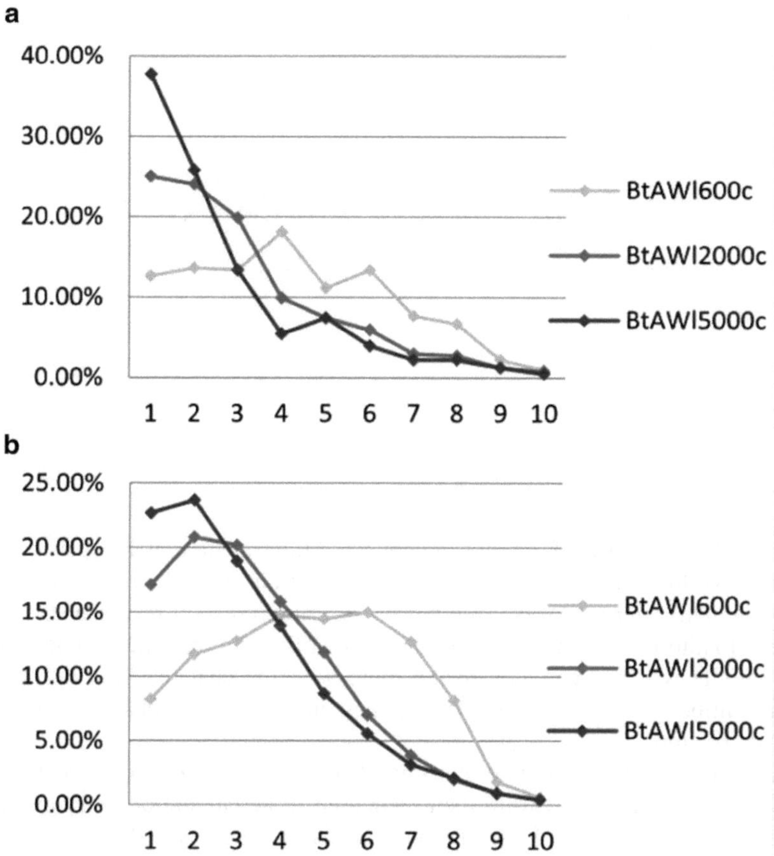

Fig. 13.6 Correlation between Betweenness Accessibility and the location of Metro Stations (**a**), Bus Stops (**b**)

We can see a pattern emerging of extreme coordination between flow potential levels commercial land use, bus stops, and metro stations locations. This coordination is best explained by circular causation presented above.

13.7 Metro Network Flow Potential Analysis

Following Derrible and Kennedy (2009) demonstrates that network topologies play a key role in attracting people to use public transit; ridership is not solely determined by cultural characteristics or city design. We now turn to the configuration of the metro lines network and evaluate the network effect on entry and exit. A network was created to represent the metro line network topology using one link between each station. For betweenness centrality, we used topological distance (one

Fig. 13.7 Topological betweenness of Shanghai Metro Network

link = 1). To model transfer between connected lines at stations level each transfer between two lines was considered as a topological distance of two thus adding two short link. This approach was taken because a limited market research showed that between to alternative route to the same destination with one line change at station level, people will be switching to the route with the change if the difference in station number was at least two. This makes the cost of line change equal to two more metro stops.

The results of flow potential analysis (Topological Betweenness centrality) of metro network (Fig. 13.7) were then compared to actual station entry/exit movement rates (Fig. 13.8a, b). A correlation analysis shows that the flow level potential value of metro network model for each station correlated with entry/exit $r^2 = 0.407$ ($p < 0.0001$) for weekend, and $r^2 = 0.497$ ($p < 0.0001$) for weekday.

The relationship between link betweenness centrality (5000 m) and the number of bus stops in the metro areas (600 m), Exit/Entry movement of metro stations are similar to its relationship with the location of metro stations (Fig. 13.9).

To recapitulate: the outer ring total area = 666,470,527 m^2; area within 600 m of Metro stations = 196,009,917 m^2 (29 %). Total Bus stop = 4042; bus stops within 600 m of Metro stations = 1619 (40 %). Total Commercial plot area = 61,889,229 m^2; Commercial plot area within 600 m of Metro stations = 27,419,968 m^2 (44 %).

13.8 Discussion

We started this paper with a review of a meta-analysis relating urban form indicators to travel behavior and identify particulars urban form indicators including non-residential land use (diversity), distance to public transport access points,

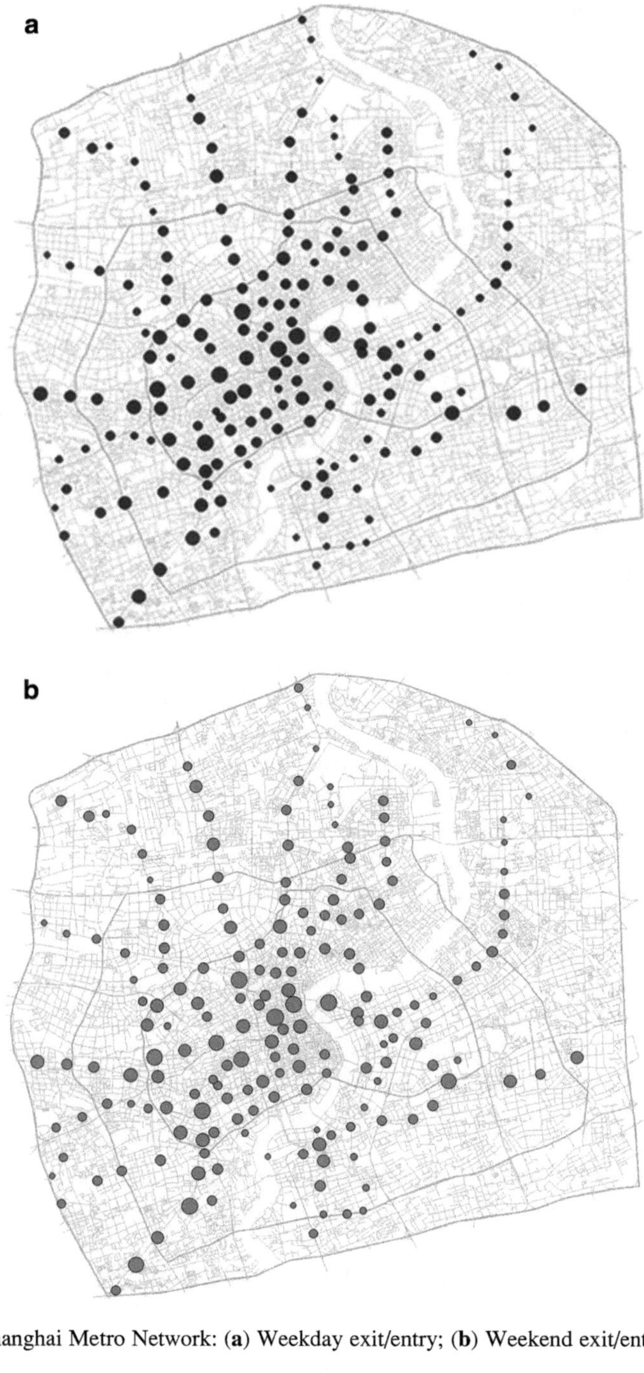

Fig. 13.8 Shanghai Metro Network: (**a**) Weekday exit/entry; (**b**) Weekend exit/entry

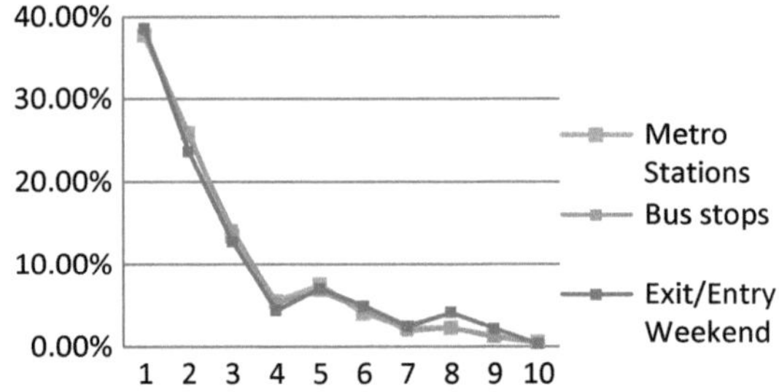

Fig. 13.9 Relationship between location of metro stations, number of bus stops in metro areas, exit/ entry movement of Metro Stations and Betweenness Accessibility (5000 m)

design of transport networks, destination accessibility related to these travel behaviors. Prior to 2010 this was known in the literature as the 5Ds (Ewing and Cervero 2001). More recently transport network composition and layout were found as important factors (Ewing and Cervero 2010; Banister 2012). One of the surprising findings was that population composition and density play a minor role once other urban form variables are taken into account. We delineated the circular causation model between accessibility and development which in part would explain such surprising finding. Urban form has evolved from being simply measured in terms of 'density' to become understood as the 3Ds (Cervero and Kockelman 1997), the 5Ds and more recently 'the spatial distribution of activities and the composition and layout of transport networks' (Bourdic et al. 2012). Reviewing the TOD model adapted in China we introduced the Node and Place model to extend and systematize Node and Place as a locus for new multi centralities. Study of centrality i.e. CBD in the US and Australia shows the important role of block size. We argued that at the structural level the understanding the "delicate design" around metro station will require the understanding of the relationship between urban block size and form in relationship to accessibility. We proposed that in order to disentangle further the land use from accessibility it will be best to use accessibility indicators that are unweighted by size effect to explore the role of accessibility mediated by urban block size and form. The advantage of these spatial accessibility indicators is that they constitute a unified multi-level analytical framework for network design analysis. Our objective was to use these indicators to distinguish between network design effect, resulting spatial accessibility and flow potential on network in relation to a range of and land use locations in Shanghai within the Outer ring. To our knowledge no such extensive analysis has been performed in such large area.

We investigated the relationship between urban block size and form and network morphology using network betweenness centrality with topological and geometric metrics that effectively discriminates the morphology of the transport network design and that can be theoretically interpreted as generic flow potential. We then investigated the empirical relationship between these multi-level accessibility indicators in relation to public transport access point (i.e. distance to transit and commercial land use, and metro station usage). In Shanghai, we found that:

- Commercial land use location is strongly distributed on the road network to maximize micro to macro high level of spatial accessibility advantage;
- Most metro stations and bus stops have location distribution that follow the same pattern: a strong location bias to macro for high level of flow potential at large radii. This is the coupling-multiplier effect for land-use surrounding metro stations identified by (Pan et al. 2007, Pan and Zhang 2008);
- Following Derrible and Kennedy (2009) we found as strong and positive relationship between metro station entry/exit usage and metro network configuration. The metro network configuration, the supply side, plays an important role in the movement generating pattern. More importantly it help to understand the Node position of each metro station in relationship to the whole metro network. It is an important planning indicator to anticipate and understand Node change interaction according to change in the whole metro network.

Overall our contribution through this exploratory cross-sectional analysis is to show how an intensive structural coordination is at work between locations of commercial land use, bus stops, metro station location identified through multi-scale accessibility and derived flow potential levels. Beyond this initial cross sectional analysis, this is the starting point of future time series analyses.

We conclude this discussion with future research and an urban design prospective.

The Node/Place model of Bertolini (1999) provides an established starting reference to develop more fully an adapted TOD model to China. The node-place model's emphasis on "conditions" is important, as it indicates a development potential that may or may not be realized, as other factors may also affect the outcome (Chorus and Bertolini 2011). Elaboration of this model (Trip 2007) emphasizes the role of urban design, place making and urban quality to maximize value capture identified in the UK literature. The node/place model distinguishes five different profiles for a station area. Each profile reflects particular relative position of a station area value on the node or place hierarchy in urban system. Future research will investigate the detailed Node/Place profile of metro station in Shanghai in relationship to travel behavior and urban quality. To this end, the development of a better understanding between block size/multi-level configurations surrounding the metro station, quality of place in relation to pedestrian usage will be of great importance. At a strategic level, a new research direction could be undertaken: how should multi networks accessibility be coordinated and distributed to balance accessibility equity, quality of place and agglomeration economies?

References

Amion Consulting, Taylor Young, Donaldsons, & University of Liverpool. (2007). *Economic value of urban design*. Liverpool: Amion Consulting.

Arefi, M., & Triantafillou, M. (2005). Reflections on the pedagogy of place in planning and urban design. *Journal of Planning Education and Research, 25*(1), 75–88.

Banister, D. (1995). *Transport and urban development*. London: Spon.

Bavelas, A. (1950). Communication patterns in task-oriented groups. *Journal of the acoustical Society of America, 22*(6), 725–730.

Bertolini, L. (1996). Nodes and places: Complexities of railway station redevelopment. *European Planning Studies, 4*(3), 331–346.

Bertolini, L. (1999). Spatial development patterns and public transport: The application of an analytical model in the Netherlands. *Planning Practice and Research, 14*(2), 199–210.

Bertolini, L., & Spit, T. (1998). *Cities on rails. The redevelopment of railway station areas*. London: Spon.

Bian, J. (2006). *Metropolitan space development and rail transit*. Beijing: China Architecture & Building Press (in Chinese).

Borzachiello, M. T., Nijkamp, P., & Koomen, E. (2010). Accessibility and urban development: A grid-based comparative statistical analysis of Dutch cities. *Environment and Planning B: Planning and Design, 37*(1), 148–169.

Boujenko, N., Marshall, S., & Jones, P. (2007). *Link and place – A guide to street planning and design*. London: Local Transport Today.

Bourdic, L., Salat, S., & Nowacki, C. (2012). Assessing cities: A new system of cross-scale spatial indicators. *Building Research Information, 40*(5), 592–605.

British Council for Offices. (2006, May 1). *Offices, value and design*. http://www.bco.org.uk/Research/Publications/Officesv2606.aspx. Accessed 1 July 2014.

CABE. (2003). *A bibliography of design value*. http://webarchive.nationalarchives.gov.uk/20110118095356/http:/www.cabe.org.uk/files/a-bibliography-of-design-value.pdf. Accessed 1 July 2014.

CABE. (2007). *Paved with gold: The real value of street design*. http://webarchive.nationalarchives.gov.uk/20110118095356/http://www.cabe.org.uk/files/paved-with-gold.pdf. Accessed 1 July 2014.

CABE Space. (2003). *The value of public space, how high quality parks and public spaces create economic, social and environmental value*. CABE Space (Commission for Architecture and the Built). https://www.designcouncil.org.uk/sites/default/files/asset/document/the-value-of-public-space.pdf. Accessed 1 July 2014.

CABE Space. (2009). *Making the invisible visible the real value of park assets*. https://www.designcouncil.org.uk/sites/default/files/asset/document/the-real-value-of-park-assets.pdf. Accessed 1 July 2014.

CABE, ODPM, & Design for Homes. (2003). *The value of housing design and layout*. http://webarchive.nationalarchives.gov.uk/20110118095356/http:/www.cabe.org.uk/files/the-value-of-housing-design-and-layout.pdf. Accessed 1 July 2014.

CABE, UCL, & DETR. (2001). *The value of urban design. Analysing commercial developments to see how excellent design adds economic, social and environmental value*. www.designcouncil.org.uk/publications/the-value-of-urban-design/. Accessed 21 Nov 2013.

Calthorpe, P. (1993). *The next American metropolis: Ecology, community, and the American dream*. New York: Princeton Architectural Press.

CBuchanan. (2008). *A programme-level evaluation of one NorthEast's quality of place activities, 2002–07*. http://www.onenortheastlegacy.co.uk/file.aspx?id=75. Accessed 1 July 2014.

Cervero, R., & Jennifer, D. (2008). Suburbanization and transit-oriented development in China. *Transport Policy, 15*, 315–323.

Chiaradia, A. J., Hillier, B., Schwander, C., & Wedderburn, M. (2012, March). Compositional and urban form effects in centres in Greater London. *Urban Design and Planning – Proceedings of the ICE, 165*(1), 21–42.

Chiaradia, A. J., Hillier, B., Schwander, C., & Barnes, Y. (2013, June). Compositional and urban form effects on residential property value patterns in Greater London. *Urban Design and Planning – Proceedings of the ICE, 166*(3), 176–199.

Chiaradia, A., Crispin, C., & Webster, C. (2014). *sDNA a software for spatial design network analysis*. www.cardiff.ac.uk/sdna/. Accessed 15 June 2014.

Chiaradia, A. J., Sieh, L., & Plimmer, F. (2015). Learning values in urban design: A studio-based approach. *Design Studies*, under review.
Chorus, P., & Bertolini, L. (2011). An application of the node place model to explore the spatial development dynamics of station areas in Tokyo. *The Journal of Transport and Land Use, 4*(1), 45–58.
Christaller, W. (1933–1966). *Die zentralen Orte in Suddeutschland – Places in Southern Germany* (C. W. Baskin, Trans.). Jena: Fisher – Prentice Hall.
CIHT. (2010). *Manual for streets 2 wider application of the principles*. London: The Chartered Institution of Highways & Transportation.
Cooper, C., Chiaradia, A., & Webster, C. (2014). *Spatial design network analysis(+)*. www.cardiff.ac.uk/sdna/. Accessed 16 June 2014.
Coupy, E. (1851). Solution d'un problème appartenant à la géometrie de situation, par Euler. *Nouvelles annales de mathématiques 1re série* 10, 106–119.
Cutini, V. (2001). Centrality and land use: Three case studies on the configurational hypothesis. *Cybergeo: European Journal of Geography [On line], Systèmes, Modélisation, Géostatistiques* 188.
Derrible, S., & Kennedy, C. (2009). Network analysis of world subway systems using updated graph theory. *Transportation Research Record: Journal of the Transportation Research Board, 2112*, 17–25.
DfT. (2007). *Manual for Streets 1*. London: Department for Transport.
Dunse, N., White, M., White, M., & Dehring, C. (2007). *Urban parks, open space and residential property values* (RICS research paper series, pp. 1–8). London: RICS.
Duranton, G., & Puga, D. (2004). Micro-foundations of urban agglomeration economies. Vol. 4, chap. 48. In J. V. Henderson & J. F. Tisse (Eds.), *Handbook of regional and urban economics, cities and geography* (pp. 2063–2117). London: Elsevier.
Ewing, R., & Cervero, R. (2001). Travel and the built environment. *Transportation Research Record, 1780*, 87–114.
Ewing, R., & Cervero, R. (2010). Travel and the built environment. *Journal of the American Planning Association, 76*(3), 265–294.
Fisher, D. R. (1999). *Waterside properties*. https://www.waterways.org.uk/pdf/restoration/trh_waterside_properties_2. Accessed 1 July 2014.
Florida, R. (2000). *Competing in the age of talent: Quality of place and the new economy*. Pittsburgh: Carnegie Mellon.
Freeman, L. C. (1977). A set of measures of centrality based on betweenness. *Sociometry, 40*(1), 35–41.
Garrod, G., & Willis, K. (1994). An economic estimate of the effect of a waterside location on property values. *Environmental and Resource Economics, 4*(2), 209–217.
Geurs, K. T., & Ritsema van Eck, J. R. (2001). *Accessibility measures: Review and applications*. Bilthoven: National Institute of Public Health and the Environment.
Geurs, K. T., & van Wee, B. (2004). Accessibility evaluation of land-use and transport strategies: Review and research directions. *Journal of Transport Geography, 12*(2), 127–140.
Giuliano, G. (2004). Land use impacts of transportation investments: Highway and transit. In S. Hanson & G. Giuliano (Eds.), *The geography of urban transportation* (pp. 237–273). New York: Guildford Press.
GLA Economics. (2003). *Valuing greenness, green spaces, house prices and Londoners' priorities*. http://www.london.gov.uk/mayor/economic_unit/docs/valuing_greenness_report.pdf. Accessed 01 July 2014.
GLA Economics. (2010). *Working Paper 42 valuing housing and green spaces: Understanding local amenities, the built environment and house prices in London*. http://www.london.gov.uk/sites/default/files/GLAE-wp-42.pdf. Accessed 1 July 2014.
Goetgeluk, R., Kauko, T., & Priemus, H. (2005). Can red pay for blue? Methods to estimate the added value of water in residential environments. *Journal of Environmental Planning and Management, 48*(1), 103–120.

Grosvenor. (2013). *Evolving cities, resilient urban form, governance and the creation of long term value*. London: The London School of Economics and Political Science.

Haggett, P., & Chorley, R. J. (1969). *Network analysis in Geography*. London: Hodder & Stoughton Educational.

Hansen, W. G. (1959). How accessibility shape land use. *Journal of American Institute of Planners, 25*(2), 73–76.

Ingram, D. R. (1971). The concept of accessibility: A search for an operational form. *Regional Studies, 5*(2), 101–107.

Jim, C. Y., & Chen, W. Y. (2010). External effects of neighbourhood parks and landscape elements on high-rise residential value. *Land Use Policy, 27*(2), 662–670.

Jones, P., Roberts, M., & Morris, L. (2007). *Rediscovering mixed-use streets the contribution of local high streets to sustainable communities*. Bristol: The Policy Press.

Kansky, K. J. (1963). *Structure of transport networks: relationships between network geometry and regional characteristics*. Research paper *4, Department of geography, University of Chicago, Chicago.

Krambeck, H. V. (2006). *The global walkability index*. http://dspace.mit.edu/handle/1721.1/34409. Accessed 1 May 2014.

Lin, L., & Moudon, A. V. (2010). Objective versus subjective measures of the built environment, which are most effective in capturing associations with walking? *Health & Place, 16*(2), 339–348.

Liu, J. (2012). *Urban multi-level development and rail transit*. Nanjing: South East University Press (in Chinese).

McIntyre, M. H. (2006). *A literature review of the social economic and environmental impact of architecture and design*. Edinburgh: Information and Analytical Services Division, Scottish Executive Education Department.

Meyer, M., & Miller, E. (2001). *Urban transportation planning: A decision-oriented approach*. Boston: McGrawHill.

Ministry for the Environment, NZ. (2005). *The value of urban design: The economic, environmental and social benefits of urban design*. https://www.mfe.govt.nz/publications/urban/value-urban-design-full-report-jun05/index.html. Accessed 1 July 2014.

Network Rail. (2011). *The value of station investment*. http://www.steerdaviesgleave.com/sites/default/files/newsandinsights/Station_Investment_Report.pdf. Accessed 1 July 2014.

Newman, M., Barabasi, A.-L., & Watts, D. J. (2006). *The structure and dynamics of networks*. Princeton/Oxford: Princeton University Press.

Pan, H., & Ren, C. (2005). Study of the spatial coupling between the metro and urban activity center—Case study of Shanghai. *Urban Planning Forum, 4*(2005), 76–82 (in Chinese).

Pan, H., & Zhang, M. (2008). Rail transit impacts on land use, evidence from Shanghai, China. *Transportation Research Record: Journal of the Transportation Research Board, 2048*, 16–25.

Pan, H., Liu, X., & Zacharias, J. (2003). Street design characteristic and green transportation choice. *Urban Planning Forum, 6*, 42–48 (in Chinese).

Pan, H., Ren, C., & Yang, T. (2007). A study on the impact on the land use of station areas brought by urban rail transport in Shanghai. *Urban Planning Forum, 4*, 92–97 (in Chinese).

Pan, H., Shen, Q., & Zhang, M. (2009). Influence of urban form on travel behaviour in four neighbourhoods of Shanghai. *Urban Studies, 46*(2), 275–294.

Pan, H., Shen, Q., & Chang, L. (2011). Transit-oriented development at the urban periphery, insights from a case study in Shanghai, China. *Transportation Research Record: Journal of the Transportation Research Board, 2245*, 95–102.

Pan, Q., Pan, H., Zhang, M., & Zhong, B. (2014). The effects of rail transit on residential property values: A comparison study on the rail transit lines in Houston and Shanghai. *Transportation Research Record: Journal of the Transportation Research Board, 2453*, 118–127.

Parks, J. R., & Schofer, J. L. (2006). Characterising neighbourhood pedestrian environment with secondary data. *Transportation research Part D: Transport and Environment, 11*(4), 250–263.

Pooler, J. A. (1995). The use of spatial separation in the measurement of transportation accessibility. *Transport Research A: Policy and Practice, 29*(6), 421–427.
Porta, S., Crucitti, P., & Latora, V. (2006). The network analysis of urban streets: A dual approach. *Physica A: Statistical Mechanics and its Applications, 369*(2), 853–866.
Reggiani, A. (1998). Accessibility, trade and location behaviour: An introduction. In A. Reggiani (Ed.), *Accessibility, trade and location behaviour* (pp. 1–16). Aldershot: Ashgate.
Reilly, W. J. (1931). *The law of retail gravitation*. New York: Knickerbocker Press.
Rogers, K., Jaluzot, A., & Neilan, C. (2012). *Green benefits in Victoria Business Improvement District*. http://www.victoriabid.co.uk/wp-content/uploads/2012/10/VBID_Green_Benefits_2012.pdf. Accessed 1 July 2014.
Rouwendal, J., Van Marwijk, R., & Levkovich, O. (2014). *The value of proximity to water in residential areas*. http://papers.ssrn.com/sol3/papers.cfm?abstract_id=2427687. Accessed 1 July 2014.
Shimbel, A. (1953). Structural parameters of communication networks. *The bulletin of mathematical biophysics, 15*(4), 501–507.
Siskna, A. (1990). *A comparative study of block size and form: (in selected new towns in the history of Western Civilization centres)*. St. Lucia, Qld: Thesis (PhD) – University of Queensland.
Siskna, A. (1997). The effects of block size and form in North American and Australian city centres. *Urban Morphology, 1*, 19–33.
Storper, M., & Venables, A. J. (2004). Buzz: The economic force of the city. *Journal of Economic Geography, 4*(4), 351–370.
The Prince's Foundation for the Built Environment. (2007). Valuing sustainable urbanism, a report measuring & valuing new approaches to residentially led mixed use growth. http://www.princes-foundation.org/sites/default/files/0707vsureport_0.pdf. Accessed 1 July 2014.
Thomas, A., & Deakin, E. (2008). Land use challenges to implementing transit-oriented development in China: Case study of Jinan, Shandong. *Transportation Research Record: Journal of the Transportation Research Board, 2077*, 80–86.
Transport for London. (2011). *Valuing urban realm*. https://toolkit.urban-realm.co.uk/. Accessed 1 July 2014.
Tribal Urban Studio, & CBuchanan. (2008). *Baseline report of the quality of place longitudinal study (capturing the impacts of quality of place)*. London: CBuchanan.
Trip, J. J. (2007, March 20). *What makes a city? Planning for 'quality of place'*. http://repository.tudelft.nl. http://repository.tudelft.nl/assets/uuid:a0ee18b2-8cf3-45f7-8e2b-45141ab28ef4/otb_trip_20070320.pdf. Accessed 1 May 2014.
Wegener, M., & Fuerst, F. (2004, January 1). *Land-use transport interaction: State of the Art*. http://ssrn.com/abstract=1434678. Accessed 1 May 2015.
Wegener, M., & Furst, F. (1999). *Land-use transport interaction: State of the art*. Deliverable 46, Institut für Raumplanung, Dortmund.
Xiao, Y., Webster, C. J., & Chiaradia, A. (2013). *How urban form shapes land use: empirical findings in Wuhan, China*. The Second International Workshop on Regional Urban and Spatial Economics in China, Beijing.
Yeung, Y. M., & Sung, Y. W. (Eds.). (1996). *Shanghai: Transformation and modernization under China's open policy*. Hong Kong: The Chinese University Press.
Zacharias, J. (2005). Non-motorized transportation in four Shanghai Districts. *International Planning Studies, 10*(3–4), 323–340.
Zhang, M. (2007). Chinese edition of transit-oriented development. *Transportation Research Record: Journal of the Transportation Research Board, 2038*, 120–127.

Chapter 14
Research on Interaction Between Traffic Improvement Around the Old Railway Station and Urban Land Utilization—A Case Study in Hohhot Railway Traffic Regulation

Linfei Han, Jianmin Guo, and Junyan Han

14.1 Foreword

As an important carrier of modern industrial society, the railway is one of the most prominently issues in the history of mankind. Since its emergence, the stations have become the important nodes in city. Taking advantage of its location, the railway station areas produced a huge agglomeration of activities along with city's development. From the initial yard to various kinds of residential buildings and commercial buildings, the land utilization, development intensity of the old railway station areas have changed significantly. Playing the role of both urban transportation hub and commercial hub, these areas have grown to be the center or sub-center of the city. However, adoption of the automobile and urban expansion made a significant impact on the traffic and functions around the old railway station. Coupled with the infrastructure obsolete, aging buildings, traffic capacity saturation, the areas have been suffering from difficulties like traffic congestion, inefficient land utilization, chaotic spatial structure, decline of environmental health conditions, increase of social instability as well as other "dirty, chaotic, and poor" issues. To make matters worse, the railway lines and stations, which are located in the inner city, fragment the city and increase traffic load. Currently the widely used way to alleviate traffic pressure is relocating the train station. However, it isn't the best way to solve the problem of development of the areas. This is because relocating the railway station can bring inconvenience to passengers, longer travel time, higher travel costs, and

L. Han (✉) • J. Han
School of Architecture and Design, Beijing Jiaotong University, Beijing, China
e-mail: usi2000@126.com

J. Guo
School of Civil Engineering, Beijing Jiaotong University, Beijing, China
e-mail: 13121717@bjtu.edu.cn

Q. Pan, J. Cao (eds.), *Recent Developments in Chinese Urban Planning*,
GeoJournal Library 114, DOI 10.1007/978-3-319-18470-8_14

other new problems. In this case, illustrating the development challenges and the relationship between traffic and land utilization in order to inspire vitality of these areas is the key to improving the appearance of the area.

14.2 The Planning Concept of Remodeling Vitality in the Old Railway Station Area

Vitality is the basis for a city, which is deeply reflected in Kevin Lynch's theory of humanistic urban planning, Jane Jacobs's criticism for the failure of urban renewal plan in the USA, and Peter Calthorpe's Transit-oriented development (TOD) theory. In fact, urban structure is constituted by people's activity, and the essence of urban vitality comes from ordered aggregation. Disordered aggregation can reduce the urban vitality as well as causing the area to be perceived as less attractive. Urban vitality has three components: economic, social, and culture (Jie et al. 2011). Kevin Lynch believes that the image of the city is mainly constituted of five most important elements, including paths, edges, district, nodes and landmarks. In order to create a pleasant environment to attract more people, urban planning and design should organize elements of urban physical environment with reason. Ordered aggregation also calls for the five elements to be arranged reasonably, which is not only closely related with traffic, but also land utilization.

Over time, as more and more residential, commercial, hotel and other buildings are filling up the old railway station surroundings, the areas have exploded in popularity. This contributes to the complex environment and socio-economic relations. However, for most cities, this aggregation appears to, even have been in disordered, leading to loss of vitality in urban areas. Therefore, to restore vitality of the old railway station areas, we must try to make the aggregation ordered on the basis of necessary population.

In the improvement plan, we ought to keep the complexity and diversity of the old railway station areas to ensure their vitality but not focus on the mechanical function zoning and unified visual space. The vital agglomeration of activities can only be orderly when all the following preconditions are met. The first condition is ensuring an uncrowded and efficient road network. Traffic is the core function of the railway station area, which should be concerned about much in the improvement plan. An uncrowded and efficient road network can not only facilitate motorized travel, but also pedestrian travel by offering safe, reliable, and green slow-walking space. Secondly, it's necessary to ensure the reasonable land use structure, for its contribution to make the city demonstrate stronger attraction and vitality by greatly facilitating citizens' life and reducing the amount of traffic. The old railway station areas has formed complete business formats, so in the improved planning, we should respect the historical heritage of the spatial layout of streets, city, workshops, lanes, houses, but avoid pursuing simple, mechanical function zoning. Thirdly, the volume rates should be appropriate (Jun 2011). Old railway station areas are generally densely build-up, resulting in no public spaces, except the station square. In improvement planning, it's necessary to (1) control the volume rates, height and density of buildings properly, (2) vacate the space after the demolition of old buildings, and (3) try to arrange public opening space or low-rise buildings instead of high-rise building. Besides, paying attention to the development and utilization of underground space, correctly handling the relationship between underground parking and the underground business is critically important (Lian and Wenzhuo 2009; Zheng et al. 2012).

In the revitalization of the old railway station area, it is important to examine the complex urban spatial pattern while also considering traffic improvement and land utilization. That is, consider both traffic improvement and land utilization to promote coordinated development of traffic and land utilization, instead of "solving the traffic problems by itself" or "solving the land utilization problems by itself".

14.3 Problems of Hohhot Old Railway Station and Its Planning

14.3.1 Problems in Development

Hohhot railway station was built in 1921, which is under the jurisdiction of the Hohhot Railway Bureau. The station issued a total of 88 trips and handles 22,000 passengers daily. The bus stations, shopping malls, large-scale wholesale markets in the surrounding areas attract almost 20 million people per day. Hohhot railway station played an important role in Hohhot's economic development, but after experiencing years of laissez-faire growth, the area is facing a lot of development problems.

Firstly, the land use structure is irrational. Now, the land use structure is "one core, two axes, five parts." The core refers to the railway station (Fig. 14.1), the two axes refer to the Station Street and Xilinguole Road, and the five parts conclude three residential areas, the city park area, and the high education area. The northern residential area is separated by the railway.

As is shown in the Fig. 14.2, there are lots of wholesale and retail stores close to the railway station, which result in a high proportion of commercial land. By contrast, the road, street and transportation land only accounts for 8.1 % of the total land. The railway lying through the city, which separates the south from the north, can bring inconvenience to citizens. There are two bus stations very near the railway, causing greater disturbance to the surrounding traffic. Besides, the green space and square land accounts for 5.5 %, far lower than the national standard (Fig. 14.3).

Secondly, the traffic of the railway station area is very crowded and the road network and the land use is uncoordinated (Fig. 14.4). On both sides of main roads high-density buildings severely weaken the adjacent road infrastructure's core function. Besides, the north-south road connectivity is poor as a result of the flawed road network. The road network density is low and uneven in general. The deformity intersections, like North Hulun Buir Road and West Station Street, reduce overall efficiency of road network. Restricted by the East-west railway, the city is

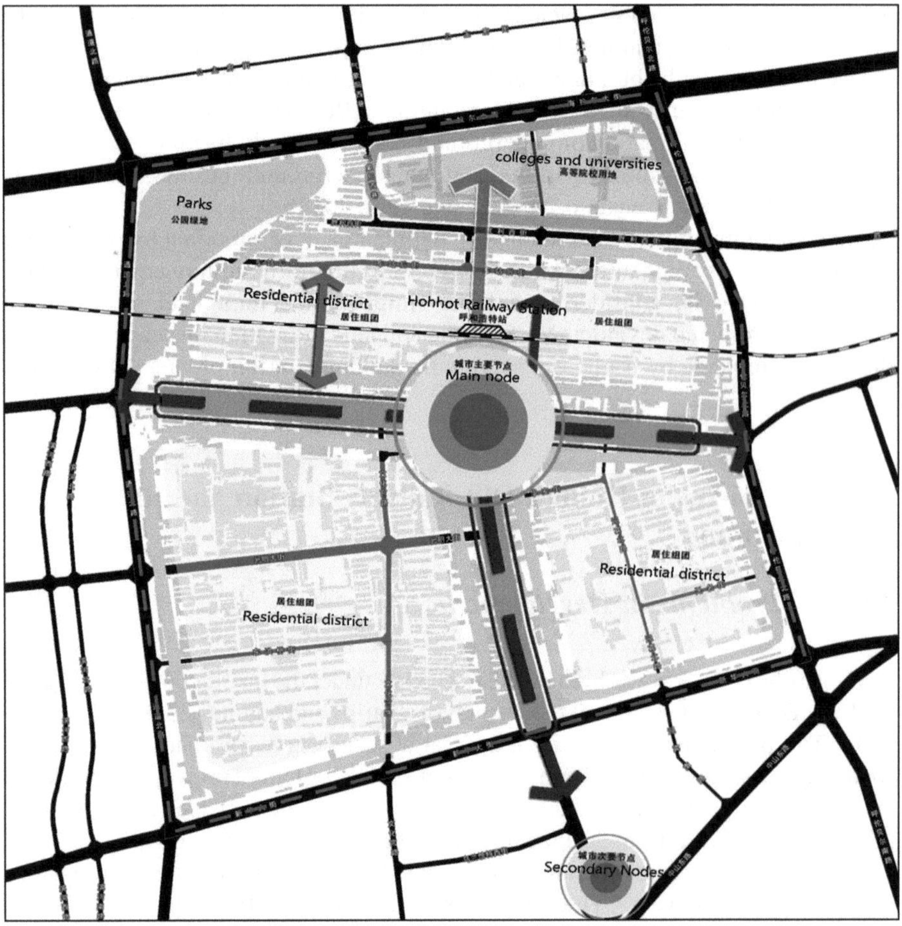

Fig. 14.1 Structure of present land use structure (Source: Study Authors)

segmented from north to south, resulting in broken roads and poor connection in the area. What's worse, the area is in a severe shortage of parking space (Fig. 14.5), and some underground parking lots are used for commercial. Inadequate supply of parking spaces leads to a large number of non-motor vehicle lanes and sidewalks parking, which seriously encroaches on the space of non-motorized and pedestrian travel. In addition, the lack of the pedestrian crossing facilities and pedestrian crossing traffic signal lamp brings some hidden safety problems (Figs. 14.6 and 14.7).

Thirdly, there is a poor management and irrational urban planning in the railway station area, which leads to the "dirty, chaotic, and poor" characteristic of the region.

Balance sheet of city construction land					
Code			Name	Land area (hm²)	Proportion(%)
High	Medium	Small			
R			Residential land	106.58	38.03
	R2		Residential land	105.99	37.82
	R3		Residential land	0.59	0.21
A			Administration and public service land	83.29	29.72
	A1		Administration land	32.81	11.71
	A2		Cultural land	3.50	1.25
	A3		Education scientific research land	36.85	13.15
		A31	Colleges and universities	16.32	5.82
		A32	Specialized secondary schools	5.81	2.07
		A33	Primary and secondary schools	6.23	2.22
		A35	Scientific research land	8.49	3.03
	A4		Sports land	3.63	1.30
	A5		Medical and health work land	6.82	2.43
B			Commercial and business facilities	36.35	12.97
	B1		Commercial land	27.13	9.68
	B2		Business land	9.21	3.29
M			Industrial land	4.03	1.44
W			Logistics and warehouse	1.11	0.40
	W1		First-rate logistics and warehouse	1.11	0.40
S			Street and transportation facilities	30.28	10.81
	S1		Road land	28.14	10.04
	S2		Rail transit land	0.00	0.00
	S3		Traffic hub land	2.14	0.76
	S4		Traffic station land	0.00	0.00
		S41	Public transportation station land	0.00	0.00
		S42	Social parking land	0.00	0.00
	S9		Other transport facilities land	0.00	0.00
U			Public facilities land	2.96	1.06
	U1		Supply facilities land	2.96	1.06
		U12	Power supply land	1.44	0.51
		U15	Communication land	1.52	0.54
G			Green space and squares	15.34	5.47
	G1		Park and green space	11.69	4.17
	G2		Green area for protection	2.00	0.71
	G3		Square land	1.65	0.59
H11			Development land	280.24	100.00

Fig. 14.2 Hohhot railway station current land use map (Source: Study Authors)

14.3.2 Target Location

According to "Hohhot City Master Plan (2011–2020) ", in the future, the railway passenger integrated hubs will be consisted of the Hohhot railway station, the north bus station, the railway public transit transfer hub, intercity railway stations, and subway stations. Together, this will work as the comprehensive hubs of intercity long-distance railway transportation, road transportation and city public traffic. The total planned land is 97,500 square meters (m^2), including 20,000 m^2 Hohhot railway station, 47,500 m^2 north bus station, 20,000 m^2 railway public transit transfer hub, and each 5000 m^2 setting aside for intercity railway stations and subway stations. According to the master plan, the old railway station area will develop into a traffic hub of the city while still playing an important role in urban development. On the basis of orientation, the improvement planning will focus on how to stimulate the urban vitality as well as the revival of the old town (Fig. 14.8).

The planning orientation of the old railway station is "traffic dominant, organic renewal". It can be accomplished by the following practice: integrating transportation networks, improving traffic systems, and suturing city "cracks." This will allow the railway station to develop into a new convenient traffic hub. By re-sorting the land use structure, balancing various types of land, and promoting all kinds of business, the area will be a booster of Hohhot's economic development. By creating public space with characteristics of Inner Mongolia, increasing walkable streets, and meeting the needs of traveling, leisure, shopping at the same time, this area will become a window to the capital's ethos. By extending the basic functions of the city and creating a complex urban spatial structure to improve its influence, the old railway station area will ultimately become the deputy center of the city.

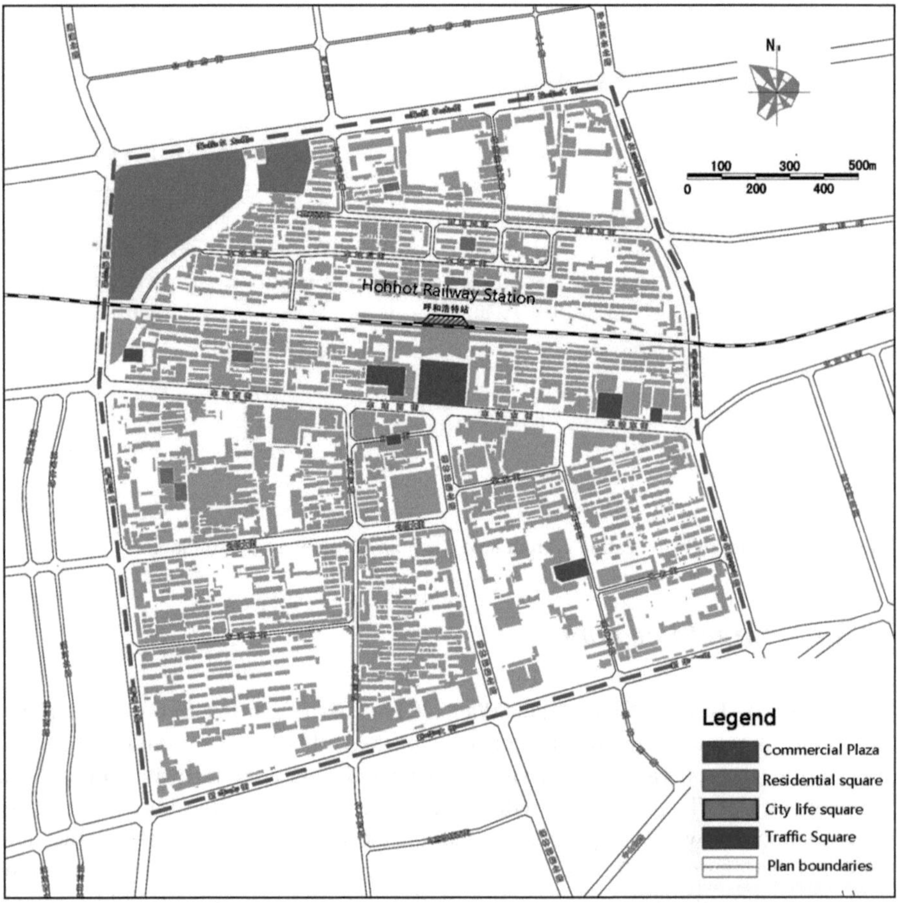

Fig. 14.3 Present square distribution map (Source: Study Authors)

14.4 Traffic Improvement Strategy for Hohhot Old Railway Station Areas

14.4.1 Creating Complex Spatial Development Structure

Some scholars believe that land utilization is a process that the market can't interfere with. Because the road network is ubiquitous and land utilization is influenced by many factors, the task of traffic planning is how to coordinate traffic with land-use trends (Zhou and Gong Rong 2010). However, this wrong idea results in causing transportation planning to be viewed as irrelevant. In fact, land utilization, which determines the volume, commuting time and the modes of transportation you choose, is the root of traffic. Moreover, the supply level of the transport

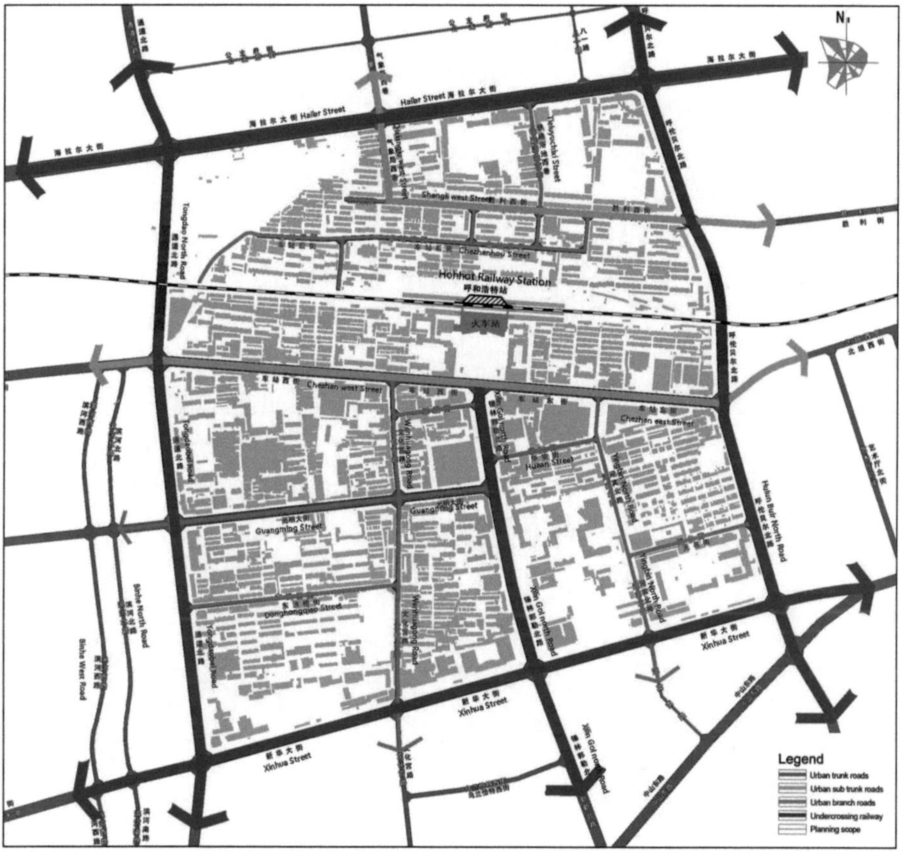

Fig. 14.4 The current status of road (Source: Study Authors)

facilities also influences the land utilization, spatial structure and scale of the city, thereby the value of land, especially the commercial land (Fig. 14.9).

Based on integrated network, temporal interleaving,[1] function aggregation and complex multiconcept,[2] the complex spatial structure is to gather the various unitary functions together to solve the contradiction between the station traffic hub and land utilization. By means of enhancing its regional statues and reconnecting the divided urban space, the railway station surrounding areas will develop from a single external transport hub into a multi-point transportation with highly concentrated complex space (Figs. 14.10 and 14.11).

[1] Three dimensional city image represent the past, now and future.

[2] functional composition and colorful life to enhance the vitality of the city.

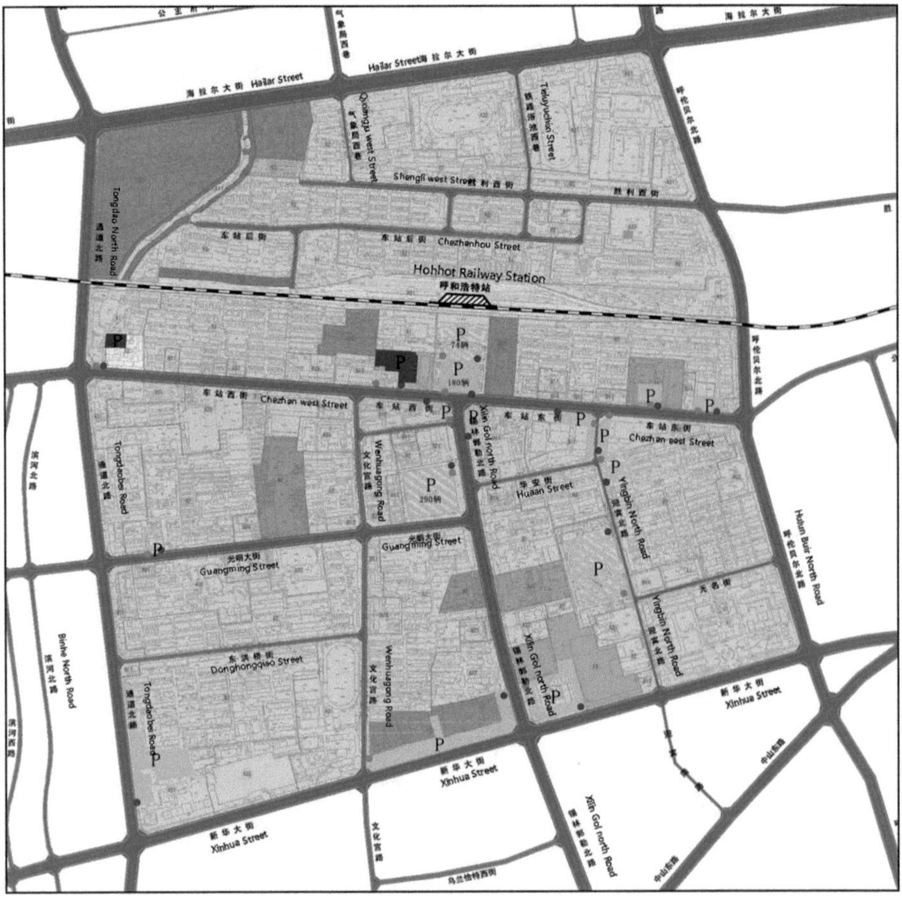

Fig. 14.5 Distribution of parking lots (Source: Study Authors)

14.4.2 Improvement Plan

First of all, adjust and optimize the land use to divert transit traffic (Figs. 14.12 and 14.13). The specific strategies are as follows:

1. Cut down the residential land use and improve building density and floor area ratio to raise the land use efficiency;
2. Integrate the present sprawling low-end commercial land into an unified and efficient modern business district;
3. Divert traffic and pay attention to the public space to enhance Hohhot's quality;
4. Increase the surrounding land use for street and transportation;
5. Build the three-dimensional transportation by building the air corridor erected on the south plaza;

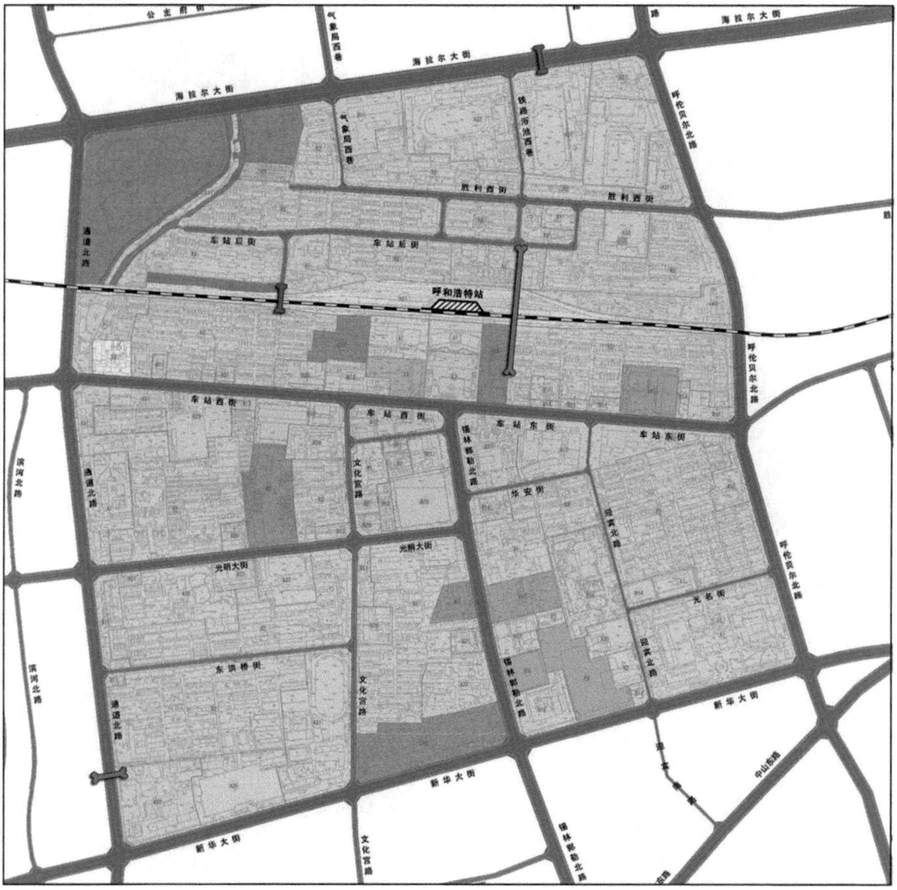

Fig. 14.6 Pedestrian crossing facilities distribution (Source: Study Authors)

6. Open up the north square to solve passengers' difficulty to the station;
7. Carry out a comprehensive traffic analysis to decrease the new traffic press.

The planning structure and layout of the land are two criss-cross axes, one core leading three points, six parts (Fig. 14.14). In the plan, Chezhan Street works as the traffic axis, Xilinguole North Road as the landscape axis; as the core, the railway station is combined with comprehensive service center, wholesale trade center and the modern business center, together to create the image of the capital; the six functional zones are: the comprehensive service center, the wholesale commercial district, the modern business district, educational center, the residential parts and the ecological landscape area. The plan means to shape the city's skyline by arranging the service facilities around the railway station. At the same time, the landmark building must be cooperative with its surrounding.

Fig. 14.7 Road congestion (Source: Study Authors' Photograph)

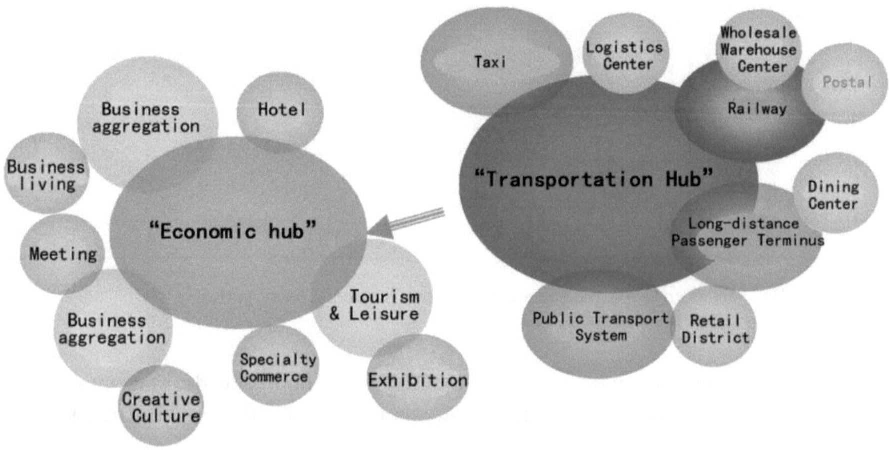

Fig. 14.8 Connection between traffic hub and economic hub (Source: Study Authors)

Secondly, improve the present traffic situation (Fig. 14.15). Specific measures should be people-oriented. The plan focuses on recent improvements in accessibility, treats facilities construction and management, static traffic and dynamic traffic with the same status.

The measures taken are as follows:

Diverting the transit traffic to reduce its impact on the city; mining the potential of urban branch roads to increase the capacity of the road system; accelerating channels construction to build a ordered network; assigning the right of way to

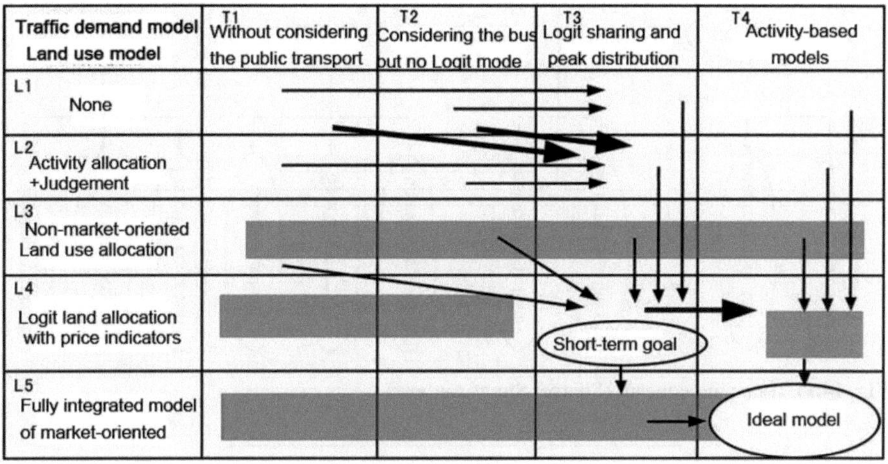

Fig. 14.9 Integrated land use and transport planning model (Source: Adapted from Miller E. et. al. 1998)

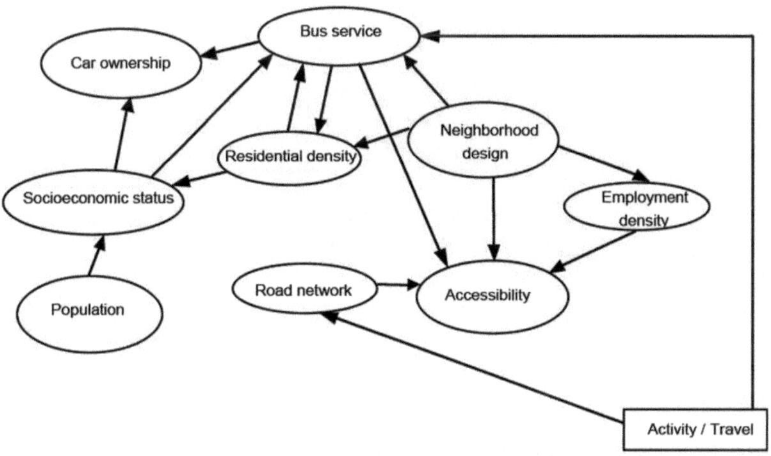

Fig. 14.10 Effect of city form on the activity and travel (Source: Adapted from Miller E. et. al. 1998)

optimize the traffic; quickening the construction of the branch roads to improve microcirculation system; optimizing the signal phase distribution to improve capacity (Figs. 14.16–14.17); implementing the policy of bus priority development to increase public transit trip contribution rates; speeding up the construction of bus station and optimizing operation management to promote the development of public transport vigorously (Fig. 14.18); enhancing traffic measures to play a role in the regulation of supply and demand requirements management (Fig. 14.19); and improving the signals and signs marking systems to regulate traffic. Specific measures taken include the following:

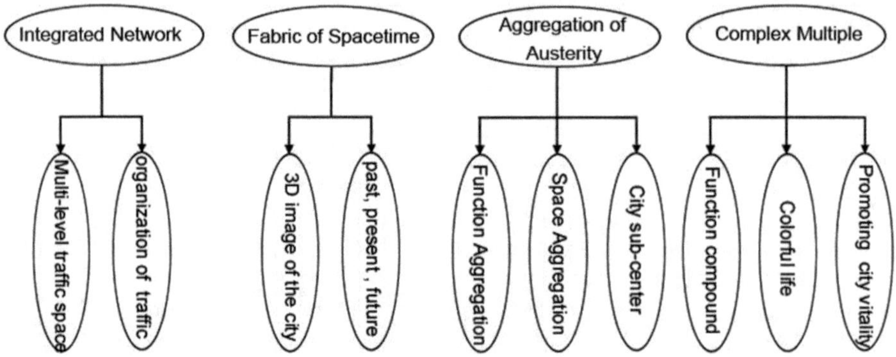

Fig. 14.11 Planning concept (Source: Study Authors)

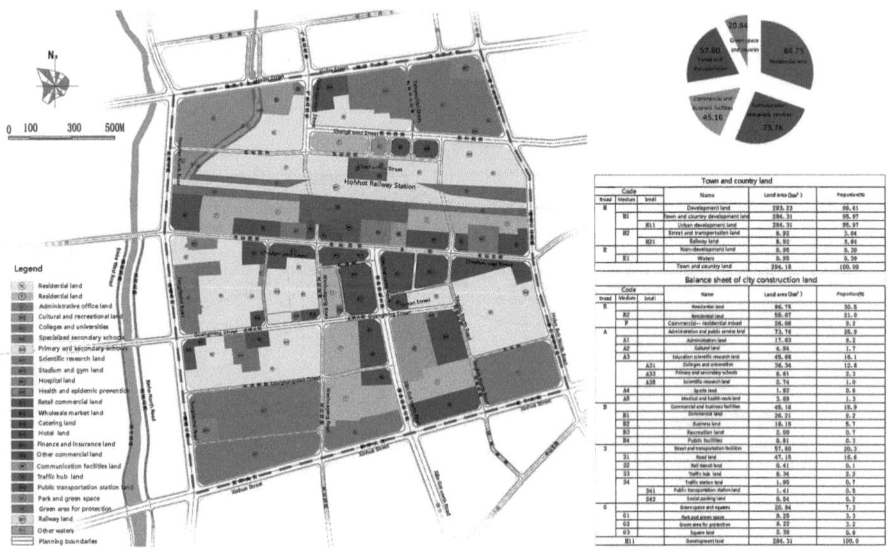

Fig. 14.12 Land use planning map (Source: Authors' Illustration)

Thirdly, coordinate the city scale, architecture scale and human scale together. City scale is closely associated with traffic and land utilization. In the improvement plan, some roads will be broadened, extended, or an underpass will be built in order to form a complete road network around the railway station. The construction of the North Plaza is also included in the plan to alleviate traffic pressure. As to adjustment of land utilization, the commercial area, which is separated from the railway station, will relocate towards south to open up a new commercial culture circle and solve traffic congestion problem (Fig. 14.20). The architecture scale mainly refers to the Station Square and the surrounding commercial buildings. The recent construction focuses on the renovation of railway station and Station Square and the development of underground space. According to the plan, the basement and the

Fig. 14.13 General layout planning map (Source: Study Authors)

ground will be converted into parking lots. Furthermore, a three-dimensional traffic network will be accomplished by the construction of a pedestrian landscape platform, which is connected with commercial buildings across the top of the road (Figs.14.21, 14.22 and 14.23). The human scale should be taken into account in the plan. Increasing public space of pleasant scale, re-organizing the urban landscape system can satisfy people's needs of interaction, leisure, playing and outdoor activities. Adding urban pedestrian system and landscape pedestrian corridor also contribute to make people's travel more safe, comfortable and convenient (Figs.14.24 and 14.25).

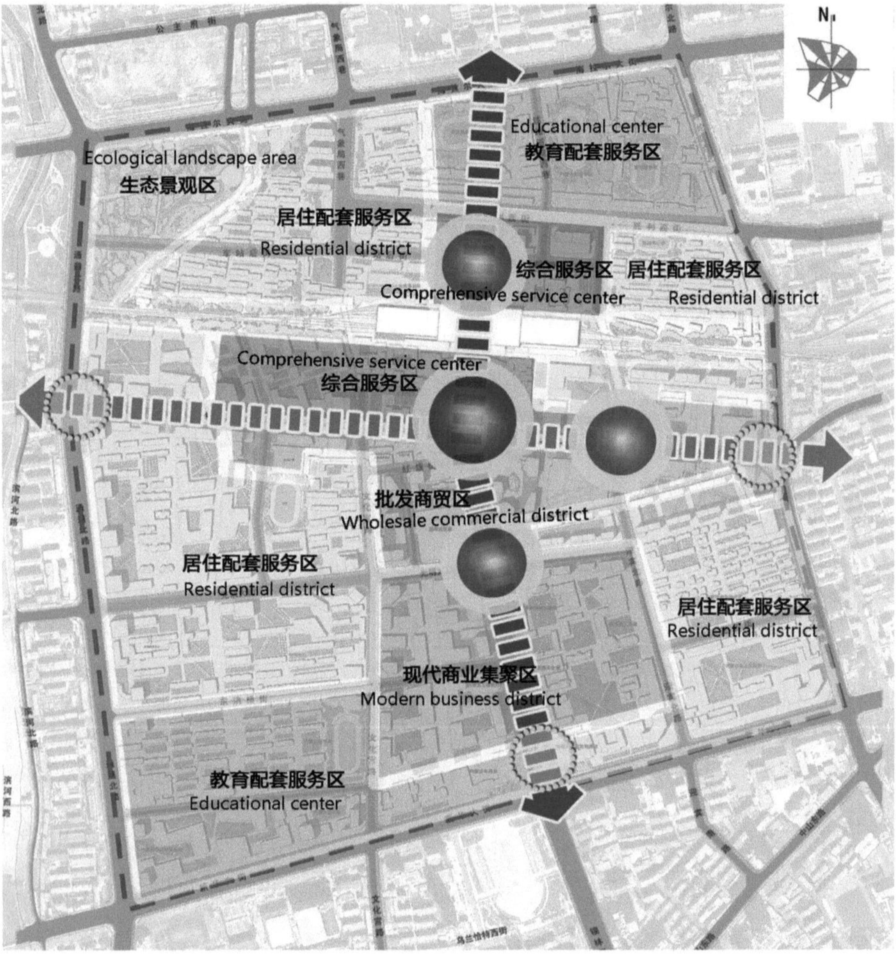

Fig. 14.14 the structure of land use map (Source: Study Authors)

14.5 Conclusion

Transformation of the old railway station and the surrounding areas are problems that lots of cities have been facing. Taking the complex urban spatial pattern combined with traffic improvement and land utilization as a solution could improve the urban transportation infrastructure, refresh the urban vitality in the old Railway Station area, improve the urban quality, and realize the coordinated development of railway station and city center.

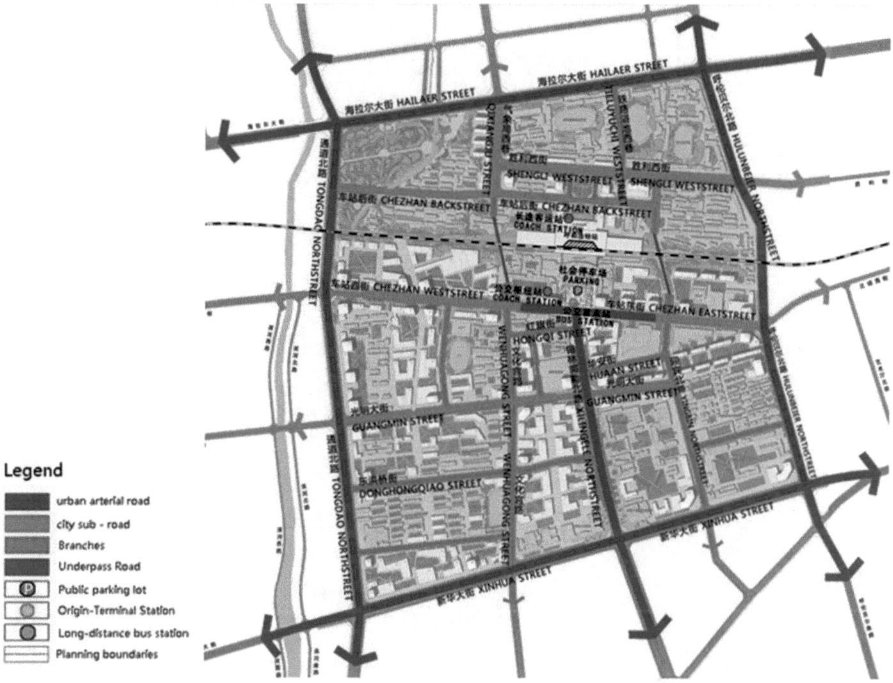

Fig. 14.15 Road and traffic planning (Source: Study Authors)

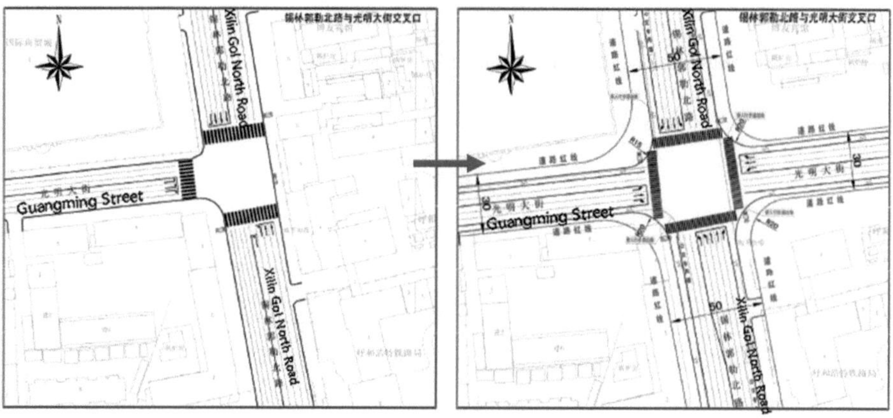

Fig. 14.16–14.17 Guangming Street-Xilinguole North Road cross improvement plan (Source: Study Authors)

Fig. 14.18 New bus stations planning (Source: Study Authors)

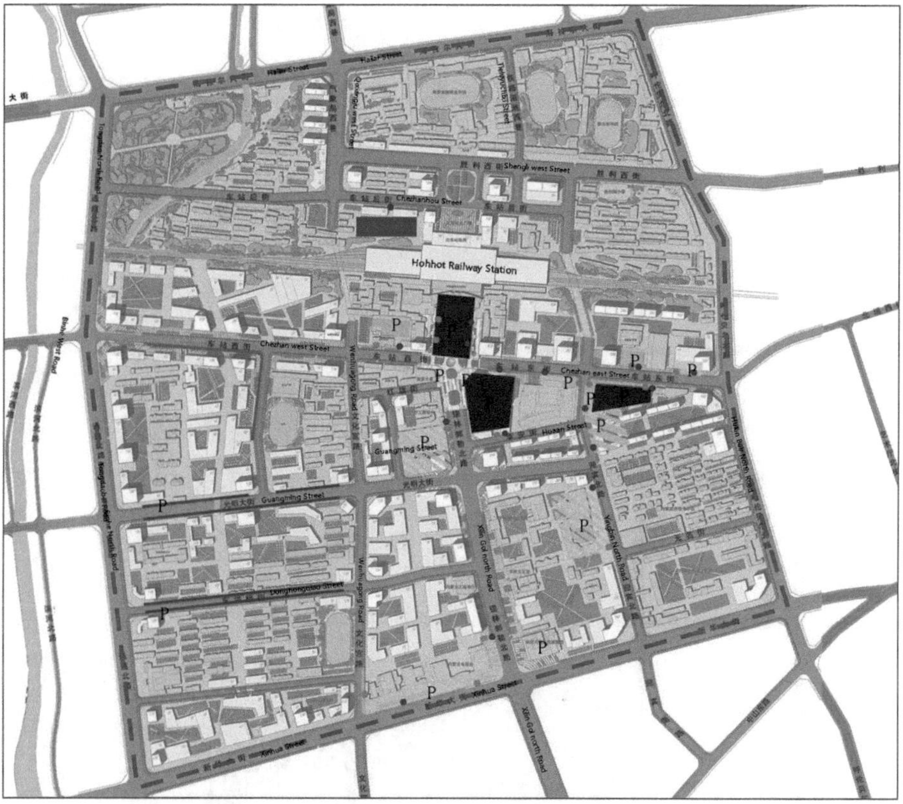

Fig. 14.19 Static traffic planning map (Source: Study Authors)

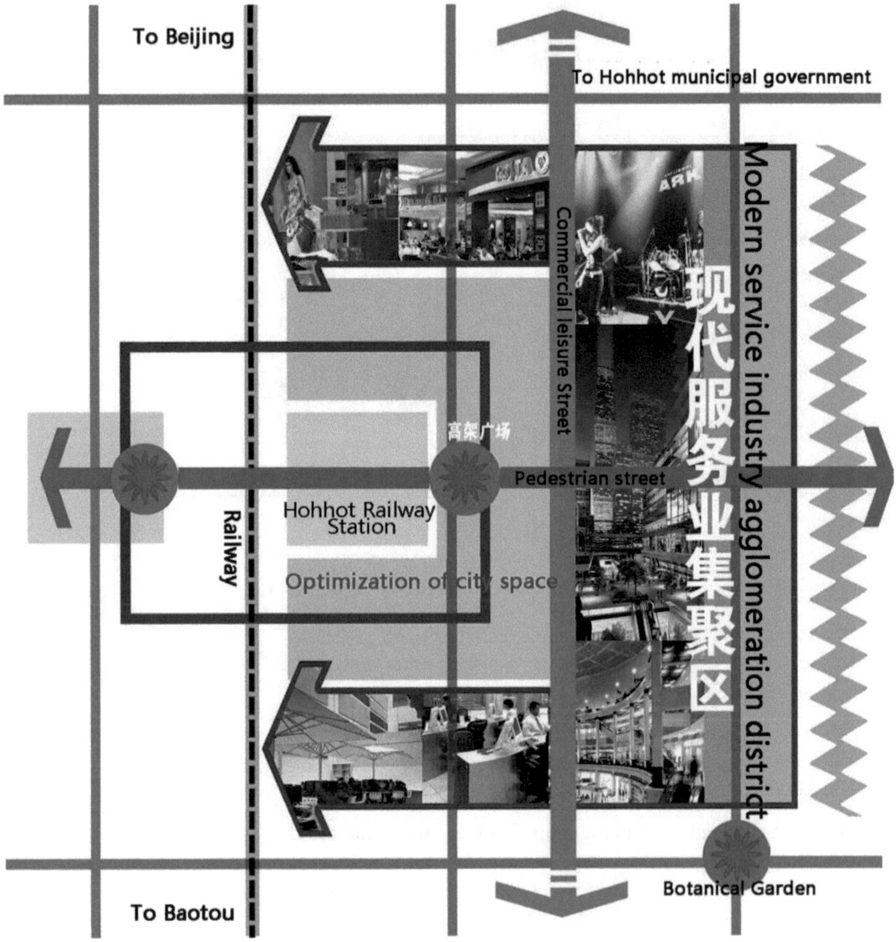

Fig. 14.20 City scale (Source: Study Authors)

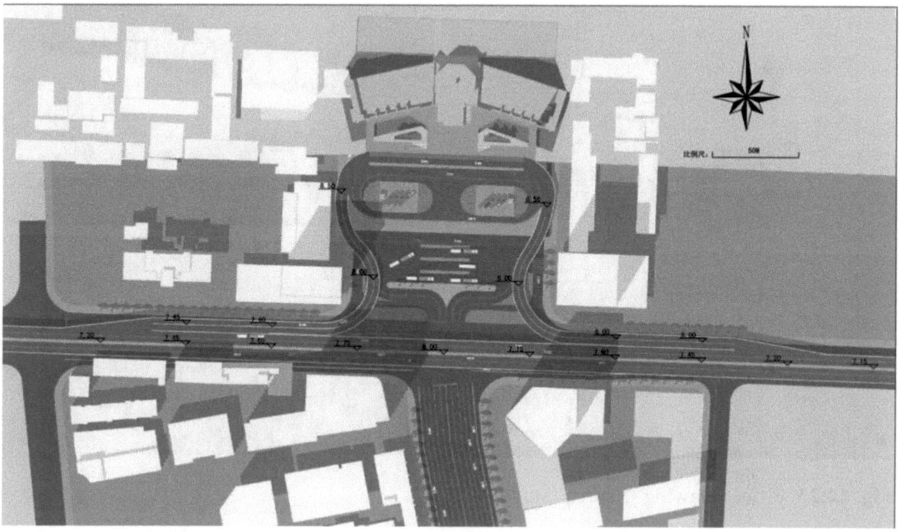

Fig. 14.21 Station square reformation plan (Source: Study Authors)

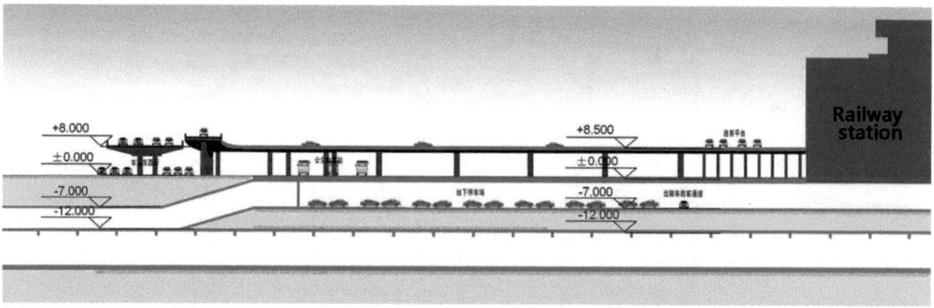

Fig. 14.22 Profile of the station square reformation (Source: Study Authors)

Fig. 14.23 Effect drawing (Source: Study Authors)

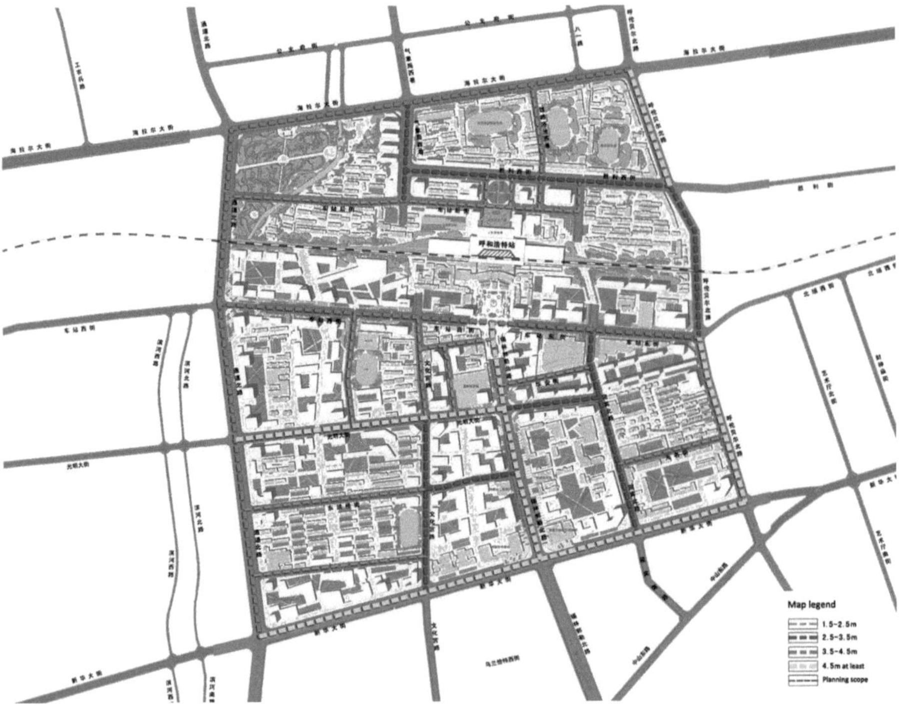

Fig. 14.24 Walking road system planning map (Source: Study Authors)

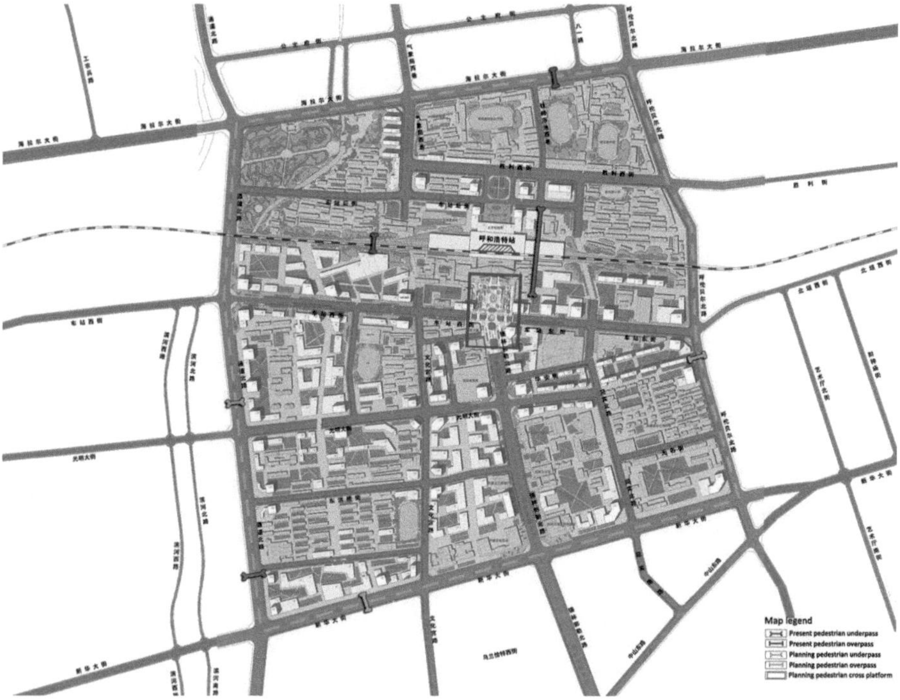

Fig. 14.25 Pedestrian crossing facility planning map (Source: Study Authors)

References

Jie, A., Zhao, W., Lei, C., & Xiaopeng, B. (2011). Generator theory's use on the urban vigor construction: Tianjin culture center design example [J]. *Planners, 27*(z1), 101–103.

Jun, Z. (2011). Urban design taking urban vitality demand as direction: Urban design in the southern city key sections of Nanjing [J]. *Urbanism and Architecture, 10*, 111–113.

Lian, W. U., & Wenzhuo, W. A. N. G. (2009). Several thoughts about the way of thinking on the urban renewal [J]. *Urban Studies, 16*(7), 51–54.

Zheng, W., Hongyuan, M., & Qiuyang, Z. (2012). Adjusting the urban vitality with architecture: Innovative architecture from and "innovative archi" [J]. *Huazhong Architecture, 30*(6), 15–18.

Zhou Bo, & Gong Rong. (2010). Renew old commercial street in the perspective of urban vitality [J]. *Journal of Hunan City University (Natural Science), 19*(2), 33–37.

Chapter 15
The Redevelopment Model of Multi-story Residences in the City Center

Hu Rui and Zhenyu Che

15.1 Introduction of the Background

15.1.1 The Development History of Multi-story Residences

15.1.1.1 Planned Economy Period (1949–1978)

In the context of the planned economy period, the development of city residences in China was directly restricted by the national industrial development policy. As a non- productive investment, the real estate industry was artificially suppressed. During this period, the urban housing construction investments and welfare housing system were implemented by the state-owned and collectively-owned enterprises (Chao Wang 2011). Considering economic factors, the fact that construction technology was not mature, and residential construction standards were relatively low, construction was mainly in the form of multi-story residence.

Because the building of the majority of urban residents' housing was the responsibility of each business unit, and city living facilities were scarce at the time, the units and enterprises built a set of service facilities in the scope of residential areas (commonly known as "enterprise-run society"). The workplace for the residents was not only the economic unit of society, but also served as the basic self-sufficient living tissue. In this period, the workers' village was a typical housing compound, such as the Kunming textile factory district and Shanghai Anshan village.

H. Rui (✉) • Z. Che
School of Architecture and Urban Planning, Kunming University of Science & Technology, Kunming, China
e-mail: lqzb19@sina.cn; czy9988@126.com

Q. Pan, J. Cao (eds.), *Recent Developments in Chinese Urban Planning*, GeoJournal Library 114, DOI 10.1007/978-3-319-18470-8_15

15.1.1.2 Since the Reform and Opening Up (1979–2000)

Since the reform and opening up, the planned economy system gradually transition to a socialist market economy system. Along with the rapid increase in urban incomes and rapid progress of urbanization, urban housing construction investment also shifted. Government-funded investment became more diversified and urban housing construction experienced rapid development (Chao Wang 2011). However, the economic capacity was limited and there was a housing shortage during this period, so a low standard of urban housing construction was an inevitable choice. Therefore, the construction of this period is also known as the low standard of residential housing, mainly in the form of multi-story residence, such as the Kunming Xinying community.

After the 1998 housing reform, the housing industry was positioned to become a forceful growth point of the national economy—by then the commercialization of housing had become the main theme of China's real estate development. After that, the quality of commercial housing construction was higher, and occurred mainly in the form of high-rise residence in the area of city center, which does not fall within the study scope of this article.

15.1.2 Update of City Center District

Large-scale development and redevelopment in the city center area has accompanied the process of rapid urbanization in China for the last 20 years. Residential districts that may have been located in the marginal zone of the city in the initial construction stage, but gradually became located in the city center district with the development of the country overall (Huizhi Geng 1999). A good location brought a variety of mature city facilities and conditions, and enhanced the land value. Thus in the process of old district transformation, there were many successful cases of replacing land uses and realizing a higher value of land. Also because of its special location, the older residential district was a key component of the city center's image, but the residential image itself was messy and the architectural appearance dull—therefore, during the process of rebuilding and image beautification, the residential districts often become the first target of reform.

Fig. 15.1 Location map of Kunming textile factory district

15.2 Cases Analysis

15.2.1 District of Kunming Textile Factory

15.2.1.1 Introduction

The Kunming textile factory was constructed in 1948. It is the largest factory in the old city of Kunming, covering an area of 56.34 acres (Fig. 15.1), and the peak number of employees reached more than 8000 people. The living area of Kunming textile factory covers an area of 25.7 acres. The worker dormitory consists of more than 60 buildings, with a total building area of 118,500 m^2, or about 3,200 households.

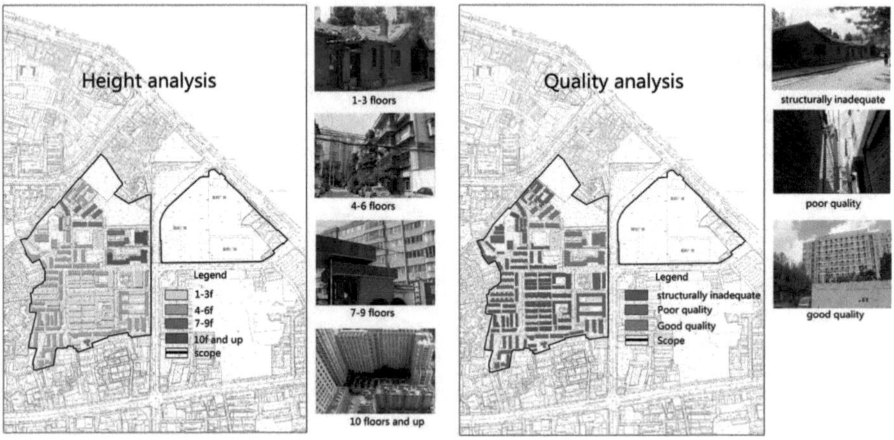

Fig. 15.2 Construction quality analysis

Since the area was completed a long time ago, the quality of the old buildings is very poor except for a new apartment built after 2000. There is internal traffic chaos and the space environment is disorderly, making this area in urgent need of improvement (Figs. 15.2 and 15.3).

The people living here are mostly older workers who worked in Kunming textile factory before—they have been living here for more than 20 years. Although the living area is small and the living condition is poor, they don't want to move out. Their education level is low, so it's hard for them to take part in modern competition (Chart 15.1).

15.2.1.2 Redevelopment Model

The Kunming textile factory announced bankruptcy in 2005. In accordance with the law, the bankruptcy property auction occurred by the end of 2006. A real estate company became the Kunming textile factory bankruptcy property buyers with 482 million RMB. Additional conditions for the purchase were that the real estate company be responsible for the reemployment of the workers and accept responsibility for the 3,200 employee households.

The real estate company made a new plan for Kunming textile district: They wanted to create a new CBD for Kunming and planned to make the largest international complex project in the Kunming core area (Fig. 15.4). The total construction was about 1 million m^2, and included international high-grade residences, 5A grade business groups, five star Hotels, 100,000 m^2 of a universal City Mall, high- grade leisure fields, international style pedestrian street, four city parks, and three centers: a “business center”, “commercial center”, “leisure and

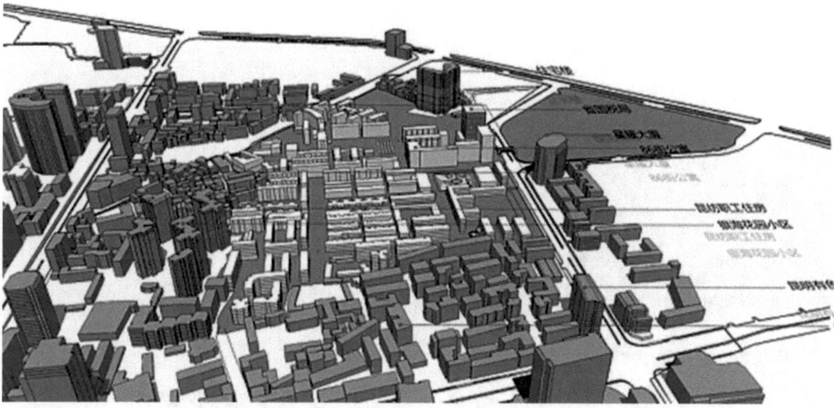

Fig. 15.3 Situation model of the site

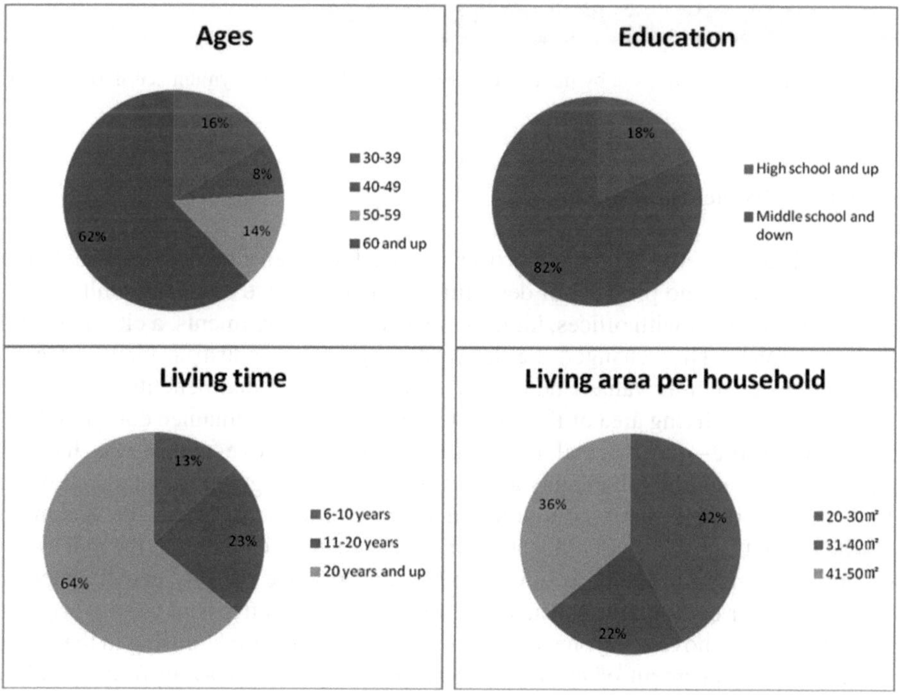

Chart 15.1 The basic conditions of the residents (total 100 questionnaires, available 86)

entertainment center." They comprehensively enhanced the overall image of the CBD and value of the region in central Kunming, and became the best international reception hall of new Kunming (http://www.ynjunfa.cn/jfcp/km_qcjy.asp).

Fig. 15.4 The plan of the site by the real estate company (http://www.ynjunfa.cn/jfcp/km_qcjy.asp)

15.2.1.3 Problems Summary

The work area of the Kunming textile district had been updated to completion because there was no pressure of demolition. The real estate company built a high-grade city complex with offices, high-grade residences, apartments, a city mall, and walking streets. They changed the decayed image of an industrial plant, released tremendous economic value, and created economic and social benefits.

However, the living area of Kunming textile district still remained comparable to its status before—the original inhabitants refused to move and it was hard to negotiate between the inhabitants and the developer. The government also did not find a solution; the social contradiction here is prominent. Since the Kunming textile district has an excellent location and great economic potential, the real estate company intended to knock it down simply, and reshape the city center space to maximize their own interests. The overall program of resettlement for the original residents was to move everyone to the new high-rise residential housing in the city's suburbs. The placement of the resettlement housing was far away from the city, which would cause a lot of difficulty for the Kunming textile workers' work and life. The original residents, then, were reluctant to move because they did not enjoy the dividends of the differential land rent. This "updating" model posed by the developers has some problems, therefore, and needs to be improved further.

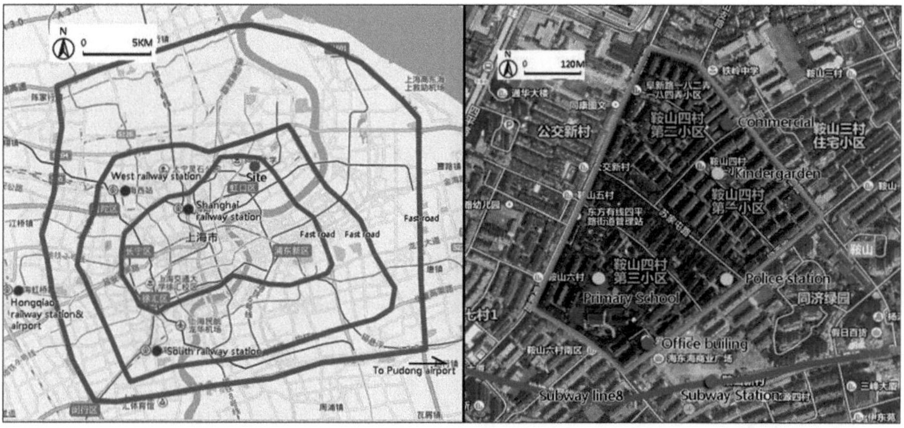

Fig. 15.5 Location map and texture of Anshan four village

15.2.2 The Shanghai Anshan Village

15.2.2.1 Introduction of Anshan Village

Anshan village is located in Yangpu district of Shanghai, close to Tongji University (Fig. 15.5). It is an old-fashioned village built in 1970s, and now, after continuous expansion, has eight villages. It is one of the earliest worker villages in Shanghai, and once had the largest area and population (it is said that in the 1970s, the population had reached one million). Anshan four village is part of Anshan village, with a total area of about 15 ha. The renewal strategy of Anshan four village was a successful case in China.

15.2.2.2 Redevelopment Model

The redevelopment model of Anshan village is different from large demolition, and instead took a more gradual approach. The specific methods consisted of adding layers, making apartment layout adjustments, and changing flat roofs to slope roofs.

1. Add layer, apartment layout adjustment

For the old worker village, there were many problems with the apartment layout—the area of every apartment is small and the area per person is far below the residential standards; additionally, the layout is unreasonable in that the function of room is disorderly and confounding with a lack of an independent kitchen and toilet system.

The transformation of Anshan four villages including taking in to account an analysis of the original dwelling size, structure, equipment and pipeline, building space and the wishes of the original inhabitants. Consideration was taken to

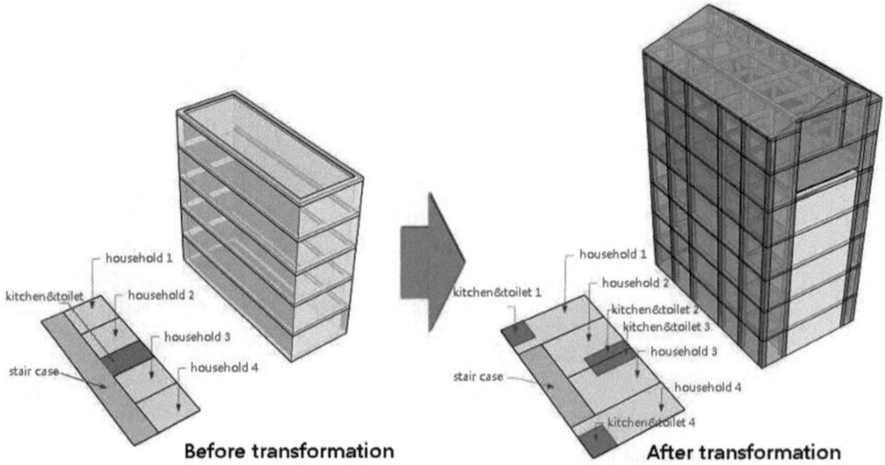

Fig. 15.6 Gradual renewal strategy of Anshan four village (Min Zhao 2010)

minimize changing the building's load-bearing structure, to ensure the premise of seismic fortification standards. The renewal project increased the area of every apartment, increased the room number, and improved equipment and facilities. Additional improvements included additional construction of the roof, increasing the attic or semi basement, and forming staggered combinations: extended plane, two households changing to one household, or three households changing to two households in a staircase (Fig. 15.6).

2. Flat roof change to slope roof

The "flat to slope" project refers to the change in residential flat roofs to the slope roofs—under the condition of the buildings' structure and what the sunshine spacing allows—and was a renovation and decoration project for the external facade to improve the building function and landscape effect. After years of development, the project moved from the architectural appearance reconstruction called "wearing new clothes, wearing new hats" at the beginning, to the comprehensive transformation project as "wearing new clothes, wearing new hats, change the inner, repair the environment, reducing energy consumption, and saving resources". The project targeted five improvements: "solve the roof leakage, improve housing performance, improve the living environment, promote community civilization, forming the new city landscape" (Min Zhao 2010), so that people could live in new homes and not need to move (Fig. 15.7).

3. Residential renovation

Anshan four village improved the residential environment by a series of effective measures, such as constructing pavement, repairing underground rain sewage drainage system, adding drainage pump, residential barrier free facilities, creating a

Fig. 15.7 The effect of "Flat roof to slope roof" project (**a**) Before transformation (photo from website), (**b**) After transformation (photo by self)

green environment, and transforming or adding convenience public facilities. These measures achieved a beneficial social effect.

15.2.2.3 Experience Study

Evolutionary redevelopment has a lot of practice in foreign countries—there are many successful cases in Japan and Germany, with a lot of research results. Professor of Tokyo University Shuichi Matsumura summarizes regeneration method diversity in European and American countries, with analysis of transformation ways and sources of funds, professional organizations, and public funds in his book "Residential District Regeneration." He also built a regeneration system in Japan (Matsumura 2008).

The range of Anshan village is wide, so it cannot be knocked down and reconstructed at one time. However, the building construction is old and the desire to improve the living conditions of the residents is very urgent. Evolutionary redevelopment improved the living conditions of residents and did not destroy the city texture or the city's context continuity. Regarding the transformation of the Anshan four villages and the renovation work, Professor Zhao Min of Tongji University identified four aspects that led to its success: "the capital management, public participation, ease the conflict between the residents, a reasonable distribution of benefits" (Min Zhao 2010). The specific contents can be found in the relevant papers.

15.2.3 Kunming Xinying Community

15.2.3.1 Introduction

Kunming Xinying community is located between No. 1 and No. 2 fast road. It was developed in the early 1990s., Including the south area and north area, it covers an area of 214 acres—it is a large community with over 12,000 households (Fig. 15.8). Once upon a time, it was known as "the rich area", but with the development of city, modern residential facilities are more desirable and the living conditions are much better. The original wealthy people who lived here have gradually moved out. Now, the Xinying community has become an ordinary member of the Kunming old district. It cannot be compared with new modern development communities from the standpoint of environment, health and security. However, the convenient traffic and desireable living facilities still attract many residents. The decline of Xinying community reflects the rapid development of Kunming city. In just 20 years, the high-grade areas have become old ordinary residences, so what strategy could be taken for revitalization?

15.2.3.2 Problem Summary

There is a large area of old multi-story residences built in 1990s, which is very common in other cities. Its main problems can be summarized by the five aspects as follows: (Hao Long 2006).

1. Plan layout

   Housing is built in rows to achieve a large volume rate under the requirements of lighting, ventilation and orientations. The rigid form leads the residential space structure to appear monotonous, and the building appearance follow the same pattern—the character of the development, therefore, is not obvious. Furthermore, the building density is close, green rate is low, and there is a lack of public activity space.
2. Apartment layout

   The types of apartment layouts are not abundant enough—there are only small and medium-size apartments with the largest area being only 120 m^2. The layout is not reasonable in that the room function is disorderly, the kitchen is small, and the facilities are simple and crude.
3. The road traffic and parking

   The internal roads in the community are narrow and the traffic is disorderly as pedestrian and vehicles are mixed. The increasing owner of private car was not considered when constructing the community, so the parking area is particularly lacking. As a result, vehicle parking is disorderly and blocks traffic which affects fire protection safety (Fig. 15.9).
4. Outdoor landscape environment

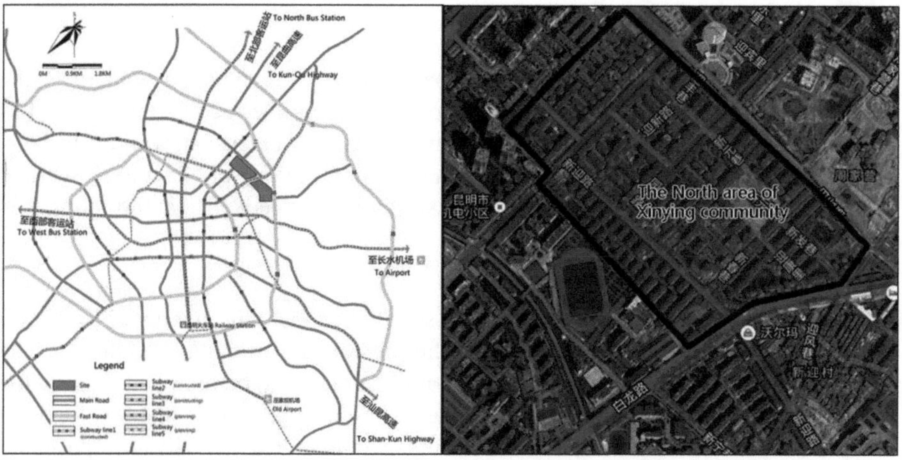

Fig. 15.8 The location and texture of Xinying community in Kunming

Fig. 15.9 The internal roads traffic and parking

Layout in the form of ranks and close building density make it so there is no large-scale green space in the community. The landscape is simple in form and located between houses at the both sides of the road—the landscape lacks features and themes.

5. Property management

   The whole community is separated into several independent groups artificially, which causes many internal broken roads and traffic inconvenience.

15.2.3.3 Redevelopment Model

The location of Xinying community doesn't have the same advantage as the Kunming textile area, so it's impossible to take large-scale reconstruction. And as 1990s commercial housing, the quality is much higher than Anshan workers village. As a high-grade residence, which was once brilliant, facing an inevitable decline, what sort of update strategy for the renaissance do we need? The author considers that:

1. City level

   At the city level, the transformation should connect with the upper-level city planning positively and combine with the subway construction in Kunming. The part of housing located in the subway station area should be removed. Taking the TOD development model as a guide, there should be high intensity development in the subway station area. These measures would increase the livable space and create large-scale public green space, making the sites around subway station dynamic hotspots.

   For the settlements located along with the main street of the city, the enclosing wall between the community and city should be pulled down. The bottom floors should be transformed to commercial and be open to the city, forming a dynamic city interface.
2. The community level

   From the community level, the man-made barriers in the community should be broken, and partial residences with the bad orientations or security risks should be demolished. The residents whose houses are demolished could move to the high-rise buildings constructed in the railway station area. This can create large-scale open spaces, and the part underground can include an underground parking garage. The part on the ground can include a community garden to form lively landscape greening in the community.

   The public service facilities and public open space should be built to improve the quality of life for the residents stimulate the vitality of the area. These measures could promote the community environment to become more sustainable.
3. The construction level

   There are a lot of successful experiences of evolutionary redevelopment from home and abroad. Specific measures include "adjustment (of) the residential apartment layout, comprehensive renovation of flat roof change to slope, facade renovation of building, green building technology applications". These effective measures should be applied to make the housing built in the 1990s to meet the modern living requirements and improve the overall quality of residential environment.

15.3 Theoretical References

15.3.1 The Land Rent Theory

Rent of land is the land owner with land ownership's ability to obtain compensation from land users. The absolute rent refers to the idea that all land use must pay the rent regardless of the condition between the lands. The differential land rent refers to different rents relying on natural conditions and land exploitation and management. Differential land rent I is due to geological conditions of the city and the geographical environment difference. Differential rent II relies on the additional investment in the same piece of land produces.

Bid rent theory was proposed by W. Alonso in 1960. According to this theory, land users, whether they be retail, office, or residence, all compete for the most accessible land within the CBD. The amount they are willing to pay is called bid rent. This can generally be shown in a "bid rent curve" (Fig. 15.10), based on the reasoning that the most accessible land, generally in the centre, is the most expensive land (Alonso 1964).

Kunming textile factory district is located in No. 1 fast road of the city, Therefore, the requirement of functional replacement and the overall upgrade is imperative. However, Anshan workers village and Xinying community are located between No. 1 fast road and No. 2 fast road, where the replacement demand is not strong.

15.3.2 TOD Development Model

"Transit Oriented Development" (TOD) model is a mixed-use residential and commercial area designed to maximize access to public transport. A TOD neighborhood typically has a center with a transit station or stop (train station, metro station, tram stop, or bus stop) (Fig. 15.11), surrounded by relatively high-density development with progressively lower-density development spreading outward from the center. TODs generally are located within a radius of one-quarter to one-half mile (400–800 m) from a transit stop, as this is considered to be an appropriate scale for pedestrians (Calthorpe 1993).

At present, Kunming is vigorously developing the city subway transportation, BRT and other large capacity public transportation. Large-scale infrastructure construction has brought new opportunities for the old residential district renewal in the center city district. Kunming Xinying community has a large volume and it should be combined with the construction of city subway transit station, taking the TOD concept as guidance, to explore a more effective and updated model.

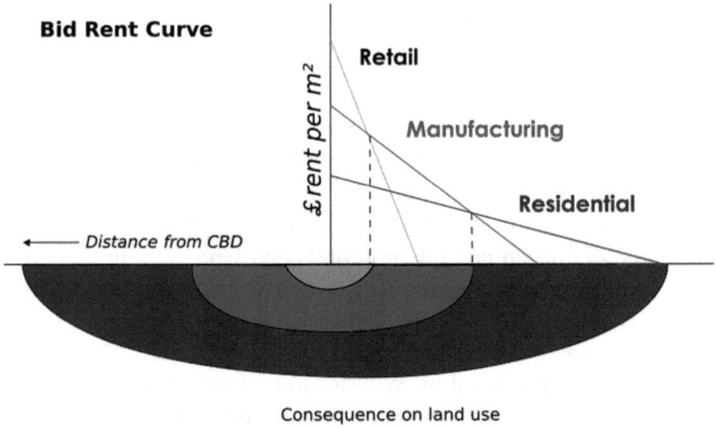

Fig. 15.10 Bid Rent Curve and land use pattern (Alonso 1964)

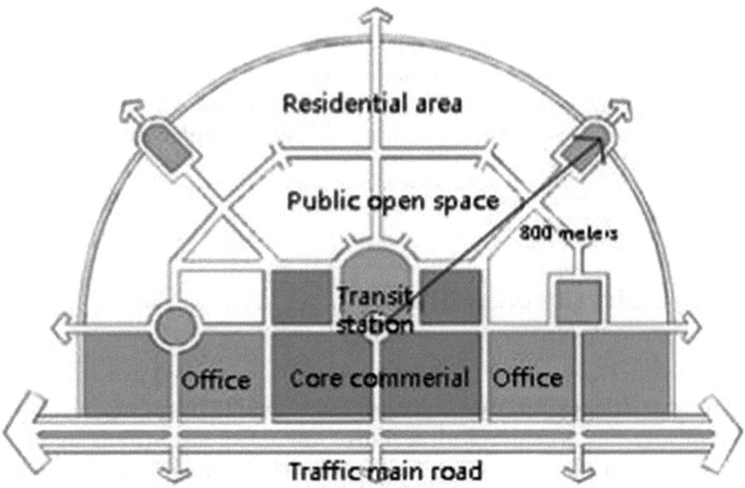

Fig. 15.11 TOD development model (Calthorpe 1993)

15.4 Model Summary

15.4.1 Upgraded Redevelopment

The location of this type multi-story residence is much advantageous, but the construction is old, the quality is poor, and the living conditions are very bad. In the reality, this kind of residential area is difficult to preserve in the city center district—they often take the upgraded redevelopment model, make the overall demolition and reconstruction. The government needed to reshape the image of city center area, the developers wanted to obtain enormous economic benefits through the replacement of function, and local residents hoped to improve their living environment.

This upgraded redevelopment model, however, is simple and crude, and often causes social contradictions as redevelopment is hard to carry on. Investigating the reason for this, local residents have to obtain the differential land rent from the relocation. The author considers that development initiatives should not simply relocate the local residents to the outskirts resettlement residential area, that rather they should be given more choice. For the young people, they should have the chance to stay in the city center area, and participation in city competition; for the elderly, redevelopment can be combined with the construction of endowment real estate, to enable them to live an easy life in a beautiful environment. In short, the upgraded redevelopment is not only for the city space to upgrade, but also for social upgrading.

15.4.2 Evolutionary Redevelopment

The old workers villages often take this model. The multi-story residence locational advantage is not very obvious, and since the volume is large, it is difficult to tear down for complete reconstruction. But the construction is old, the quality of housing is low, the room layout doesn't meet the needs of modern life, and the ruined appearance hinders the city image.

Evolutionary redevelopments in China are often dominated by the government—the city's low-income residents have had improved living conditions through initiatives such as "apartment layout adjustment, flat roof change to slope, (and) residential renovation." More mature and effective technical means have been applied, and the residents' investment is relatively little. In the western developed countries, residential evolutionary redevelopment has already formed during industrialization. Our country should attempt to set up progressive organization to promote the old residential district renewal with this type of industrialization, and guide the promotion and gradual implementation of the old residential district renewal industrialization.

15.4.3 Revitalization Redevelopment

With the development of the times, the early construction of commercial housing has begun to decline. The quality of this period's multi-story residences is better than the workers village, but also is faced with many similar problems. How to get rid of decline and revitalize the residences would become an urgent issue in future years.

The paper discussed the possible methods of revitalization redevelopment at the city level, community level, and construction level. The revitalization redevelopment should take advantage of the successful experiences of both upgraded redevelopment and evolutionary redevelopment. The residential buildings located in the land with high value (like in the transit station area) should be demolished and changed to commercial or office building. The majority of the residential building should be preserved, and use the natural, evolutionary ways to improve the living quality. Finally, revitalization needs to make the fading residential community to keep pace with the new. The specific operating mechanism between local resident, government and developers need further study.

15.5 Conclusions

Multi-story residence has played an important role in the residential development of our country. With the development of the times, the pressure of redevelopment in the city center area is increasing. The multi-story residence is facing the fate of being renovated or even being demolished.

Through the analysis of three different types of the multi-story residence cases, the author summarized three different redevelopment models such as upgraded redevelopment, evolutionary redevelopment, and revitalization redevelopment. These models give some guidance to the redevelopment of the multi-story residence in the city center. It answers the questions of "which part should be demolished and which part should be preserved?" and gives some effective measures to improve the preserved living condition. These models should be applied to the multi-story residence in the city center according to the specific situation.

Acknowledgement This study was supported by a grant from the National Natural Science Foundation of China (NNSFC), No. 51168019 and No. 51368023.

References

Alonso, W. (1964). *Location and land use: Toward a general theory of land rent* [M]. Cambridge, MA: Harvard University Press.

Calthorpe, P. (1993). *The next American metropolis: Ecology, community, and the American dream* [M]. Princeton: Princeton Architectural Press.

Chao Wang. (2011). *The study of urban design strategy of old residential renovation in cities.* Tsinghua University master of architecture professional degree thesis, Beijing.

Hao Long. (2006). The research on reforming methods of old urban residence [J]. *City Architecture*, 21–23.

http://www.ynjunfa.cn/jfcp/km_qcjy.asp

Huizhi Geng. (1999). Impetus mechanism for renewal of urban centers in China [J]. *Urban Planning Forum, 1999*(3):27–31.

Matsumura, S. (2008). *Residential district regeneration* [M]. Beijing: China Machine Press.

Min Zhao. (2010). On the gradual urban renewal: Theory and practical strategies [J]. *International City Planning*, *25*(1):24–32.

Chapter 16
Elementary Analysis of the Impact of Large-Scale Sports Events on Space Regeneration of the Hosts: Cases from the 2008 Beijing Olympic Games and 2012 London Olympic Games

Ying Gu and Ze Zhang

Large-scale sports events provide a strategic means to market a city because they bring excellent opportunities for the hosts to promote their value and show themselves to the world. If cities host large-scale sports events successfully, they will enjoy plenty of benefits, such as increasing economic development, publicizing their culture, and raising their profile. Therefore, a large number of cities seek to hold large-scale sports events. At the same time, large-scale sports events can influence the urban spatial pattern development. After large-scale sports events, cities must return to their normal state—both in social and economic aspects. To comprehensively analyze the impact of large-scale sports events on hosts, urban spatial patterns of development should be considered along with factors, such as the economy, culture, environment, and society.

This paper focuses on the impact of large-scale sports events on the space regeneration of the host. Space regeneration means restoration of some old parts of the city to turn them into prominent, vivacious urban spaces. Space regeneration also creates opportunities for potential urban sub-centers to develop. To provide a thorough inquiry into the influence of large-scale sports events, this paper examines four areas: urban spatial pattern regeneration, land value increment and change of land use, infrastructure construction due to large-scale sports events, and subsequent utilization of relative facilities. To detail the impact of large-scale sports events on hosts' space regeneration, this paper discusses two cases—the 2008 Beijing Olympic Games and 2012 London Olympic Games—both of which are typical cases in which governments used large-scale sports events as space regeneration tools. While 2012 was not the first time London held the Olympic Games in 2012, the differences in London compared with cities that made their first attempt were not in areas that affected the topics discussed in this paper.

Y. Gu (✉) • Z. Zhang
Department of Urban Planning, Tongji University, Shanghai, China
e-mail: 931782016@qq.com

Q. Pan, J. Cao (eds.), *Recent Developments in Chinese Urban Planning*, GeoJournal Library 114, DOI 10.1007/978-3-319-18470-8_16

16.1 The Spatial Arrangement of Large-Scale Sports Facilities Influences Urban Spatial Pattern Regeneration

Because of external effects, large-scale sports events influence their hosts' development. Hence, when selecting locales for large-scale sports events in a city, plenty of factors should be taken into consideration (Roche 2000a). First, the locales must meet the events' demands for area, terrain, transportation, and services. Next, they should be consistent with the area's urban development strategy and assist in the economic, spatial, and social aspects of urban development. In general, sites of large-scale sports events are typically located in key districts ready for regeneration or new districts important to urban spatial expansion. Therefore, stadiums, infrastructure, and matching facilities for large-scale sports events would be a part of the urban basic space structure. The land development on the periphery of the stadiums, infrastructure, and matching facilities would prompt the formation of comprehensive urban areas with full functionalities.

16.1.1 Overall Spatial Arrangement of Stadiums for Large-Scale Sports Events

In general, different cities take different approaches to locate stadiums for large-scale sports events. For example, they may take city conditions and scale and influence of sports events into account. Hence, no particular pattern of spatial arrangement exists. Nevertheless, the overall spatial arrangement of stadiums can be divided into three patterns: distributed, centralized, and a combination of distributed and centralized patterns (Metropolis Commission 2002).

A distributed pattern usually takes advantage of existing stadiums across the city, which reduces the costs and time needed for large-scale stadium construction. As existing stadiums are distributed across a city, a distributed pattern is better for relieving transportation stress than a centralized pattern. A distributed pattern also requires less land for construction. On the one hand, by using existing stadiums across a city, a distributed pattern can select stadiums flexibly and easily integrate them into an urban development structure. On the other hand, a distributed pattern has only a limited impact on the urban spatial pattern transformation. Therefore, the city has few opportunities to take advantage of the assembling effects of large-scale sports events to give rise to urban public center formation.

A centralized pattern tends to concentrate stadiums in an urban center or marginal area of an urban center. In this pattern, cities always intend to promote urban regeneration of run-down areas and improve the urban function layout. The centralized pattern has the distinct advantage of offering high-efficient organization and administration, as most of stadiums are located in one region, and therefore, travel time between facilities is reduced. The facilities can also serve as large-scale public spaces after events. Furthermore, because of the enthusiasm for sports after such events, these spaces could add to the urban center's cultural facilities. Run-out districts may also experience accelerative regeneration.

However, a centralized pattern has many challenges. The pattern requires construction of many new stadiums to provide the infrastructure and matching facilities in a short period of time in a concentrated area. As a result, a commensurate amount of land must first be acquired. However, after the construction area is selected, the value of land nearby the facilities starts to increase. Additionally, traffic becomes concentrated in the development areas, which can cause traffic jams and lead to inconveniences for residents and tourists. As well, after the events, many stadiums suffer deficits due to lack of use and questions of how to operate and use the new large stadiums after the events arise. This paper examines two cases with the centralized pattern: the 2008 Beijing Olympic Games and 2012 London Olympic Games. The 2008 Beijing Olympic Games took place in three districts and the center, which was the Beijing Olympic Center. The 2012 London Olympic Games was located in the main district in urban center (Trafalgar Square) and three districts, which were all within 10 km from the sight of the urban center.

A combination of distributed and centralized patterns has the advantages of the distributed and centralized patterns. It concentrates the main facilities in the core area, which is more convenient for operators in the process of construction and management. The core area also becomes the characteristic urban image. The combination pattern also includes other facilities distributed across city. This allows people who attend or watch large-scale sports events to be distributed across stadiums or courts in different areas, relieving traffic jams in the core area. Taking advantage of the existing facilities in city, this pattern actively interacts with the city and promotes an organized, unitary, and dynamic urban space structure. Examples of when the combination development pattern was used include the Olympic Games in Seoul, Athens, and Barcelona.

16.1.2 Large-Scale Sports Facilities Influences Urban Spatial Pattern Regeneration

Large-scale sports events drive run-out area regeneration and new district construction, namely a "1 × 1" pattern and a "1 + 1" pattern, leading to urban spatial pattern regeneration. A "1 × 1" pattern concentrates on improving existing urban spaces and facilities. Existing facilities, such as stadiums and hotels, may fall short of the demands of large-scale sports events, so necessary maintenance and infrastructure construction are indispensable in the development process. A "1 + 1" pattern concentrates on new facility construction in new urban districts, which impels urban spatial patterns to transform (Lin 2002). Though large-scale sports events have apparent commercial and public intent, they can influence urban space development. Resulting from improvements in existing urban centers or the formation of new urban centers or sub-centers, both patterns contribute to multi-center city structures and new functional center construction.

A “1×1” pattern usually affects the run-out area regeneration that has a government lead and investment from developers. Improving the quality of public spaces and organizing urban public center systems, a “1×1” pattern emphasizes certain locations for attracting public recreational activity. As well, stadiums essential for the sports events, hotels, restaurants, shopping malls, and other auxiliary facilities are constructed at the same time. With fully equipped facilities, growing numbers of people want to return to the inner city to live or entertain, which makes the inner city vibrant. In turn, more commercial activities in inner city and run-out areas contribute to a more prosperous economy. When the land in the inner city is renovated, less undeveloped land is needed and urban sprawl is alleviated. Take the 2008 Beijing Olympic Games for example. By holding the Olympic games, the city aimed to become a “green city” with less pollution and regeneration of traditional industrial districts. To meet the goal of being the “green” Olympic games, the Beijing government took a series of anti-pollution actions. After 5 years, about 150 factories were close or moved away. With the traditional factories out, the functional replacement had a positive effect. It has provided development space for cultural and creative industries and the entertainment industry in the Shi Jing Shan District. By 2010, the Shi Jing Shan District, a formerly industrial district, had become a comprehensive public center for the west district of Beijing. Recently, the Shi Jing Shan District undertook comprehensive service functions, including administration, commerce, industrial heritage conservation, and tourism with an industry theme. Similarly, Stratford, England used to consist of slums and an industrial wasteland. However, it became the primary regeneration target for the 2012 London Olympic Games. Before the LOCOG (The London Organizing Committee of Olympic Games) took over Stratford, 75 % of its soil was badly polluted by gasoline, petroleum, and heavy metals. Before constructing the Olympic Park, many projects were undertaken to restore the land, including pushing over all buildings, purifying polluted soil, and dredging rivers. As a result 80 % of soil where many of the Olympic stadiums, public facilities and houses were built underwent remediation. Consequently, Olympic Park has become the largest city park in Europe and many companies and residents have settled there.

A “$1 + 1$” pattern always has some effect on urban spatial pattern development by infusing more investment and construction in new focus areas. The 2008 Beijing Olympic Games took this approach. The overall plan was developed before Beijing’s successful Olympic Bid, and it did not take the Olympic games into consideration when conforming the plan to city’s projected development demand. In 2004, to maintain consistency with the “new Beijing new Olympic games” strategy, the *Beijing Municipal Institute for Economic and Social Development* (BMIESD) recommended readjusting the layout of urban functions. BMIESD promoted a plan to evacuate certain central functions and reconstruct space structures. The plan had particular measures, such as advocating for industry and housing in new functional zones by preferential policy and high-quality public service facilities.

The Beijing Olympic Center includes 600 ha of a city park located alongside the central axis in the north section of Beijing. Thus far, Beijing has been developed based on the existing city structure and following the symmetrical layout radiating from the Forbidden City. Judging from this plan and the realistic construction action, the layout of the 2008 Beijing Olympic Games continued Beijing's original urban spatial pattern and contributed to a new cross urban space structure. The Beijing Olympic Center was a new comprehensive urban public center and the following steps were taken to fulfill the plans. First, the Beijing Olympic Center has various functions. Sports facilities including a stadium, natatorium, and training hall are part of the Beijing Olympic Center's core. The core also includes the China National Convention Center and other conference centers and exhibition centers. Cultural facilities, entertainment facilities, and shopping malls exist around the sports center as well. Second, transportation is readily available. The Beijing Olympic Center is close to other Olympic facilities so it is convenient to travel from one to the next. Third, the infrastructure around the center is fully equipped because the neighboring areas were built up around it. Fourth, the area around the facilities is densely populated. After the games ended, the residents became the potential consumers of the facilities constructed for the games. Fifth, new communities, dormitories, and office buildings were developed, which help make the area more suitable for living (Fig. 16.1).

Nevertheless, the plan was not completely successful because some key functions remained in the old city during the 2008 Beijing Olympic Games (Fig. 16.2).

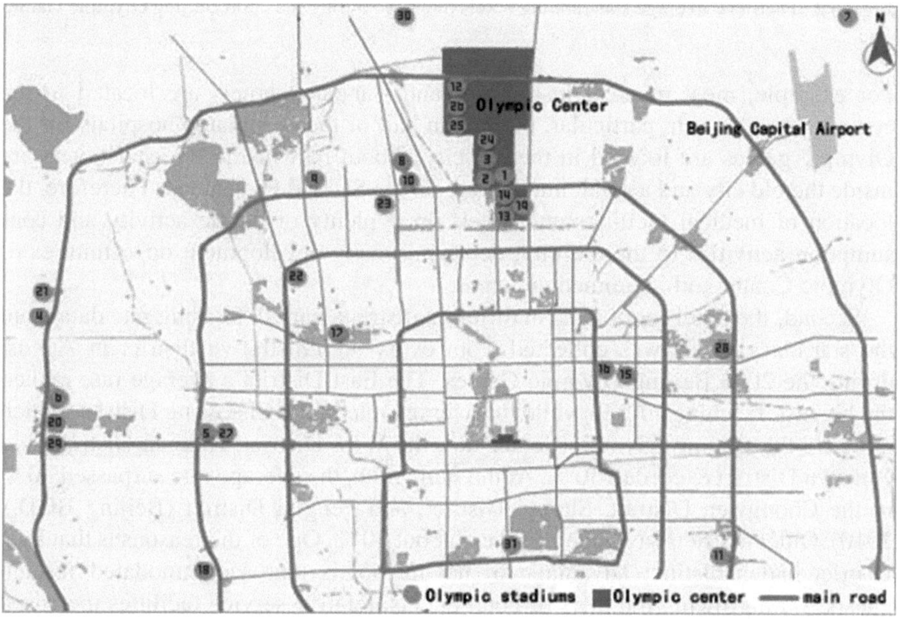

Fig. 16.1 The layout of Beijing Olympic Games stadiums (http://www.onegreen.net/maps/HTML/25743.html)

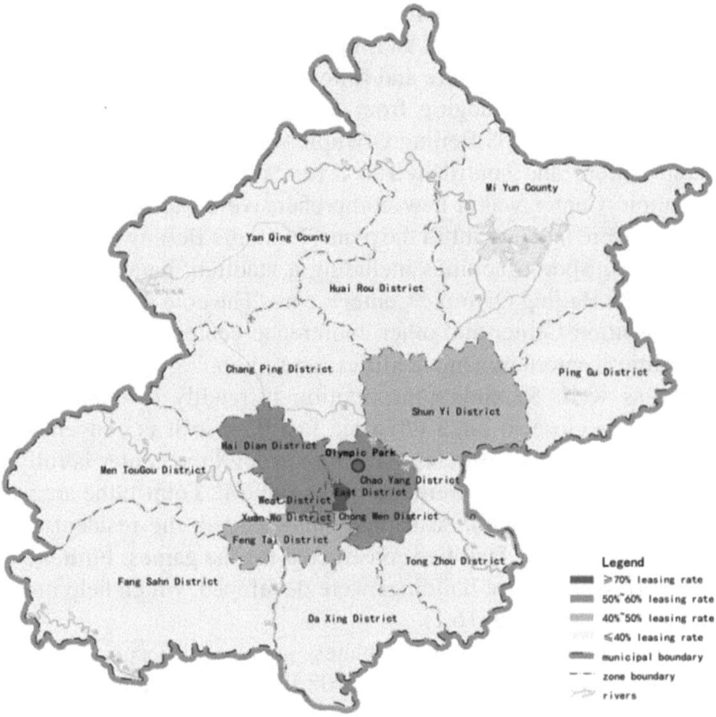

Fig. 16.2 Districts' average star-ranking hotels rent ratio during the 2008 Beijing Olympic Games

For example, most medical institutions and star-rated hotels are located in the central urban area. In particular, more than half of the designated hospitals for the Olympic games are located in the old city. About half of the five-star hotels are inside the old city and a small number are on the Second Ring Road. Therefore, the location of medical facilities and hotels drew plenty of public activity and consumption activities to the old city, depriving many development opportunities of Olympic Center and its immediate areas.

Second, the hotel room rates in different districts varied. Leasing rate data from the star-class hotels was collected from every administrative district in August during the 2008 Beijing Olympic Games. The East District's average rate ranked the highest, reaching 77.2 % while the average rate in the Chaoyang District, which is where the Beijing Olympic Center lies, the West District, Haidian District, and Xuanwu District exceeded 50 %. At the same time, the average rate surpassed 40 % in the Chongwen District, Shunyi District, and Fengtai District (Beijing BODA 2010). Other districts' average rate were about 40 %. One of the reasons is that East District had a distinct advantage of having hotels that accommodated foreign guests. With growing numbers of hotel clients, relative service facilities increased demand, such as entertainment facilities, restaurants, and shopping malls. Such spots attracted customers from other districts; therefore, customers were distributed around the various districts with some districts having fewer total customers in 2008 than the same time in 2007.

The amount of customers between the old shopping areas and new ones differed. For instance, as the first Commercial Street, Wangfujing Commercial Street accommodated increasing numbers of customers, many who were viewers of the 2008 Beijing Olympic Games. In contrast, other shopping areas had flat levels of business, some worse than in 2007. Additionally, Olympic stadiums tourists mainly focused on famous resorts, especially the resorts in the old city, as evidenced by traffic data. The Imperial Palace, Tiananmen, and Shicha Lake were still their primary choices, and most of the classic resorts are located in the old city.

In conclusion, the 2008 Beijing Olympic Games did not prompt a focus on the new districts. Thus, the expected results of plans were not achieved. The old city must be fully analyzed and evaluated, which will lead to more effective measures.

16.2 Land Value Increment and Land Usage Changes

The impact of large-scale sports events on urban land use, which includes land value increment and land usage changes, is immediate and intrinsic. The reason large-scale sports events cause land value increments is that they increase the demand for land. Because large-scale sports events are held immediately and occasionally, capital flows to certain spots to be exploited immediately, causing local land value to climb (Roche 2000b). Holding large-scale sports events also requires a large amount of capital. For example, Fig. 16.3 shows Beijing's total investment in fixed assets and real estate development investment over time. Due to the financial crisis, investment declined in 2008. The local areas attracted capital from more regions to their neighboring districts. Then local infrastructure

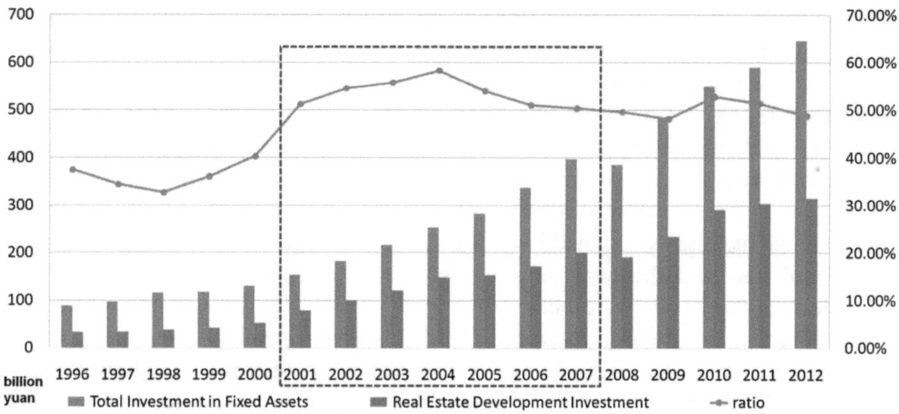

Fig. 16.3 Beijing's total investment in fixed assets and real estate development investment (Reproduced from Beijing Statistical Yearbook 2013)

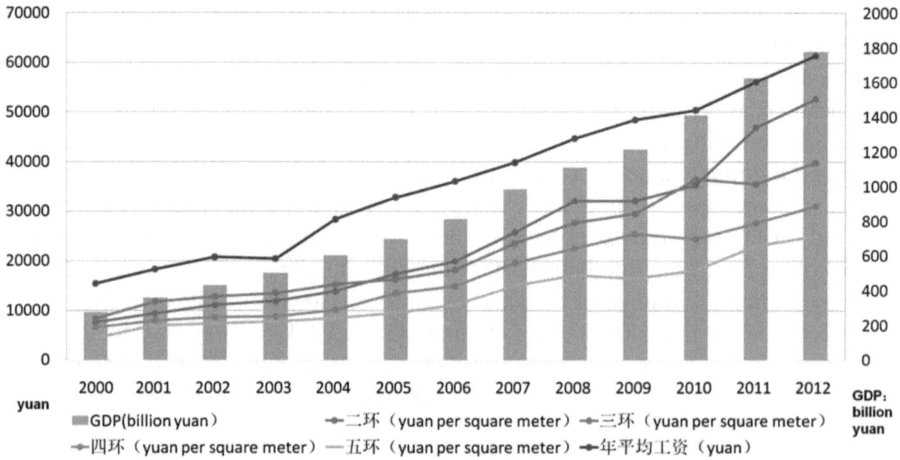

Fig. 16.4 Beijing Housing Price (http://house.qq.com/zt2013/lsbnb/index.htm)

construction and economic development led to differential rent changes, resulting in direct land usage changes. For instance, the increasing local land values in residential areas turned into commercial usage and farmland became residential areas. Thus, it is significant to explore a way to use land sustainably.

The 2008 Beijing Olympic Games influences urban land use as described above. On the one hand, Beijing's property values increased continually before and after the games, which was partly a result of the games (Fig. 16.4). First, after Beijing's 2001 Olympic Bid, its average property values soared 25.8 %. At that time Beijing's average housing price per square meter was ¥ 1,500 higher than Shanghai's, while Beijing's income per head was less than Shanghai's. Before Beijing's 2001 Olympic Bid, the average housing price per square meter in the Olympic District ranged between ¥ 4,000–¥ 5,000. However, after the bid it climbed to ¥ 6,000–¥ 7,000. Even in 2006, the average housing price per square meter in the Olympic District was above ¥ 9,000 with a portion ranging between ¥ 12,000–¥ 15,000, which was about 30 % more than Beijing's average housing price per square meter. In 2009, it reached ¥ 18,000. Prompted by the government and the housing market, urban land usage around the Olympic stadiums changed almost completely. At first, many residential areas and office buildings around the Olympic stadiums were forced to move away. A removal allowance was paid to compensate them, but many people had to leave the places they had lived for a long time. The residents experienced the emotional loss and the costs of moving. Yet, after moving the residents, new upscale neighborhoods were built.

In contrast, the London Olympic District was not developed as planned. The London Olympic District was to be a comprehensive district including housing, sports facilities, commercial facilities, and entertainment facilities to meet consumers in the Docklands Financial District. Because the London Olympic District is close to Stratford, traveling between Stratford and Docklands Financial District was convenient. However, the construction process was barely completed. At first, it was difficult to persuade original residents to move away. Worse yet, the financial crises happened, which brought financial pressure to the London Development Agency (LDA). As a result, the construction of affordable housing that LDA promised was unachievable. To some extent, land usage did not change according to plans.

16.3 Urban Infrastructure Construction Accelerates Urban Space Regeneration

To support large-scale events, most of the infrastructure has to be ready to use. As well, it needs to be built according to special standards and to be constructed for efficiency, particularly for the urban transport facilities. The urban transportation system both outward and inward has an important impact on the urban structure. An improved urban transportation system offers a prominent advantage to large-scale sports events because it makes it convenient for viewers and participants to reach their destinations. As well, urban space expands along the urban transportation system. In the urban transportation system plans, both large-scale events and city future development plans should be considered.

Large-scale sports events allow cities to regenerate mainly through considerable urban infrastructure construction. They also demand urban service facilities including stadiums, transportation facilities, communication facilities, attractions, and municipal facilities. Stadium construction is only a small part of the regeneration; the main goal is infrastructure construction. For example, Fig. 16.5 shows that in many years, the infrastructure construction expenses of the annual Olympic games has exceeded more than 50 % of total expenses.

The total investment for the 2008 Beijing Olympic Games was 295 billion, which included 2 billion for operation, 13 billion for stadium construction, and 280 billion for urban construction including infrastructure improvement, transportation construction, and environmental improvement (Fig. 16.6). During 2001–2007, infrastructure investment grew annually, especially the ratio of infrastructure investment to total investment in fixed assets (Fig. 16.7).

Financing helped fund many different upgrades in Beijing to prepare for the Olympic Games. Financing was used to enhance the Beijing Capital Airport. By 2008, a new runway was built and the airport terminal expanded to 600,000 m^2. Thus, Beijing Capital Airport's capacity increased notably. In 2000, its passenger throughput was 35 million and airfreight was 774,000 tons annually. In contrast, in 2008, its passenger throughput increased 37 % reaching 48 million, and its airfreight increased 68 % reaching 1,300,000 tons annually (Fig. 16.8). Financing also helped fund railway construction or improvement, which included the Beijing-Shanghai High-Speed Railway, the Beijing-Qinhuangdao Railway, and the passenger line of Beijing Capital Airport and Tianjin Airport. Additionally, financing was

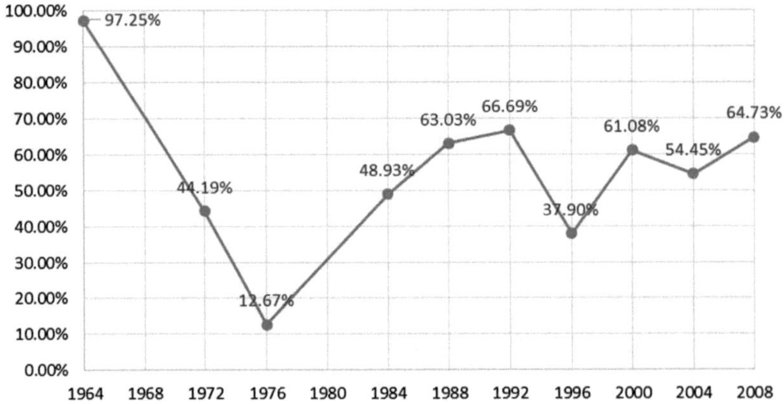

Fig. 16.5 Ratio of infrastructure construction expense/whole expense of Olympic Games (Liao and Pitts 2006)

Fig. 16.6 Total expense composition of Beijing Olympic Games (http://house.qq.com/zt2013/lsbnb/index.htm)

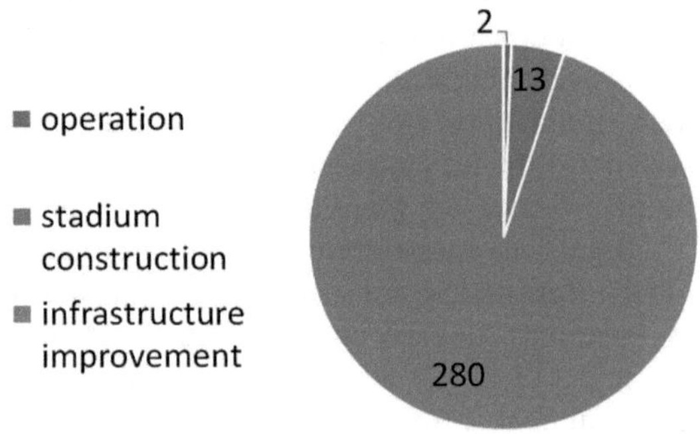

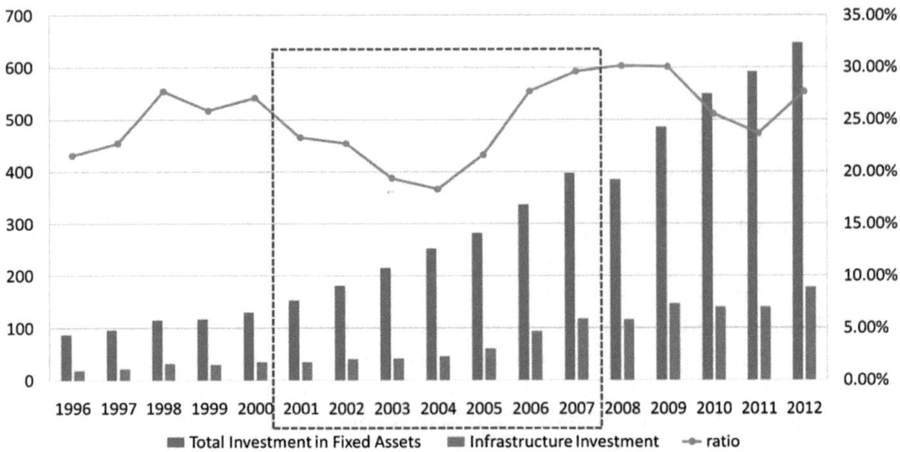

Fig. 16.7 Beijing's infrastructure construction investment (Beijing Statistical Yearbook 2013)

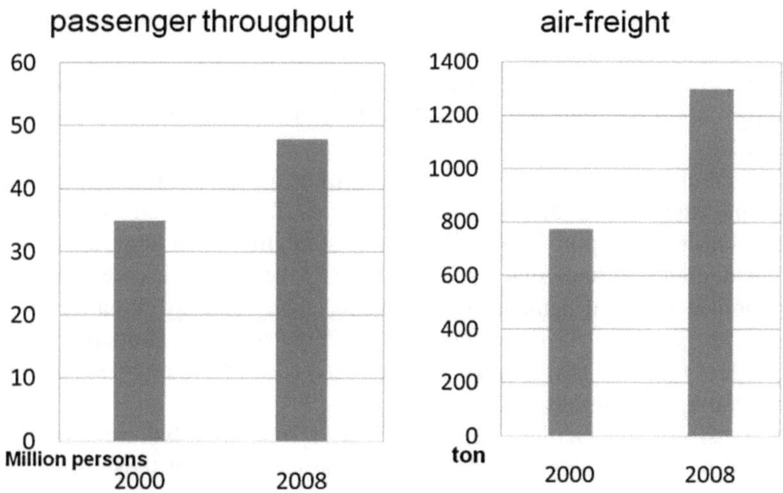

Fig. 16.8 Beijing Capital Airport passenger throughput and air freight 2000 and 2008. http://house.qq.com/zt2013/lsbnb/index.htm

used for highway and road construction. Beijing's highway increased 6,449 km from 2001 to 2008, namely 46.4 % length. Similarly, Beijing's urban road increased 51.9 % length. As well, financing was invested to construct metro lines. Estimating the numbers of people who would attend the games, seven metro lines were built by 2008 for a total length of 200 km. Metro lines played a significant role in transporting passengers conveniently during the games, and now transport 13.3 % of the total 1.28 billion passengers annually. Finally, some financing was used for communication network improvements (Beijing Government 2009).

The typical intention of hosting the Olympic games is to regenerate a city through urban infrastructure construction. However, this does not typically work as well in smaller cities. Financing for urban infrastructure construction was a relative small part of the total spent in some smaller cities, such as the 1972 Munich Olympic Games, 1984 Los Angeles Olympic Games, and 1996 Atlanta Olympic Games (Liao and Pitts 2006). Therefore, construction scale should depend on the urban economy, environment, social conditions, and urban scale.

16.4 Follow-up use of Facilities and Surrounding Development Influence Urban Space Regeneration

The impact of different large-scale sports events on urban development depends on various factors, such as urban developmental stages, specific measures, and financial conditions. Similarly, the follow-up effects promote urban space regeneration to different degrees (Kaplanidou and Karadakis 2010). In general, facilities' follow-up usage and surrounding development mainly affect urban space regeneration, both of which require scrupulous strategies and smart policies. Many facilities are neglected after the games because maintenance charges are expensive. Their operating conditions depend on whether they are financially successful. Thus, finance plans are necessary before and after large-scale sports events. Additionally, builders should make allowances for the market demand for facilities' capacity before and after events. Of course, during events, facilities such as hotels, restaurants, amenities, transportation facilities, and stadiums are in great demand. However, the demand declines sharply after events. If these facilities are constructed to accommodate the total demand during events, many end up idle afterwards. For instance, 96 % of the 85 large stadiums in China were not profitable and 40 % received less than 500,000 through lease rentals. As a result, many stadiums relied on financial appropriations.

To transform the Beijing Olympic District into an urban center, operators developed positive strategies that increased the chances that they would be used after the games. First, they cooperated with famous international games including the China Open Tennis and Beijing central international road cycling race. Second, the stadiums were able to generate some profits from tourism annually. Third, such large stadiums as the National Aquatics Center and Bird's Nest were used for large entertainment events such as concerts.

Nevertheless, those Olympic stadiums are still confronted with financial pressures. After the 2008 Olympic games, tourism in Beijing declined and costs increased. Operators faced great pressure to generate sources of income. For example, in 2011, the National Aquatics Center's self-operated business costs were as high as 99.299 million Yuan, including energy consumption, maintenance cost, assets depreciation, marketing expenses, administrative expense, and fiscal expenditures. As well, its labor cost reached 57.563 million Yuan, because of its large size. Conversely in 2011, its self-supporting income was only 88,000 Yuan (Xianpeng Liao 2012). As a consequence, it relies on financial support from the government to continue to operate.

16.5 Concerns About Using Large-Scale Sports Events to Regenerate Urban Space

It is necessary to conform to urban spatial pattern evolution rules to have a positive effect on urban spatial regeneration. Large-scale sports events' influence has been far-reaching for a long time. Thus, it is necessary to take into account sustainable urban long-term strategies and structures. It is also important to track the influence afterwards. The following are factors that should be considered when determining whether a large-scale event would be beneficial.

16.5.1 Identify Sustainable Ways to Plan Large-Scale Sports Events

First, analyze the situation with a broader vision and comprehensive perspective that examines factors from the city to the metropolitan area and from space to the economy, society, and environment. Second, before making final plans, ask opinions from the government, enterprises, and citizens. Third, have a forward-looking perspective to lead the city toward a better development trend.

16.5.2 Choose Stadium Locations to Improve Urban Spatial Pattern

There are two kinds of location choices. One is based on existing facilities and the other is to build new facilities. Different decisions are made by different cities when considering the financial budget and urban scale. If a city has held large-scale sports events more than once, it typically chooses a combination of the two approaches.

16.5.3 Pursue Projects in Accordance with the Overall Urban Over Plan

Projects to be constructed include infrastructure, stadiums, and supporting facilities. Ostensibly those projects are aimed at large-scale sports events. However, those projects are mainly used after large-scale sports events. The projects should be consistent with urban development demands. Furthermore, the most important criterion for agreeing to host an event is whether the city has an opportunity to improve itself.

16.5.4 Combine Commercial Exploitation and Public Space Construction in Follow-Up Usage

On the one hand, holding large-scale sports events are effective ways to maneuver all resources. Public and private financing are offered synchronously to build facilities for large-scale sports events. Thus, it is easier to carry out certain projects by holding large-scale sports events because many cannot be implemented through regular policies. On the other hand, conflicts of interest between commercial exploitation and public space construction often exist. Holding large-scale sports events is a multi-faceted game, in which the public interest should be put first. For example, the Beijing workers' stadium is not just one of the Olympic stadiums. It also plays an important role as a vibrant public space.

References

Beijing BODA. (2010). http://www.beijing2008.cn/live/pressconference/pool/bimc/n214495372_1.shtml

Beijing Government. (2009). *Beijing Data*. http://www.bjpc.gov.cn/fzgh/guihua/11_5_zx/11_5_yb/200612/t148049.htm

Kaplanidou, K., & Karadakis, K. (2010). Understanding the legacies of a Host Olympic City: The case of the 2010 Vancouver Olympic Games. *Sport Marketing Quarterly, 19*, 110–117.

Lin, G. C. S. (2002). The growth and structural change of Chinese cities: A contextual and geographic analysis (J. Malysis, Ed.). *Cities, 19*(5), 55–76.

Liao, H., & Pitts, A. (2006). A brief historical review of Olympic urbanization. *The International Journal of the History of Sport, 23r*, 1232–1252.

Metropolis Commission. (2002). *The impact of major events on the development of large cities*. Barcelona: World Association of the Major Metropolises, Metropolis.

Roche, M. (2000a). *Mega-events and Modernity: Olympics and exposing the growth of global culture* (M, Ed., pp. 276–277). London: Routledge.

Roche, M. (2000b). *Mega-events and Modernity: Olympics and expos in the growth of global culture* (M, Ed.). London: Routledge.

Xianpeng Liao. (2012). *Research on Olympic stadiums' follow-up utilities after Olympic Games*. State General Administration of Sport. http://www.sport.gov.cn/n16/n1152/n2523/n377568/n377613/n377763/390128.html

Chapter 17
Study on Planning of Healthcare Facilities in High-density Urban Residential Areas—Taking Three New Communities in Beijing as an Example

Jing Li and Tian Chen

17.1 Background and Awareness

A large portion of urban population currently live in high-density residential areas, and because of continuous urbanization, this is likely to be the most common living pattern in the future. Most Chinese people have accepted high-density living habits; however, the health and livability issues are catching more and more attention and have become the issues of most concern in housing selection.

According to the definition by the World Health Organization (WHO) in 1994, a healthy city is one that is continuously creating and improving the physical and social environment and expanding community resources which enable people to mutually support each other in performing all the functions of life and developing to their maximum potential. Fu Hua, a professor from the School of Public Health at Fudan University, brought up a more understandable definition: "a healthy city is one that focuses on human health in urban planning, construction and management. It provides healthy living and working conditions for urban citizens, and is necessary in the development of a human society with healthy people and a healthy environment.

Healthcare facilities in residential areas, usually in the form of community health stations, are important nodes responsible for the health of residents. Especially in

This paper is funded jointly by Youth Fund of Humanities and Social Science Research Projects in Ministry of Education (**13YJCZH080**) and the Beijing Excellent Talent Training Project of Beijing Municipal Committee Organization Department (**2013D005002000007**).

J. Li (✉)
School of Architecture, Tianjin University, Tianjin, China

College of Architecture, North China University of Technology, Beijing, China
e-mail: debbylj@sina.com

T. Chen
School of Architecture, Tianjin University, Tianjin, China

Q. Pan, J. Cao (eds.), *Recent Developments in Chinese Urban Planning*,
GeoJournal Library 114, DOI 10.1007/978-3-319-18470-8_17

the large cities such as Beijing, which serves as a microcosm of the national health service system, community health stations will complement large general hospitals and various special hospitals to form a comprehensive healthcare system. However, at present, community health stations are still under development, with some deficiencies in spatial design. Resident satisfaction with community health station is not high enough.

This paper analyzes the rational arrangement of healthcare facilities in high-density urban residential areas through field research and questionnaires in three high-density residential areas in Beijing. The questionnaires and open-ended interviews focused on two aspects: firstly, basic information, personal health, and treatment conditions of interviewees, including age, gender, living condition, revenue source, medical insurance, current health self-assessment, and utilization of healthcare services; secondly, assessment of healthcare facilities, including coverage and radius of services and base conditions of community health station, building assessment, and external environment.

17.2 Data and Research Methods

17.2.1 Research Data and Sample Community

This investigation focuses on the utilization of community health stations in high density populations. The samples in this study are taken from three new communities which are located between the fourth and fifth ring roads in Beijing and built after 2005. The communities include Yongjing Siji Community at Western Fifth Ring Rd., Xishan Fenglin Community at Western Fifth Ring Rd., and Yuanyang Qinshanshui Community at Western Fourth Ring Rd. Each community is required by the Beijing government to be provided with a community health station. These three communities are typical urban residential communities with dense population, appropriate scale, relatively complete infrastructure facilities and occupancy rate of over 70 %. They are relatively large integrated urban residential areas, with the floor area ratio of greater than 2.2 and over 2000 households settled. Table 17.1 shows the summary information of these three samples.

These data are sourced from a survey of living conditions and health station utilization of about three similar residential areas in Beijing. The survey was composed of open-ended interviews and questionnaires in order to determine the utilization of health centers by the residents through observation, assess the scale, range and services of community health station through open-ended interviews, and assess the utilization of current health station by the residents through questionnaires.

The three communities are typical mid-size residential areas in Beijing. They have various supporting facilities found in newer communities and have been provided with community health stations, which are becoming the basic health

Table 17.1 Basic conditions of three sample communities

Items	Yongjing Siji Community	Xishan Fenglin Community	Yuanyang Qinshanshui Community
Location	Located at Western Fifth Ring Rd., adjacent to Badachu Rd in the west, neighboring Fushi Rd in the south and bordering on Tiancun Rd. in the north.	Located at the east of Badachu Rd in Shijingshan District, north of diversion canal of Yongding River, south of shooting yard, and west of Western Fifth Ring Rd.	Located between Western Fourth and Fifth Ring Rd., neighboring the west extension line of Chang'an to the north, and adjacent to the new planned expressway outside west postern in the south.
	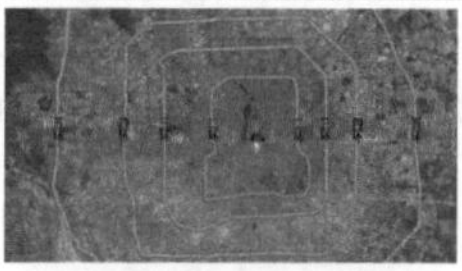		
	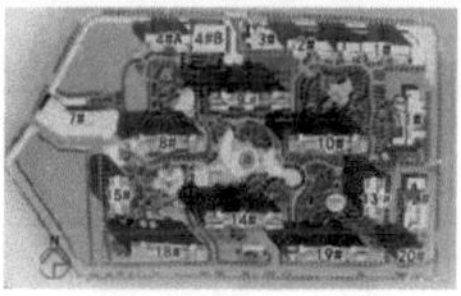		
Floor area	112,700 m^2	260,000 m^2	113,600 m^2
Households	2012 households	5026 households	3980 households
Floor area ratio	2.71	2.58	2.92
Gross floor area	380,000 m^2	700,000 m^2	320,000 m^2
Residence area	280,000 m^2	580,000 m^2	290,000 m^2
Settle-in time	In 2009	In 2005	In 2010

organization and node serving the citizens of Beijing. Through Beijing healthcare reform, such community health stations have been included in the medical security system, becoming an important urban station for medical care, life care, and chronic disease treatment.

17.2.2 *Field Survey on Healthcare Facilities in Sample Communities*

17.2.2.1 Analysis on Base and Current Conditions

The three health stations are respectively located inside communities or outside the commercial district on the first floor of the external urban road. The building entrances of two communities face the street, and one of the two is situated inside the community, which means that any person cannot enter the building from the outside.. Moreover, accessibility and the amenity of overall outdoor space are not considered when selecting community location, so there are many special constrains and no reserved space for development. The basic information of health stations are generalized in the following Table 17.2.

The floor area of the three health stations also only meets temporary medial demands. It is insufficient for future development and has no connections with the surrounding environment. The floor area is about 300 km^2, only considering the most basic medical treatment demands. The three health stations are medical-insurance designated divisions, so the residents with Beijing social insurance can benefit from these services. The three health stations are built in new communities, which boast of excellent building conditions, complete equipment and reasonable staffing.

17.2.2.2 Summary of Utilization of Health Stations in Three Sample Communities

1. **Structure of population with medical care**

According to our research, in the three communities, the number of people requiring medical care at the health stations within 1 week (five working days) respectively is 105, 110 and 132. Of these numbers, Table 17.3 shows that 75 % were elderly (ages over 55 years old) and about 15 % were children (ages under 12 years old). The proportion of youth is relatively lower. Most adult and young people work during the week, so the groups the health stations serve are the elderly and children. Most adult and young people come to the health station for infusion therapy or other specific therapy if they are sick. Therefore, the population seeking conventional or basic treatment is less prevalent in the health stations.

2. **Medical activity analysis**

We assessed the medical activity in 1 week (five working days) in Yongjing Siji Community as an example, shown in Table 17.4. In this time period, medical activities in the station mainly included blood pressure taking, ECG test, and blood glucose monitoring. This reflects the high proportion of elderly people with chronic diseases in users of the health stations. In this medical care environment, the mainstream of serving group and services of health station can be relatively controlled, and the charge is also affordable, and thus it is extremely important to add healthcare and medical care consulting. The field survey results show that the basic information of other community health stations is basically the same as in Yongjing Siji Community.

3. **Analysis of time on medical service**

We again looked at medical activity in 1 week (five working days) in Yongjing Siji Community as an example, shown in Fig. 17.1. On a typical working day

Table 17.2 Basic information of health stations in three sample communities

Items	Yongjing Siji Community	Xishan Fenglin Community	Yuanyang Qinshanshui Community
Location	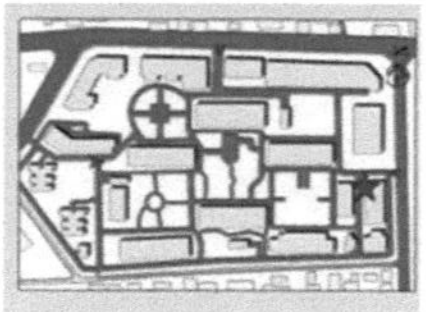	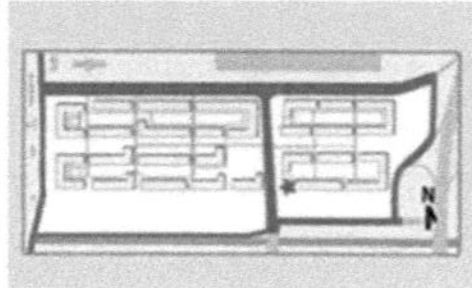	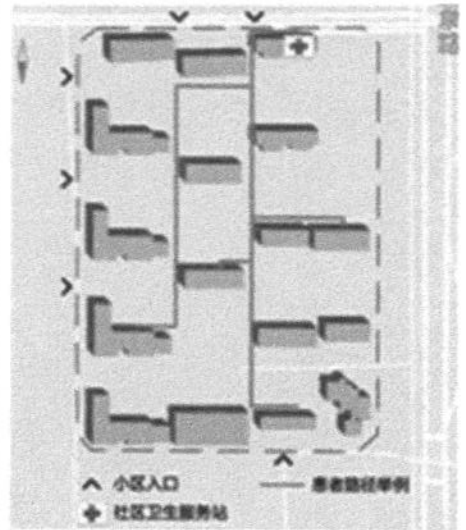
	Located at a proper place, close to the commercial district on the first floor at one side of urban road.	Located at the commercial district on the first floor at one side of the urban road dividing the community due to large floor area of the community	Located at commercial district on the first floor at one side of the urban road close to the community
Internal layout	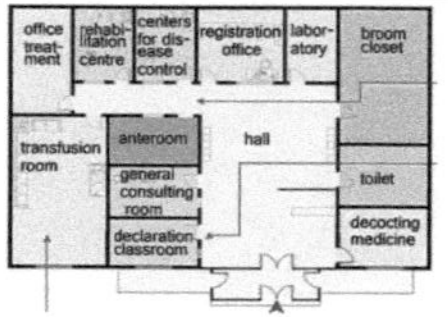	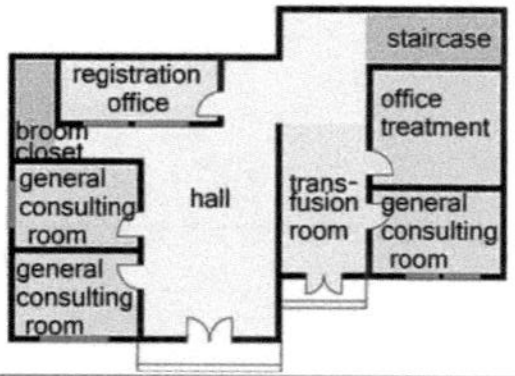	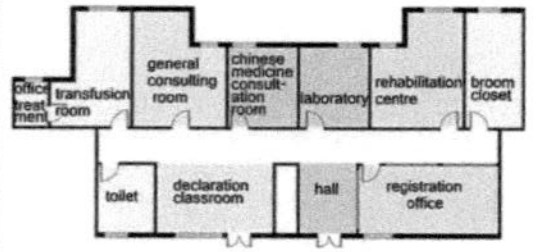
Floor area	About 370 m^2	About 240 m^2	About 310 m^2

(continued)

Table 17.2 (continued)

Items	Yongjing Siji Community	Xishan Fenglin Community	Yuanyang Qinshanshui Community
Sector layout	General Consulting Room, TCM Consulting Room, Rehabilitation TCM Consulting Room, Medicine Room, Laboratory Room, Toilet, Decocting Room, Treatment Room, Lounge, Infusion Room, Maternity Guidance Room, Health Promotion Room	General Consulting Room, Medicine Room, Laboratory Room, Toilet, Treatment Room, Lounge, Infusion Room, Health Promotion Room	General Consulting Room, TCM Consulting Room, Rehabilitation TCM Consulting Room, Medicine Room, Laboratory Room, Toilet, Treatment Room, Lounge, Infusion Room, Health Promotion Room
Medical insurance dedicated point	Yes	Yes	Yes
Number of staff	7	6	5

Table 17.3 Structure of population with medical care

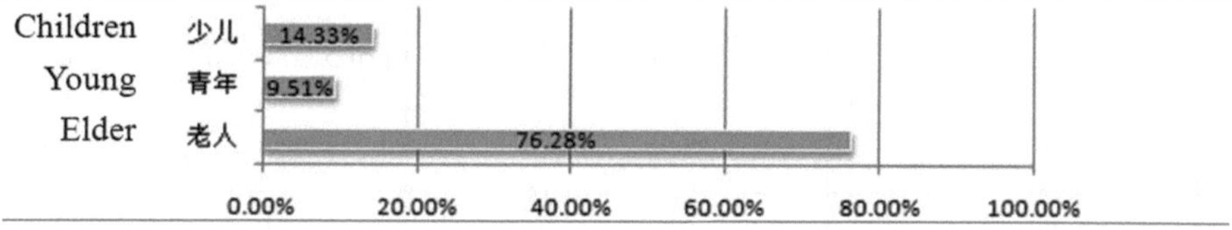

Table 17.4 Statistics of medical activities in sample communities (taking one as an example)

Service	Blood Pressure Taking	Auscultation	ECG	Weighing	BG	Prescription
Times per week	42	32	12	43	7	65
Equipment	Sphygmomanometer	Stethoscope	ECG instrument	Scale	BG meter	Medicine room
Qty.	4	4	1	1	4	1
Quality	Common	Common	Common	Common	Common	Slightly poor
Charge	Free	10¥	8¥	Free	4¥	Different

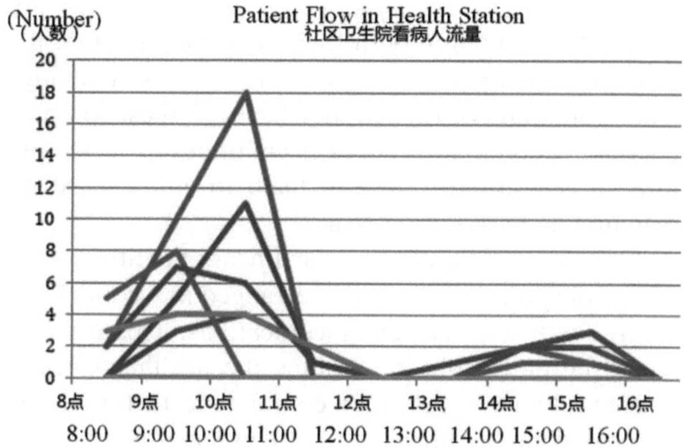

Fig. 17.1 Patient flow in health station

(Table 17.5), most of the treatment sought by residents occurred in the morning rather than in the afternoon. According to this information, it can be said that the serving time and services of the health station should be adjusted according to actual demands, to make it more convenient and to increase the utilization, thereby gradually increasing the sense of trust and improving the overall efficiency and utilization of health care.

Table 17.5 Statistics of time on medical care (taking one for example)

Time	Date						
	April 1	April 2	April 3	April 4	April 5	April 6	April 7
8: 00—9: 00	0	0	5	2	2	3	Vacation
9: 00—10: 00	5	3	8	10	7	4	Vacation
10: 00—11: 00	11	4	0	18	6	4	Vacation
11: 00—12: 00	1	2	0	0	1	0y	Vacation
12: 00—13: 30	Lunch hour	Lunch hour	Lunch hour	Lunch hour	Lunch hour	Vacation	Vacation
13: 30—14: 30	1	0	0	0	0	Vacation	Vacation
14: 30—15: 30	2	2	2	1	2	Vacation	Vacation
15: 30—16: 30	3	3	1	1	2	Vacation	Vacation
16: 30—17: 00	0	0	0	0	0	Vacation	Vacation
Total population	23	14	16	32	20	11	0

17.2.2.3 Summary of Sample Community Conditions

High-density residential areas typically suffer from having a shortage of space and resources per capita. However, in our survey of sample communities, the resources seem sufficient. In fact, various healthcare & fitness centers and health-related commercial facilities are emerging in an endless stream around the community, making it difficult for large, public and community hospitals to engage in registration and consultation with local residents. Despite the Beijing municipal government continually proposing various services including "health file, "family doctor" and "community-oriented healthcare," these problems of commercial services overtaking community services continues to occur. However, community residents lose an opportunity close at hand to enjoy these services due to management, planning and promotion issues.

In most residential areas, healthcare facilities are not included at the beginning of planning and are randomly arranged in a commercial district on the first floor in an area with the relevant specifications. This often leads to traffic congestion. It would be preferable to combine the healthcare facilities with open space in the residential area or to form a complete residential health system.

Healthcare facilities are simple in functions and services but they fail to meet the demands of the residents in high-density residential area during working hours. Though the cost is low, the technologies are inadequate and fail to keep up with the pace of the times. In contrast with the services for consulting, health care and nursing are significantly inadequate. Furthermore, the staff turnover at health stations is high, which results in lack of resident trust in the health station.

17.3 Research on Planning of Healthcare Facilities in High-density Urban Residential Areas

17.3.1 Residential Area Morphology, Healthcare Facilities Distribution and Utilization Mode

In our survey on sample residential areas, the residents put high importance on geographical location and the environment surrounding the health station. Residents also favor increasing the overall utilization of health clinics. Due to the high density of the residential areas, the community can be divided according to living conditions around the residential area so as to determine the serving radius and geographical location of the health station. Health stations should be located at the geographic center of the residential area to ensure that residents can walk to the station in under 10 min, (walking distance not more than 800 m).

1. Our research shows that residents are likely to trust health stations more with an increase in services and serving radius of the station. The residential area form and community management mode tend inward, the healthcare facilities have inherent outward features. Under the guidance of current medical and health concepts, healthcare facilities have some attractiveness. The attractiveness to surrounding residents will greatly affect its own utilization.
2. In crowded residential areas, future healthcare facilities will need to be diverse in their uses. In addition to medical conditions, health care and health consulting will become the main work of health stations. With an increasingly aging population, health stations should be constructed together with various faculties. The treatment of chronic diseases will become an important function of community health stations.
3. It is necessary to add some fitness facilities combined with healthcare facilities in high-density residential areas. Community health stations should evolve into community health centers, and open space around the facilities should be included in the design to form key fitness and health care places for residents. For example, the health station should provide fitness tools and opportunities for exercise to improve the overall health level of the community and to improve the utilization of public places. This will alleviate space shortages in high-density areas while simultaneously promoting the healthy development of high-density residential areas.

17.3.2 Optimization and Planning Policies of Healthcare Facilities

Because of the current expansion of mega-cities, new residential areas are likely to develop continuously. In order to achieve sustainable operation of residential area space and healthcare facilities, excellent micro-level spatial design should be used. Healthcare facilities should be continually optimized on the existing conditions of the built-up area, facilitating integration with the new community and forming a sound organizational structure of healthcare facilities in the urban community.

According to the optimization orientation of spatial organization, the policies in this study have the following implications:

1. Improve the effectiveness of healthcare facilities—The effectiveness of healthcare facilities should be improved with the development and expansion of high-density residential areas. Especially inside the residential areas of megacities, how can we improve the utilization of limited space with effective planning and design, strengthen residents' sense of belonging to the residential area, and create good relationships in the neighborhood? This can be achieved by spreading health concepts and healthy living to the residents, and putting forward the improvement points. Moreover, improving the effectiveness of healthcare facilities will have a positive effect on the establishment of the overall urban medical network.
2. Change in management of healthcare facilities—One important reason for the low effectiveness of healthcare facilities is because managers pay more attention to the overall form rather than the details. For example, beauty shops, spas, and health centers around the residential area are booming, but the healthcare facilities are deserted. Management, operating time, and other specific works of healthcare facilities are excessively institutionalized and formalized, failing to meet the actual demands of the residents. The management of healthcare facilities should play a more effective role, and the combination of dual management of maker and government is an important way to create flexible and practical high-density healthcare facilities.
3. Changes in design ideals of the public environment in residential areas—The design of public space in residential areas has been greatly changed since the residential areas have become open to the market, where the emphasis on appearance is of greater importance. However, it is important to combine the design of public spaces with the utilization of healthcare facilities, thereby improving the health of the whole residential area. High-density population, less per capita green space, and less per capita open space may adversely affect residents of high-density residential areas. Thus, it is important to integrate healthcare facilities, outdoor public space, overall slow-down system of the residential area to improve the health level of the high-density residential areas.

17.4 Conclusions

With a high population density, sustainable development of residential space in large cities is strongly related to health level and environmental sustainability. The health issues caused by high-density population and limited outdoor activity space are unique issues encountered by Chinese people that cannot be solved through foreign experience. Our desired goals are to improve the effectiveness of healthcare facilities, continually meet the demands of the residents, determine services, standards and management rules of healthcare facilities, prepare associated management regulations, and create long-term, effective healthcare facilities through community construction and market joint construction with the financial support of the government. This will help us build a truly harmonious, healthy, sustainable, high-density residential environment.

Bibliography

Beijing Community Health Service. http://www.bjchs.org.cn/

Chen Yi. (2007). *Preliminary study of high-density development pattern of residential areas in large cities*. Master's thesis of Southeast University [D].

Corburn, J. (2004). Confronting the challenges in reconnecting urban planning and public health [J]. *American Journal of Public Health, 94*.

Dong Chunfang. (2012). *Architecture to high density* [M]. Beijing: China Architecture & Building Press

Huang Yong, & Liu Zhiqiang. (2010). Promote the public health by landscape planning and designing [J]. *Forestry Survey and Design, 1*, 37–40.

Li Qing, & Zhong Liqun. (2011). High density and livability—Duxton plain public housing [J]. *Times Architectural, 4*, 70–75.

Liu Binyi, et al. (2006). Promoting health through design: A brief introduction of "Active Living by Design", an American national program [J]. *Foreign Urban Planning, 2*, 60–65.

Loonj, V., & Frank, L. (2011). Urban form relationships with youth physical activity implications for research and practice [J]. *Journal of Planning Literature, 26*(3), 280–308.

Wang Puqing. (2007). *Hustle and Gap—The research of the city morphology under the high density state*. Master's thesis of Hefei University of Technology [D].

Zou Jingyu, & Zhang Huizai. (2004). Research on ecological settlement of high density city–discussion about Hong Kong mode [J]. *New Architecture, 4*, 51–54.

Chapter 18
Jobs-Housing Balance: The Right Ratio for the Right Place

Qian Wu, Ming Zhang, and Daniel Yang

18.1 Introduction

The concept of Jobs-housing balance (JHB) has attracted many city and transportation planning agencies for the interest of increasing place quality and reducing travel demand. Many studies have revealed the inter-relationship between land use pattern and travel behavior. The spatial location of jobs and residence has a strong impact on travel patterns (Cervero 1989; Ewing et al. 1996; Zhao et al. 2009, Loo and Chow 2011), especially commute patterns (Bento et al. 2005; Cervero 1989; Downs 2004; Horner 2008; Wang and Chai 2009). For example, jobs-housing ratio has been adopted as the measurement of JHB in Washington, California, and Georgia to guide land use development. It also has been incorporated into either comprehensive land use plans or land use regulations, such as the Sacramento region of California (SACOG 2012) and Atlanta Regional Commission in Georgia (Miller 2010).

Austin is a fast growing area, which is facing increasing travel demands, traffic congestion, suburban expansion, along with air pollution and environmental deterioration. The CAMPO (Capital Area Metropolitan Planning Organization) is responsible for developing and updating the Regional Transportation Plan for the

---

Q. Wu (✉)
Kunming Urban Design & Planning Institute, China, 213 Shangyi Road, Kunming, Yunnan 650041, China
e-mail: wuqian8119@126.com

M. Zhang
The University of Texas at Austin, 1 University Station, Stop 7500, Austin, TX 78712, USA
e-mail: zhangm@austin.utexas.edu

D. Yang
Capital Area Metropolitan Planning Organization, 1616 Guadalupe Street, Ste. 4.202, Austin, TX 78701, USA
e-mail: daniel.yang@campotexas.org

Q. Pan, J. Cao (eds.), *Recent Developments in Chinese Urban Planning*,
GeoJournal Library 114, DOI 10.1007/978-3-319-18470-8_18

Austin region every 5 years. In the CAMPO 2035 Plan, it aims to improve jobs-housing proximity by incorporating JHB to guide future land use development. The plan employs the concept of activity centers connected by high capacity tránsit corridors, targeting the accommodation of 31 % of the population and 38 % of employment in Central Texas by 2035 (CAMPO 2010). At the time when CAMPO defines the appropriate ratios of JHB for the activity centers, critical questions arise: what is a good ratio? Further, how should JHB be quantified for guiding land use development? And to what extent could jobs-housing ratio be effectively used as an intervention instrument?

This study attempts to provide theoretical and empirical evidence of JHB and examine the applicability of jobs-housing ratio for different planning purpose in local context. Based on a rich literature review, the report removed the "deceptive simple concept" (Cervero 1991, p. 10) of JHB on the surface and gathered insights on JHB from existing exemplary studies. Absent a single consensus of a balanced jobs-housing ratio, the goal of this report is to present the possible ways of measuring and defining JHB in complex urban development. This study analyzed existing JHB of the Austin region, presenting the limitation of jobs-housing ratio in guiding the distribution of employments and housing. Local municipalities might consider more detailed factors in terms of the application of jobs-housing ratio in local context.

18.2 Literature Review

18.2.1 Definition of Jobs-Housing Balance

The concept of (JHB) stems from a balanced or self-contained community initiated by Howard, Purdom, and Munford (Giuliano 1991, pp. 305 & 312). Giuliano (1991) defines JHB is "the distribution of employment relative to the distribution of workers in a given geographic area" (p. 305). Real balanced communities should be "self-contained" or "self–reliant", which means a spatial match of jobs and workers and residence within a given area (Giuliano 1991). It implies the attainment of jobs within a reasonable commute distance to homes. The degree of balance is determined by the proportions of locally employed residents or locally residing workers in a defined area. Achieving true balance requires a 'match-up' between jobs opportunities and workers' skills as well as housing price and worker's income (Cervero 1989, p. 137).

The concept of JHB was firstly translated into planning practice in Margolis's discussion on the relationship between tax rate and land use pattern (residential and industry) in 1957. He primarily provided the concept of balanced community with a ratio of employment to resident labor force ranging from 0.75 to 1.25. A city is seemed as a "dormitory city" if this ratio is lower than 0.75 and as an "industrial enclave" if the ratio is larger than 1.25 (Margolis 1957, p. 227). Some researchers claimed that, however, true JHB should be "self-containment" in defined regions. Evidence from Cervero's study (1996) in the San Francisco Bay Area shows that a small share of locally reside workers in some perfectly balanced cities, though jobs-employed residents ratios of 20 cities have been improved during the 1980s. Cervero concluded that there is a weak relationship between the ratio of JHB and self-containment. Giuliano also noted that real balance needs a "broad mix of housing type to accommodate households of a range of income categories" (Giuliano 1991, p. 305). Nevertheless, jobs-housing ratio is the most popular and easiest way to measure JHB because a quantified number is easy and feasible to be translated into land use planning and housing policy.

18.2.2 Significance of Jobs-Housing Balance

Traffic congestion is the main factor that initiates concerns about JHB (CRP 2008). According to 2009 National Household Travel Survey, commute trips take up 27.77 % of total vehicle miles traveled (VMT) in America (FHWA 2011). A rich literature has proved that balanced growth in jobs and housing has a substantial impact on commuting and VMT (Bento et al. 2005; Cervero 1989; Cervero and Duncan 2006; Downs 2004; Ewing et al. 1996; Frank and Pivo 1994; Peng 1997; Weitz et al. 2003).

Along with reduction in VMT and alleviation of traffic congestion, JHB offers benefits to travel cost, which is the second highest expenditure for most households after housing (Home + Transportation cost % Income). In addition, by improving jobs-housing proximity, associated social, economic and environmental benefits could also be gained: better job accessibility, greater family stability, lower public expenditure on infrastructures, more housing choices, better air quality, more open space preservation, and more travel choices (CPR 2008, p. 5, Frank 1994; Giuliano 1991; Levine 1998; Macek et al. 2001; Zhan et al. 1999). Another marginal benefit of JHB is improved personal health through encouraging non-motorized travel (Ewing and Cervero 2001; Frank 1994; Handy et al. 2006).

The extent of commuting benefits of JHB varies from region to region. Extensive studies have proved that there is clear correlation between spatial distribution of work place and residence and commuting distance or time (Bento et al. 2005; Cervero 1989; Cervero and Duncan 2006; Frank 1994, Giuliano 1995, Levinson and Kumar 1994; Peng 1997; Wang and Chai 2009). Conversely, a number of empirical studies show that relatively more balanced jobs-housing ratio has little impact on shortening commuting distance or time (Downs 1992, 2004; Miller and Ibrahim 1998; Giuliano 1991; Giuliano and Small 1993). Zhao et al. (2009) concluded that the differences in the correlation between JHB and commuting result from four theoretical and methodological limitations: the variations in the definition and measurement of JHB, the different geographic scales of study areas, the thresholds of jobs-housing ratios to detect the commuting impact of JHB, and specific local settings, such as institutional influence, culture difference, demographic characteristics, work shifts rate, etc.

Overall, there is a majority of experts believe that improving JHB helps to alleviate traffic congestion and reduce travel cost as well as social equity and environmental issues.

18.2.3 Measurement of Jobs-Housing Balance

How can JHB be quantified and qualified in the planning field? Peer-reviewed studies have explored the correlation between the distribution of workers and residents and commute duration. Although many transportation scholars are more interested in quantifying JHB by estimating theoretical commuting time and actual commuting time, in the land use planning field JHB has widely been quantified through calculating jobs-housing ratios.

Expressing JHB by ratio has been widely accepted in planning practices. Yet scholars and researchers have different calculation of the jobs-housing ratio. Some studies use jobs/housing units (Ewing et al. 1996; Cervero 1991; Weitz et al. 2003) while others use the jobs/ household ratio (CPR 2008). Some studies use jobs/ employed-residents ratio (labor force) (Cervero 1989, 1996) and others use jobs/ resident workers ratio (Giuliano 1991). Giuliano (1991) also roughly calculated the ratio of jobs and population of Orange County as a way of evaluating longitude change in JHB. Some people argue that housing units include vacant housing units and might distort housing supply in communities. However, jobs to housing units are still a simple and adequate way to measure jobs-housing ratio if the average number of workforce per housing unit could be estimated (Weitz et al. 2003, pp. 4–5). Using jobs/housing units has merit in revealing the real supply of housing market if the ratio is used as an indicator of housing relocation. American Planning Association and CPR (2008) recommend the ratio of jobs-employed residents as the best expression because it's easier to evaluate the degree of balance if a ratio of 1.0 is signified as perfect balance (Weitz et al. 2003; CPR 2008).

In addition to disparity in denominator of the ratio, the variation in terms makes quantifying JHB more complicated. For example, "jobs" may only include wage and salary workers (Giuliano 1991, p. 305); it may also contain home-workers, or farmers (Miller 2010, p. 24). CPR (2008) also pointed out that incongruence of "jobs" counts could occur if data is provided by different entities. The Table 18.1 lists terms used in previous studies.

Comparing the terms above, simplifying the term as jobs/employed residents (number of residents in the community who are employed) makes it easier for planners to review and assess JHB in different regions or cities. As Cervero (1989) pointed out, real balance should rely on the percentage of residents who live and work locally (p. 137). However, due to unavailable data and frequent job-shift in the U.S., few studies have calculated jobs-housing ratio by the percentage of local-employed residents. Jobs–employed resident ratio (JER), nevertheless, is still valuable in directly grasping job opportunities and housing distribution (Yang and Ferreira 2005).

Table 18.1 Interpretations of variables of jobs-housing ratio

Jobs	
Weitz	Employed residents = Labor force = Residents workers
Cervero	Workers = Jobs
	Resident workers = number of workers who reside locally
	Employed residents = Number of residents in the community who are employed
Giuliano	Resident workers = Labor force
CPR	Employed residents = those in the labor force who are currently working

18.2.4 Where to Measure?

The primary question is to what extent jobs and housing could be balanced. Many scholars have discussed the geographic size of JHB measurements. Numerous studies on the commuting impact of JHB were conducted in different regions/cities at different geographic levels from different perspectives, such as county level (Giuliano 1991; Cervero 1996, Miller 2010), census tract level (Frank 1994; Yang and Ferreira 2005; Zhao et al. 2009), etc.

Although there are disparities of geographic size in previous studies, the consensus is that measuring JHB at the metropolitan/region level is meaningless. According to Cervero, "the larger the size, the more likely the balanced-at the extreme, planet earth has a perfect balance of jobs and employed residents." (Cervero 1996, p. 495) Because the spatial distribution of employment locations and different types of employment generate commuters within the region (Giuliano 1991; Peng 1997), jobs and housing are usually imbalanced at the counties, blocks and census tract level. Thus, the measurement at micro level is more likely to reflect relationship between travel behavior and land use pattern. At macro geographic level (i.e., county and city level), large amounts of internal work trips are neglected.

However, Peng (1997) argues that census tract data based on population density cannot illustrate the reality of work-trip travel behavior. Because work trips across neighboring census tracts or travel analysis zones (TAZs) tend to be categorized as jobs-housing mismatch. He suggests a "meso level" of geographic unit defined by average commuting time in the region. Nevertheless, there is a lack of consensus as to the most appropriate size of circular area. Levingston recommended 6–8 miles as the home to work trip distance by driving, while Deakin thinks a work trip with 3–10 miles of driving distance is acceptable (Peng 1997). Cervero (1989) uses a 3-mile radius of suburban job center to assess JHB. Giuliano suggests using 3-mile radius circle rather than an estimated commute distance of 9-mile within suburban areas across the nation. Peng argues that a commute shed should be aggregated zones built upon the average or median actual commute distance because actual average commute distance reflect the real impact of housing and employment locations on commuting. Thus, he recommended aggregated zones partially or totally covered by a 5-mile buffer as the basic units of measurement (Peng 1997).

Since JHB is utilized as either transportation policy or guidance for land use development, the factors that outline measured units should consider current urban spatial structure, growth policy, land capacity for development, administrative limitations of jurisdictions, the goal and the scope to measure JHB, as well as what kind of measures to be used. Without clear standards to follow, a good assessment of JHB comes from comprehensive understanding of the local context.

18.2.5 Defining a Good Jobs-Housing Ratio

Based on a review of the literature, the only consensus is that the appropriate ratio should be a range rather than an absolute and arbitrary number. The jobs-housing ratio only indicates the "potential for better balance". Jobs-housing ratio could vary from region to region and is a changing process from imbalance to balance, then back to imbalance (Cervero 1991, p. 12). Existing studies have provided recommendations of good jobs-housing ratio in three measurements (see Table 18.2).

18.2.5.1 Jobs to Household Ratio

Cervero defined jobs to household ratio at 1.5 as the threshold in a city, based on the assumption that 90 % of workers in the U.S. share housing with others and 70 % of cohabitant households have two or more workers. As noted that in the literature review, his study focused on the San Francisco Bay Area with high economic connection and inter-work trips among cities, thus this assumption might not be suitable for low-urbanized cities, such as many cities in Central Texas. Ewing's estimate that is built upon internal work trips captured within a 3–5 miles area is more practical.

Frank and Pivo's recommendation for a balanced ratio by census tract is identified through relatively reduction of work trips distance along with changes in jobs-household ratios. Balanced census tracts are trip destinations with average work trips length (6.9 miles) 28 % percent shorter that others (9.6 miles). Thus, the recommended ratio might fluctuate because of the change in travel distance, travel mode, and census tract size.

SCAG (Southern California Association of Governments) established target jobs-household ratio within 14 miles of job centers through comparing with the regional average ratio of 1.25. The recommended ratio was controlled within "the middle 20 % of the SCAG region." As noted by SCAG (2001), this definition might be updated with variation in job center size, spatial distribution of job centers through the region. More importantly, the 14-mile commute shed might change over time because congestion and average commute time differ from region to region, and will change year to year. Moreover, the ratio from 1:1 to 1:1.29 based on one wage earner per household (SCAG 2001) might be arbitrary. An increase in wage earners per household is a demographic trend in most metropolitan areas in the U.S. Thus, it is necessary to estimate local average wage earner per household when calculating jobs to household ratio in different localities.

Table 18.2 Recommended jobs-housing balance ratios

Measurement	Study	Definition of JHB	Equation	Recommended ratios and scale
Jobs to household ratio	Cervero (1989)	"the share of jobs in a community actually filled by residents, and conversely the share of workers finding a place to live in that community… a match-up between the skill levels of local residents and job opportunities as well as between the earnings of workers and the cost of local housing"	1.7 jobs/cohabitant household*0.9 cohabitant household	Ceiling ratio 1.5 (multiple workers) at city level
Degree of self-containment			Workers residing locally/workers	1:1 at City level
Jobs to household ratio	Ewing et al. (1996)	Not mentioned		1.3:1–1.7:1 for 3–5 miles area around a development site
	Frank and Pivo (1994)	Relationship between the distribution of employments and households in an urban area.	Jobs/household	0.8:1–1.2:1 for census tracts
	Peng (1997)	"Spatial relationship between the number of jobs and housing units within a given geographical area"	Jobs/household	1.2:1–2.8:1 for TAZs covered by 5-mile radius of a central TAZ
	SCAG (2001)	"A provision of an adequate supply of housing to house workers employed in a defined area"	Jobs/household	1:1–1:1.29 Commute shed within 14 miles radius of job centers
Jobs to population ratio	Giuliano (1991)	"The distribution of employment relative to the distribution of workers within a given geographic area"	Jobs/population	NA for County level
Resident-workers to jobs ratio			Resident workers/jobs	
Jobs to housing units ratio	Cervero (1991)	Workers have opportunities to live within a reasonable distance to workplace	Jobs/housing units	1.4:1–1.6 for Medium-sized community
Jobs to housing units ratio	Sultana	An equal distribution of residential and employment	Jobs/housing units	0.75:1–1.5:1 for TAZs within 7 miles radius of the zone centroid

(continued)

Table 18.2 (continued)

Measurement	Study	Definition of JHB	Equation	Recommended ratios and scale
Jobs to employed residents ratio	Cervero (1996)	Not mentioned	Jobs/employed residents	0.8:1–1.25:1 at city level
	Yang and Ferreira (2005)	Jobs-housing proximity and qualitative and quantitative match between jobs and housing	Jobs/employed residents	NA for floating areas with census tracts "whose centroids are within 10 miles radius of target census tract"

Peng's study defines commute shed according to the mean commute distance of Portland metropolitan area. The catchment area is rational but cannot be universally applied to other metropolitan areas. Portland is a medium-sized region. The spline function used in Peng's research might not fit to large metropolitan areas. The range of balanced ratio is vulnerable in travel pattern and the size of TAZ and metropolitan area.

18.2.5.2 Jobs to Housing Units Ratio

Jobs-housing unit ratio seems to be deceptive because it usually conceals vacant housing units in communities. However, housing unit data at different geographic unit is easily obtained from census. Cervero's recommendation on jobs to housing units ratio should be reconsidered according to local employment rate and housing occupancy rate since his assumption is that there are two or more workers per household. In Miller's Virginia metropolitan study, he noted that jobs-housing unit ratio explains lower degree of imbalance than jobs-employed residents ratio for the same community (Miller 2010, p. 25). Cervero also noticed that perfect jobs-housing unit ratio distort real low percentage of residents who worked locally and workers residing locally (Cervero 1996).

18.2.5.3 Jobs to Employed Residents Ratio

The variable of employed residents is more reliable than housing units or households because household size varies in localities. Cervero's definition (1996) is reasonable because he considers the variations in land use patterns and travel patterns. He identifies the balanced ratio between 0.80 and 1.25 by applying a standard deviation of 0.5 based on the average jobs-employed residents ratio of 1.02 in the region.

In general, disparities in defining good jobs-housing ratios result from variations in measurements and measured units. However, it is clear that jobs-employed residents ratio is the most accurate way to quantify JHB. Several studies have failed to explain changes in commute patterns based on the calculation of jobs-population ratio and jobs-housing units. Even the ratios had moved forward to balanced range, commute time or VMT increased (Cervero 1989, 1996; Giuliano 1991; Miller 2010). In addition, no absolute balanced jobs-housing ratio can be achieved since JHB is a dynamic process. It is effective and reasonable to identify appropriate ratio range relative to regional average ratio.

18.3 Case Study of Austin MSA

18.3.1 Planning Context

The CAMPO 2035 Plan for the Austin MSA developed a comprehensive multi-model transportation system that incorporates the concept of activity centers to improve JHB and reduce VMT. The designated activity centers are expected to catch 31 and 38 % of the total population and employment respectively in CAMPO region by 2035 (CAMPO 2010, p. 22). The targeted growth for each center aims to achieve a balance growth in housing and jobs locally. CAMPO's new version of the Regional Transportation Plan suggests a rough ratio of population and employment ranging from 1:4 to 4:1 depending on the size of hierarchical activity centers in regional context (see Fig. 18.1). Defining appropriate jobs-housing ratio is CAMPO's desire for future guidance on the regional distribution of housing and jobs, which will affect proportionally funding investment. This study is going to analyze the applicability of jobs-housing ratio as an indicator for JHB in the Austin region.

18.3.2 Current Research Limitations

Although many states have adopted jobs-housing ratio as guidance for land use development, there is limited research on activity centers. Current planning practices have not provided theoretical evidence or systematic analysis in terms of the catchment area and targeted ratios of JHB for activity centers. The definition of catchment area in existing practices is general and lacking of specific consideration on the characteristics of activity centers. For example, the regional transportation plan of the Sacramento Area measures JHB within four miles of regional job centers (SACOG 2012). Similarly, there are limitations in CAMPO 2035 Plan:

1. The catchment area of JHB with circles of 2 mile, 1 mile, and 0.5 mile radius for large, medium, and small centers respectively is conceptual. There is a lack of consideration of the characteristics of activity centers in terms of location, function, transit service, current land use pattern, roles and size at city and regional level. For example, some small centers with rail stops and bus stops might measure jobs-housing ratios on a larger scale than those without transit

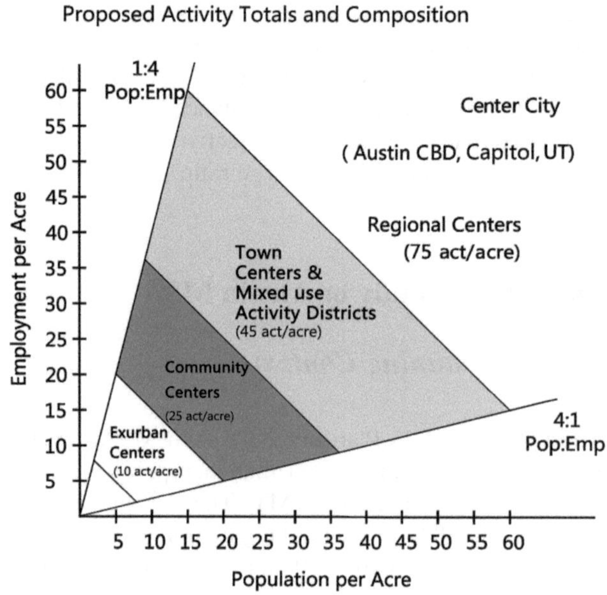

Fig. 18.1 Estimated job-housing ratios (Source: Capital Area Metropolitan Planning Organization, 2013a)

availability, such as the Northwest Center. Some small centers might have no land capacity for targeted population and jobs within 0.5-mile radius areas, such as Highland Mall. In addition, the conceptual boundaries of the activity centers are vague to extract data for measuring existing JHB.

2. The measurement of jobs-housing ratio is too rough by calculating the ratio of employment to population. This method conceals the unemployed group in communities and thus cannot directly reveal jobs distribution and housing demands in given areas. In addition, because of variation in household size, it is not appropriate to evaluate JHB by jobs-population ratio if one job per labor force is signified as a baseline in a self-contained community.

18.3.3 Ways to Evaluate Jobs-Housing Ratio

The key step to evaluate JHB is to define the study area. The geographic scale of catchment area plays a key role in measuring JHB. When JHB is used as policy tool, as stated before, the measuring scale depends on the planning goals, the urban structure, administrative limitations of jurisdictions, land capacity, as well as the role of catchment area in region. Activity centers defined in the CAMPO 2035 Plan are designated for improving land use efficiency, increasing transit ridership, and reducing VMT. Thus, the measuring scale should consider JHB in relation to commuting. Drawing from methods from precious studies (Peng 1997; Sultana 2002; SCAG 2001), it is rational to define measured units by a commute shed. According to a 2004 Austin Commuter Survey report, home-to-work trip distance

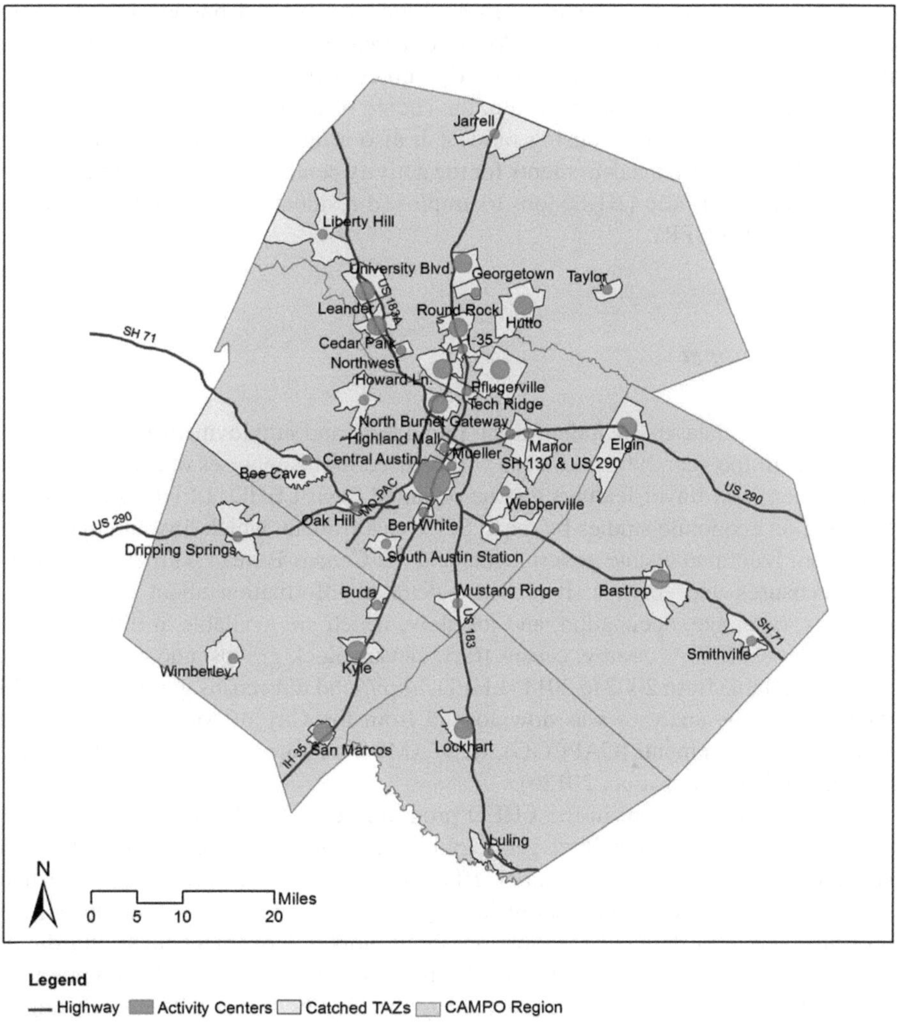

Fig. 18.2 Catchment areas of activity centers

ranges from a quarter mile to 70 miles. The average commute distance is 12.3 miles in the Austin area. Seventy-two percent of workers commute within 15 miles to the workplace and only 28 % of workers travel more than 15 miles (Bhat et al. 2004, p. 41). If this study used average commute distance as commute shed of JHB analyses, all the activity centers will join together since activity centers are designated very close to each other. Therefore, this study redefined the catchment areas by TAZs (Traffic Analysis Zones) that are totally or partially covered by defined circles of each activity center in the CAMPO 2035 Plan (see Fig. 18.2). Because the essential goal of improving JHB is to reduce VMT, the catchment areas by TAZs with travel data can help planners and local municipalities define appropriate ratios in relation to commuting benefits. In addition, the defined boundaries are more accurate to incorporate with "OntheMap" tool created by the U.S. Census Bureau in the Longitudinal Employer-Household Dynamics (LEHD).

In addition to the measuring scale, the choice of measurements and the dataset used for calculating jobs-housing ratios can also affect outcome. To examine the applicability of these measurements for the activity centers, this study will calculate jobs to household ratio (JHR), jobs to employed residents ratio (JER), and jobs to population ratio (JPR).

18.3.4 Dataset

CAMPO TAZ datasets including basic population and employment information of each TAZ in the year 2005 and 2010 will be used in the analyses of JHB. Details of employment and travel features can be obtained from LEHD. LEHD is a program focusing on economic studies by the U.S. Census Bureau, which link administrative data from federal and state governments and the Census Bureau with data extracted from censuses and surveys. It provides detailed information about employment, earnings, race, age, occupation, and job flow, which are available at different geographic levels (MSA, county, census tract, census block groups, and census block). Data is available from 2002 to 2011 (LHED, http://lehd.did.census.gov/led/). The GIS data used in the analyses was downloaded from the City of Austin, Capital Area Council of Governments (CAPCOG), and CAMPO websites (Capital Area Metropolitan Planning Organization, 2013b).

The "OntheMap" tool in the LHED program describes the distribution of jobs and workers by user specified geographies that can be imported from the GIS shapefiles. The maps provide labor efficiency (residents employed locally) and employment efficiency (workers reside locally) by defined areas. In addition, the tool also provides detailed information about worker movement regarding direction, destination, and travel distance of home-to- work trip and work-to-home trip. The LHED data is a good source to analyze the distribution of jobs and housing at local level. It generally outlines the degree of self-containment in the defined areas.

18.3.5 Analysis of Jobs-Housing Ratios

18.3.5.1 Jobs-Housing Balance Profile in the Region

The jobs-housing imbalance has been a critical issue in Central Texas. By 2010, almost half of all workers commute across county lines to reach workplaces (CAPCOG Central Texas Regional Data, 2013). The share of work trip distance greater than 50-miles increased 5 % from 2005 to 2010 in the Austin MSA (see Table 18.3). Travis and Williamson Counties are the largest job pools, attracting large amounts of commuters from the north and southwest of the CAMPO region

Table 18.3 Austin MSA work to home travel profile

	2010		2005	
Travel distance	Count	Share (%)	Count	Share (%)
Less than 10 miles	306,265	40.60	306,212	46.70
10–24 miles	199,660	26.50	173,759	26.50
25–50 miles	52,927	7.00	43,278	6.60
Greater than 50 miles	195,607	25.90	133,043	20.30
Total Primary Jobs	754,459	100.00	656,292	100.0

Data source: Longitudinal Employer-Household Dynamics, http://lehd.ces.census.gov/

Table 18.4 Changes in jobs-housing ratio and commute distance

County	Year	Local workers	Jobs	Local workers (%)	Employed residents	JER	Share of trip distance >10 miles (%)
Travis	2005	314,591	560,377	56.1	406,810	1.38	26.6
	2010	312,729	621,672	50.3	455,766	1.36	33.1
Williamson	2005	39,005	76,965	50.7	146,841	0.52	52.0
	2010	49,916	109,649	45.5	175,448	0.62	56.0
Hays	2005	15,680	37,967	41.3	44,309	0.86	69.9
	2010	15,355	40,631	37.8	59,971	0.68	77.5
Caldwell	2005	2,242	4,894	45.8	12,261	0.4	58.3
	2010	2,313	5,512	42.0	15,921	0.35	64.6
Bastrop	2005	6,267	10,638	58.9	25,070	0.42	58.0
	2010	6,570	12,489	52.6	31,763	0.39	62.5

Data source: OntheMap http://onthemap.ces.census.gov/

(Table 18.4). Figure 18.3 shows the general profile of travel distance and direction in the CAMPO region. The darkest gray color and the lightest gray color in radar charts represent shortest (less than 10 miles) and longest travel distance (greater than 50 miles) respectively, with the number of commuters in direction. Table 18.4 shows the changes in the share of local workers (workers residing locally), jobs-employed residents ratios and work-to-home (all travel modes) trip distance in five counties of the CAMPO region between 2005 and 2010. Travis and Williamson Counties experienced slight increases in jobs-housing balance. In contrast, the jobs-employed residents ratios of Bastrop, Hays, Caldwell Counties reduced between 2005 and 2010, but commuters with trip distance greater than 10 miles increased greatly. Williamson and Hays Counties have relatively high jobs-employed residents ratios but the shares of local workers are much lower than other counties. In other words, local job type cannot satisfy local residents. Conversely, Bastrop County is the highest self-contained community though it shows an extreme employed-resident ratio imbalance. Therefore, it seems that the jobs to housing ratios did not explain the exact spatial distribution of jobs and workers.

However, the jobs-housing ratios at county level can provide a basis of reference for the activity centers within each county. For instance, SCAG defines the

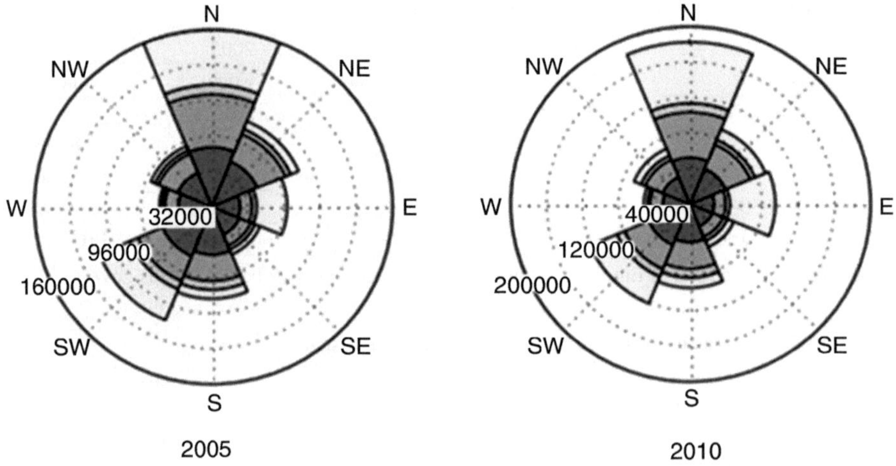

Fig. 18.3 Jobs counts by distance/direction in 2005 and 2010 (all workers in the CAMPO region) (Source: OntheMap http://onthemap.ces.census.gov/)

appropriate ratio by using 20 % fluctuation of regional ratio (SCAG 2001). Cervero applies a standard deviation of 0.5 based on the regional ratio of 1.01 (Cervero 1996). Similarly, Fairfax County identifies the ratios for TOD centers based on empirical ratios in D.C. region (Fairfax County Department of Planning and Zoning 2012).

18.3.5.2 Jobs-Housing Ratios of Activity Centers

The study aggregated data from TAZs that are covered within activity centers and calculated jobs to household ratios (JHR) and jobs to population ratios (JPR) in 2005 and 2010. Jobs to employed-residents ratios (JER) were calculated by using the LEHD data. Particularly note that JPR is vulnerable in household size that varies from 1.7 to 3.3 in the activity centers. Thus, JPR is not a good measurement of JHB (see Table 18.5).

As stated before, good jobs-housing ratios differ in geographic units, urban pattern, and metropolitan size. Since the CAMPO region is a medium-sized metropolitan area like Portland, Oregon (Table 18.6), this study drew from Peng's results, assuming that JHRs within 1.2–2.8 are balanced ratios. JERs and JHRs were grouped based on Peng's definition (see Figs. 18.4 and 18.5).

Figure 18.4 shows activity centers with improved or reduced jobs-housing balanced ratios from 2005 to 2010; Fig. 18.5 displays activity centers that kept balanced ratios in 2005 and 2010. Plots shapes of JER and JHR in two figures show inconsistent trend, implying the variations of these two measurements in explaining the degree of jobs-housing balance. Differences in plots shape of JHR and JER indicate the variations in demographic characteristics among these centers, such as

Table 18.5 Jobs-housing ratios in activity centers (2005–2010)

	Name	Defined growth type	Average household size 2010	JER 05	JHR 05	JER 10	JHR 10	JPR 10
1	Georgetown	Medium	2.4	1.84	1.46	2.02	2.31	0.82
	Bastrop	Medium	2.5	1.64	1.76	1.6	1.95	0.75
	Taylor	Small	3.3	1.15	1.24	0.96	1.4	0.48
	RM 2222 & RM 620	Small	2.5	2.48	1.69	1.19	1.82	0.75
	Oak Hill	Small	2.6	2.05	2.12	1.19	1.88	0.72
	Bee Cave	Small	2.6	2.34	1.81	2.65	1.79	0.69
	Buda	Small	2.4	1.21	2.19	0.56	2.02	0.71
2	Leander	Medium	2.9	0.6	1.11	0.4	1.45	0.5
	Round Rock	Medium	2.5	0.64	1.1	0.59	1.25	0.43
	San Marcos	Medium	1.7	1.94	2.39	1.83	3.26	0.94
	Tech Ridge	Small	2.6	0.42	1.13	0.43	1.24	0.48
	Mueller	Small	2.7	0.76	0.9	0.59	1.67	0.64
	Jarrell	Small	2.8	0.37	1.12	0.71	1.22	0.43
	SH 130 & SH 71	Small	3.2	0.31	0.44	0.38	0.54	0.16
3	Central Austin	Large	2.8	5.03	2.99	5.67	3.19	1.49
	North Burnet Gateway	Medium	2.2	2.79	3.57	2.45	3.7	1.67
	Howard Ln.	Medium	2.6	1.17	1.04	0.96	1.18	0.47
	Lockhart	Medium	2.7	0.37	0.55	0.45	0.75	0.28
	Cedar Park	Medium	3	0.59	0.62	0.32	1.14	0.38
	Kyle	Medium	2.8	1.36	0.32	0.32	0.39	0.12
	Pflugerville	Medium	3	0.11	0.39	0.65	0.39	0.13
	Elgin	Medium	2.9	0.84	0.65	0.83	0.75	0.26
	Hutto	Medium	3.1	0.52	0.78	0.43	0.64	0.21
	Northwest	Small	1.8	2.51	4.03	2.03	5.73	2.17
	Highland Mall	Small	2.3	5	3.43	4.95	3.67	1.49
	Smithville	Small	2.5	0.49	0.64	0.39	0.79	0.3
	Liberty Hill	Small	2.9	0.43	0.55	0.34	0.54	0.18
	I-35 & SH 45 N	Small	2.8	3.4	6.02	7.56	6.69	2.38
	Luling	Small	2.8	0.04	0.75	0.05	0.93	0.34
	South Austin Station	Small	2.6	0.46	0.29	0.2	0.27	0.1
	Mustang Ridge	Small	3.1	0.2	0.28	0.23	0.46	0.15
	SH 130 & US 290	Small	2.6	0.16	1.6	0.18	1.07	0.34
	Manor	Small	2.7	0.33	0.33	0.37	0.36	0.12
	Ben White	Small	2.6	2.03	3.11	1.55	2.95	1.22
	Dripping Springs	Small	2.8	0.62	1.18	0.7	1.13	0.4
	University Blvd.	Small	2.3	1.49	0.01	9.63	0.76	0.27
	Webberville	Small	3.5	0.12	0.14	0.12	0.14	0.04
	Wimberley	Small	2.4	0.52	0.82	0.55	0.88	0.37

Source: Authors' calculation based on datasets from CAMPO and LEHD

Table 18.6 Region profiles of Portland and Austin

Statistics	Portland-Vancouver-Hillsboro	Austin-Round Rock-San Marcos
Total population	1,789,580	1,716,289
Population density (per sq. mile)	406.8	406.7
Land area (sq. miles)	4,481.3	4,280.1

Data source: American Fact Finder 2012

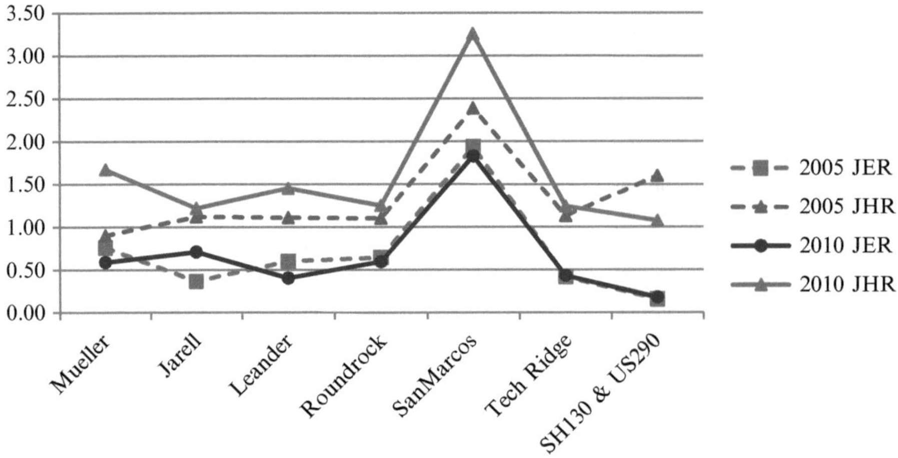

Fig. 18.4 Ratios portraying jobs-housing balance in activity centers (ratios out/in the range of 1.2–2.8) (Data source: TAZ dataset from CAMPO)

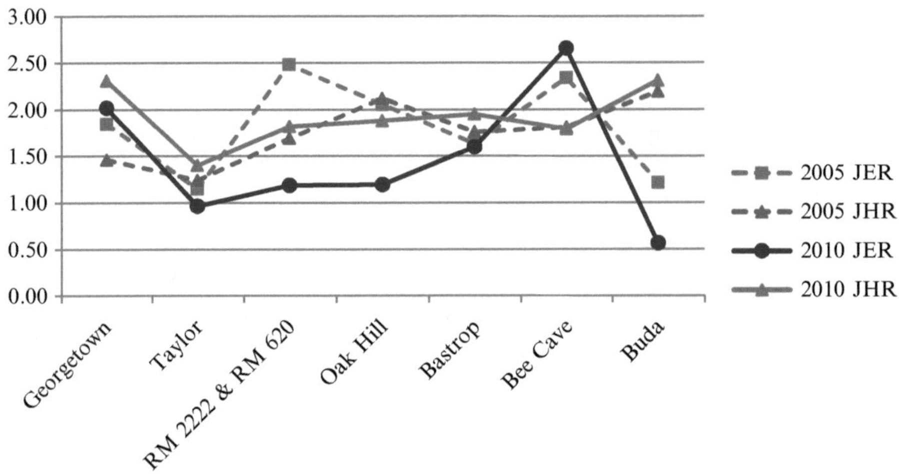

Fig. 18.5 Ratios portraying jobs-housing balance in activity centers (ratios kept in the range of 1.2–2.8) (Data source: TAZ dataset from CAMPO)

average wage earners per household and employment rate. If the application of JHB focuses on commuting, JHR is not a good indicator because it fails to convey real commuters.

In order to test the reality of jobs-housing ratios, more information about distribution of jobs and housing within each center was extracted from LEHD by using the U.S. Census Bureau's OntheMap tool. Table 18.7 presents the JERs, self-containment, and travel distance (all travel modes) larger than 10 miles. Currently, all the activity centers have pretty low employment efficiency (share of workers reside locally below 20 %) and labor force efficiency (share of residents employed locally below 20 %). Some activity centers with assumed balanced jobs to household ratios (based on Peng's definition) have low degrees of self-containment and larger amounts of long distance commuters, such as Georgetown, Taylor, Bastrop, Bee Cave, and Buda. Over half of workers in all the activity centers endure long distance work-to-home trips (longer than 10 miles). Base on the data collected, no clear relationship was found between jobs-employed residents ratios and self-containment, jobs-employed residents ratios and commute distance. Some activity centers with perfect ratios also show very low self-containment, such as Oak Hill, RM 2222 & RM 620, Taylor, and Howard Ln centers. There might be a severe mismatch between jobs opportunities and workers' skills, the salary of workers and housing prices.

Overall, existing jobs-housing ratios of activity centers cannot fully explain the real JHB based on the data applied in this study. Activity centers with high ratios share similar poor self-containment and long distance commutes as those with low ratios. More balanced jobs-housing ratios are not associated with reduction in commute distance. Beyond the ratios, more factors that influence JHB should be explored. Nevertheless, there is a positive correlation ($r = 0.3$) between jobs to employed-residents ratios and local employed residents, indicating that improvement in jobs to employed-residents ratios could increase self-containment in communities because of more jobs opportunities in given areas.

18.4 Discussions and Suggestions

There are researchers that have traditionalist attitudes to JHB as a policy tool to relieve traffic congestion and VMT (Downs 2004; Levine 1998; Peng 1997). As Giuliano stated, jobs-housing ratio can be adopted for solving transportation problems in two contexts: (1) JHB can be only accomplished by government intervention, and (2) there is a significant impact of JHB on travel patterns (Giuliano 1991, p. 305). Yang and Ferreira (2005) asked a more primary question: how can existing urban patterns be characterized with regard to spatial distribution of jobs and housing? (p. 171)

The first issue in the application of jobs-housing ratio is the limitation of government intervention. Longitudinal evidence from many cities has revealed that JHB is a dynamic process dominated by market force in urban development.

Table 18.7 Jobs-employed residents ratios and travel distance

Activity center	Growth type	Job-employed resident ratio (JER)		Local workers		Share of work-home trip >10 miles	
		2005	2010	2005 (%)	2010 (%)	2005 (%)	2010 (%)
Central Austin	Large	5.03	5.67	8.6	7.3	53.2	59.0
Highland Mall	Small	5.00	4.95	1.2	1.0	57.6	63.9
I-35 & SH 45 N	Small	3.40	7.56	1.5	1.1	41.5	44.1
North Burnet Gateway	Medium	2.79	2.45	3.8	3.2	45.6	55.1
Northwest	Small	2.51	2.03	0.5	1.0	57.2	54.1
RM 2222 & RM 620	Small	2.48	1.19	1.8	1.8	52.6	61.0
Bee Cave	Small	2.34	2.65	2.3	2.7	60.1	59.9
Oak Hill	Small	2.05	1.19	2.7	2.6	58.0	57.4
Ben White	Small	2.03	1.55	2.0	2.2	55.8	59.0
San Marcos	Medium	1.94	1.83	14.5	10.9	53.8	60.9
Georgetown	Medium	1.84	2.02	7.8	5.7	58.3	50.3
Bastrop	Medium	1.64	1.60	17.4	13.5	56.8	60.3
University Blvd.	Small	1.49	9.63	0.0	0.1	68.2	69.6
Manor	Small	1.36	0.32	2.1	1.3	48.9	71.2
Buda	Small	1.21	0.56	4.0	2.8	60.7	64.0
Howard Ln.	Medium	1.17	0.96	2.8	2.6	49.4	46.2
Taylor	Small	1.15	0.96	17.0	12.7	49.0	56.1
Luling	Small	0.84	0.83	28.8	23.8	64.1	68.8
Mueller	Small	0.76	0.59	2.4	2.3	61.3	63.5
Roundrock	Medium	0.64	0.59	14.4	9.8	44.4	53.3
Dripping Springs	Small	0.62	0.70	7.1	7.4	65.6	69.2
Leander	Medium	0.60	0.40	7.9	7.8	62.8	59.2
Smithville	Small	0.59	0.32	20.1	27.7	60.1	51.4
Wimberley	Small	0.52	0.55	23.0	18.4	49.7	55.2
Hutto	Medium	0.52	0.43	8.0	6.2	52.8	61.5
Elgin	Medium	0.49	0.39	27.0	20.9	53.0	62.9
Kyle	Medium	0.46	0.20	5.7	6.4	64.2	65.4
Lockhart	Medium	0.43	0.34	30.1	25.0	51.5	58.9
Tech Ridge	Small	0.42	0.43	3.0	3.3	52.6	60.3
Cedar Park	Medium	0.37	0.45	12.5	9.2	53.4	58.2
Jarell	Small	0.37	0.71	6.3	5.7	76.8	80.2
Pflugerville	Medium	0.33	0.37	18.5	15.4	37.0	42.2
Liberty Hill	Small	0.31	0.38	12.4	10.0	60.4	69.3
South Austin Station	Small	0.20	0.23	7.6	7.4	54.2	63.0
SH130 & US290	Small	0.16	0.18	0.0	0.4	61.2	70.0
Mustang Ridge	Small	0.12	0.12	0.0	1.3	78.1	76.9
SH 130 & SH 71	Small	0.11	0.65	2.8	0.8	59.7	55.0
Webberville	Small	0.04	0.05	4.8	5.0	69.0	65.0

Data Source: CAMPO and LEHD

On one hand, the choice of location of residence and firms are the simultaneous process. Employers tend to locate firms close to labor pool and resident tend to choose "utility-maximizing location" where extra commuting cost can be traded off with the marginal saving of housing and services. Under this colocation of employment and residence, free market forces will push jobs-housing toward a general balance. On the other hand, the assumption that suburban residents will move to more balanced areas might be unrealistic unless an extreme imbalance has caused high commuting cost. Personal preferences and tastes (e.g., services, environment, housing size, etc.) also strongly influence the locations of homes. Residents' self-selection indicates that even if planning interventions achieved balanced jobs-housing ratio, there might be workers who move to imbalance neighborhoods (Peng 1997, pp. 1231–1233).

Planners and policy makers should note that many factors behind jobs-housing ratio could affect commute pattern. Peng's (1997) study also found population density and income greatly influence commute patterns. Other factors play important roles in housing selection process, including housing price, housing type, school quality, neighborhood characteristics, transportation availability, and environmental amenities. Notwithstanding this, the potential to achieve a better balance in jobs and housing exists in extremely imbalanced communities. Jobs-housing imbalance implies the failure of the market force in adjusting jobs and housing supply. Self-selection of home location or employment location is restricted or misguided by market forces. Thus, the task of planning is to understand free market forces and eliminate obstacles to ensuring a jobs-housing balance.

The numerical balance or imbalance might conceal the truth of commute patterns. Improving JHB, however, is not simply adjusting jobs-housing ratio. For example, counties in the Central Texas show the numerical balanced jobs-employed residents ratios, but the percentage of long distance commuting have increased, and the self-containment has decreased between 2005 and 2010. This finding is consistent with Cervero (1989) and Giuliano's (1991) studies in larger metropolitan areas. Cervero (1996) tracked the changes in jobs-housing ratios in the 20 largest cities in the San Francisco Bay Area in the 1980s, and the evidence shows that some cities have moved forward to a better balance, but frequent internal commuting trips generated an increasing VMT in total. Simply adding more jobs into the labor pool might decrease commuting distance but simultaneously increase commute time to workplaces. Thus, introducing job opportunities to dormitory communities should consider the match-up between job qualifications and residents' skills and education level.

Another dilemma of using jobs-housing ratio as a policy instrument is defining a good ratio of JHB. Jobs-housing ratio is sensitive depending on the geographic scale used. The ratio of jobs-employed residents at 1:1 is achievable at the metropolitan level but might not be a feasible goal for sub-regions. Determining the appropriate geographic units of jobs-housing ratio depends on local needs, the economic structure, and land use pattern of metropolitan areas. Primarily, planners should clarify the main purpose of defining jobs-housing ratio at different geographic scopes. JHB cannot be multiple solutions for metropolitan problems caused by urban sprawl or suburbanization, such as congestion, air pollution, and segregation (Giuliano 1991).

When JHB policy is applied at the regional level in the context of traffic mitigation and air pollution, the targeted ratios should be redefined in a regional context. For instance, with the goal of relieving interregional long distance commutes, the Sacramento Area Council of Government (SACOG 2012) pursues a better JHB in a sub-region area within 4 mile-radius of regional job centers. A realistic target ratio of 1.2 is redefined for sub-regions based on the existing even balanced ratios in the six counties of SACOG region (pp. 3–28). While JHB policy is applied at the local level, defining good jobs-housing ratio involves more specific information and local needs, such as the current degree of land use mixture and job supply, the capacity of land use and infrastructure and the economic incentives of measured area. For example, in Fairfax County, VA, with an emphasis on providing housing in areas around transit stations and 13 regional activity centers, the target ratios of these centers were reduced based on the average ratio of all regional activity centers in the Metropolitan Washington area (Fairfax County Department of Planning and Zoning 2012, pp. 7–10).

Whatever the geographic unit is used, it is essential to note that the size of the metropolitan area and the degree of urbanization affect the usefulness of jobs-housing policy. Highly urbanized areas with high homogeneity of industrial clusters in a certain subarea, like the San Francisco Bay Area and Southern California, are more likely to generate large amount inter-regional work trips. In contrast, small communities tend to generate more internal work and non-work trips. Hence, for a small metropolitan region with rare cross-county work trips, it is meaningless to measure JHB by county level. It is more practical to target the balance at micro units (census tracts or TAZs) in order to reduce internal work trips. Defining jobs-housing ratio by commute shed should consider the characteristics of travel patterns. A rational and reasonable commute shed should consider commuters' acceptable travel distance. For example, Southern California initiated a survey to define jobs-housing balance areas within about 30-min travel distance at an average speed of 28 mph (Weitz et al. 2003). Furthermore, acceptable travel distance differs in travel modes. If JHB policy attempts to encourage transit ridership in a given area around a transit station, it is helpful to set JHB goal within an accessible transit service area. If multimodal transportation means are considered, the catchment area may include areas within reasonable driving or biking distance to transit stations. In addition, the acceptable distance might be affected by other factors, such as culture and weather. For example, 15 min of walking to a transit station might be common in the northern cities, such as New York and Boston, but might be unacceptable in hot Texas.

When jobs-housing ratio is applied as land use policy regarding measured units, administrative limitations should be taken into account. For instance, in some states, counties have no legislative power over zoning, so it is feasible to measure JHB by city rather than by county. In addition, a good jobs-housing ratio should meet the local needs. Local municipalities should have knowledge of existing conditions about jobs-housing balance. The applicability of jobs-housing ratio strongly relies on communities' expectations. Some communities may prefer "semi-rural lifestyles" and keep bedroom status, while some may pursue non-residential development because of tax needs (Koh 2012, p. 30).

In general, choosing a good ratio for a given area involves comprehensively understanding the impact of market force on jobs-housing proximity, the relationship between jobs-housing ratio and urban pattern, commuting impact of existing urban pattern in local settings, as well as future growth trends. Planners should keep in mind that improving JHB is a dynamic process.

18.5 Conclusions

This paper has explored the application and applicability of the JHB policy through a thorough literature review of the definition of JHB, ways to measure JHB, and effectiveness of jobs-housing ratio. The study aims to answer the core question: what is the appropriate way of defining a good jobs-housing ratio in the local context?

Jobs-housing ratio is the easiest and most widely used method to measure JHB in many states. Beyond numerical balance, JHB are more influenced by other factors, including match-up between jobs type and the qualification of workers, worker's wage and housing price, market force, and residents' self-selection. Thus, defining a good ratio of JHB involves specific factors, depending on measured geographic units and planning goals. The appropriate geographic units of jobs-housing ratio are determined based on local needs, commute patterns, the economic structure and land use patterns of metropolitan areas. The size of metropolitan area and the degree of urbanization strongly affects the usefulness of jobs-housing policy because large communities and small communities usually have different land use patterns and travel patterns.

This study analyzed existing jobs-housing ratios in the defined activity centers in Central Texas. An extreme mismatch exists between jobs and housing distribution based on estimated commute distance from LHED. However, existing jobs-household ratios can hardly explain this mismatch. Poor relationship was found between commute distance, self-containment, and jobs-employed residence ratio. The outcome of the case study, nevertheless, indicates the potential to improve JHB through the adjustment of the distribution of employment and residential in detailed local settings.

As the initiative of improving JHB is reducing traffic congestion and VMT, a future extension of this study is to explore the impact of jobs-housing ratio on commuting. In order to define more appropriate geographic scale related to acceptable commute shed, more detailed information about travel patterns will be considered, such as work VMT by activity center, external trips and internal trips for activity centers. An alternative way is conducting a survey for acceptable travel distance as commute shed. Beyond pursuing good jobs-housing ratio, future studies should be aimed at promoting JHB through improving the match-up between jobs and housing and more socioeconomic factors should be included.

References

Bento, A. M., Cropper, M. L., Mobarak, A. M., & Vinha, K. (2005). The effects of urban spatial structure on travel demand in the United States. *Review of Economics and Statistics, 87*(3), 466–478.

Bhat, C. R., Sivakumar, A., Sen, S., Guo, J., & Copperman, R. (2004). *Austin commuter survey: Findings and recommendations*. Austin: The University of Texas at Austin.

California Planning Roundtable. (2008). *Deconstructing jobs-housing balance*. National Park: California Planning Roundtable.

CAPCOG (Capital Area Council of Governments). *Commuting patterns*. Retrieved June 31, 2013, from CAPCOG website: http://www.capcog.org/data-maps-and-reports/central-texas-regional-data/#commutingpatterns

CAMPO 2010. (2010, May). *CAMPO 2035 Regional Transportation Plan*. Austin: Capital Area Metropolitan Planning Organization.

Capital Area Metropolitan Planning Organization. (2013a, April). *Revised CAMPO Centers Criteria*.

Capital Area Metropolitan Planning Organization. (2013b, June). *2005 to 2035 demographics* [Geospatial data]. Retrieved from http://www.campotexas.org/programs_gis.php

Cervero, R. (1989). Jobs-housing balancing and regional mobility. *Journal of the American Planning Association, 55*(2), 136–150.

Cervero, R. (1991). Jobs housing balance as a public policy. *Urban Land, 50*(10), 10–14.

Cervero, R. (1996). Jobs-housing balance revisited: Trends and impacts in the San Francisco Bay Area. *Journal of the American Planning Association, 62*(4), 492–511.

Cervero, R., & Duncan, M. (2006). Which reduces vehicle travel more: Jobs-housing balance or retail-housing mixing? *Journal of the American Planning Association, 72*(4), 475–490.

Downs, A. (1992). *Stuck in traffic: Coping with peak-hour traffic congestion*. Washington, DC: The Brookings Institution Press.

Downs, A. (2004). *Still stuck in traffic: Coping with peak-hour traffic congestion*. Washington, DC: The Brookings Institution Press.

Ewing, R., & Cervero, R. (2001). Travel and the built environment: A synthesis. *Transportation Research Record: Journal of the Transportation Research Board, 1780*(1), 87–114.

Ewing, R., DeAnna, M., et al. (1996). *Best development practices: Doing the right thing and making money at the same time*. Urban Land Institute; Florida. Dept. of Community Affair.-Washington, DC: American Planning Association.

FHWA (U.S. Department of Transportation Federal Highway Administration). (2011, June). *Summary of Travel Trends: 2009 National Household Travel Survey (2009 NHTS)*. Washington, DC.

Frank, L. D. (1994). An Analysis of relationships between urban form (Density, mix and jobs: Housing balance) and travel behavior (mode choice, trip generation, trip length and travel time). *Transportation Research Part A: Policy and Practice, 34*, 171–205.

Frank, L., Pivo, G. (1994) *Relationships between land use and travel behavior in the Puget sound region*, 397. Seattle: Washington State Transportation Center.

Giuliano, G. (1991). *Is jobs-housing balance a transportation issue?* (Transportation research record, no. 1305, pp. 305–312). San Francisco: The University of California Transportation Center.

Giuliano, G., & Small, K. A. (1993). Is the journey to work explained by urban structure? *Urban Studies, 30*(9), 1485–1500.

Giuliano, G. (1995).The weakening transportation-land use connection.ACCESS Magazine, 1(6).

Handy, S., Cao, X., & Mokhtarian, P. L. (2006). Self-selection in the relationship between the built environment and walking: Empirical evidence from Northern California. *Journal of the American Planning Association, 72*(1), 55–74.

Home + Transportation cost % Income. (n.d.) Retrieved June 15, 2013, from H + T Affordability Index website: http://htaindex.cnt.org/map/

Horner, M. W. (2008). Optimal accessibility landscapes? Development of a new methodology for simulating and assessing jobs—housing relationships in urban regions. *Urban Studies, 45*(8), 1583–1602.

Jobs-Housing Ratios: National Perspectives and Regional and Local Benchmarks (Fairfax County Department of Planning & Zoning, Comp.). (2012, December).

Koh, J. (2012, April). *To what extent do municipalities in metro Vancouver support a jobs-housing balance? An evaluation of official community plans in metro Vancouver*. Vancouver, Canada: The University of British Columbia.

Levine, J. (1998). Rethinking accessibility and jobs-housing balance. *Journal of the American Planning Association, 64*(2), 133–149.

Levinson, D. M., & Kumar, A. (1994). The rational locator: Why travel times have remained stable. *Journal of the American Planning Association, 60*(3), 319–332.

LEHD, http://onthemap.ces.census.gov/. (n.d.). Retrieved July 25, 2013, from Longitudinal Employer-Household Dynamics http://lehd.ces.census.gov/

Loo, B. P., & Chow, A. S. (2011). Jobs-housing balance in an era of population decentralization: An analytical framework and a case study. *Journal of Transport Geography, 19*, 552–562.

Macek, N. M., Khattak, A. J., & Quercia, R. G. (2001). What is the effect of commute time on employment?: Analysis of spatial patterns in New York metropolitan area. *Transportation Research Record: Journal of the Transportation Research Board, 1780*(1), 43–52.

Margolis, J. (1957). Municipal fiscal structure in a metropolitan region. *The Journal of Political Economy, 65*(3), 225–236.

Miller, J. S. (2010). *Feasibility of using jobs/housing balance in Virginia statewide planning* (No. FHWA/VTRC 11-R1). Charlottesville: Virginia Transportation Research Council.

Miller, E. J., & Ibrahim, A. (1998). Urban form and vehicular travel: Some empirical findings. *Transportation Research Record: Journal of the Transportation Research Board, 1617*(1), 18–27.

Peng, Z. R. (1997). The jobs-housing balance and urban commuting. *Urban Studies, 34*(8), 1215–1235.

SACOG (Sacramento Area Council of Government). (2012, February). *Metropolitan transportation plan sustainable communities strategy (MTP/SCS 2035)*. http://www.sacog.org/2035/draft-final-mtpscs/

SCAG. (2001). *The new economy and job-housing balance in Southern California* (pp. 19–20). Los Angeles: Southern California Association of Governments.

Shen, Q. (2000). Spatial and social dimensions of commuting. *Journal of the American Planning Association, 66*(1), 68–82.

Sultana, S. (2002). Job/housing imbalance and commuting time in the Atlanta metropolitan area: Exploration of causes of longer commuting time. *Urban Geography, 23*(8), 728–749.

Wang, D., & Chai, Y. (2009). The jobs–housing relationship and commuting in Beijing, China: The legacy of Danwei. *Journal of Transport Geography, 17*(1), 30–38.

Weitz, J., A. P. Association, et al. (2003). *Jobs-housing balance*. Chicago: American Planning Association Chicago.

White, M. J. (1988). Urban commuting journeys are not "wasteful". *The Journal of Political Economy, 96*(5), 1097–1110.

Yang, J., & Ferreira, J. R. (2005). Evaluating measures of job–housing proximity: Boston and Atlanta, 1980–2000. *Access to Destinations, 7*, 171–192.

Zhang, M., Shen, Q., & Sussman, J. (1999). Strategies to improve job accessibility: Case study of Tren urbano in San Juan metropolitan region. *Transportation Research Record: Journal of the Transportation Research Board, 1669*(1), 53–60.

Zhao, P., Lu, B., & Linden, G. J. (2009). The effects of transport accessibility and jobs–housing balance on commuting time: Evidence from Beijing. *International Planning Studies, 14*(1), 65–83.

Chapter 19
The Implication of City Form Transformation in Rongcheng Area of Jieyang City

Jian-wen Huang and Ying Xu

19.1 Background

Rongcheng is an area located in the center of Jieyang, a medium-sized city in eastern Guangdong province (Figs. 19.1 and 19.2). Rongcheng is a large inner city that is more than 800 years old (Fig. 19.3). The inner city dates from the Shaoxing period of the South Song Dynasty, when local authorities picked Yujiao village (site of the current Rongcheng inner city) as the county seat. In those early days, only south and north markets and large structures, such as county Yamen (administrative offices) and Confucian temples, existed and there was no block structure. The area underwent a series of changes in subsequent decades and centuries, including reconstruction after it was destroyed by fire in the Yuan Dynasty, expansion during the Ming and Qing Dynasties to increase security, and redevelopment in the Republic era. It gradually developed into a traditional inner city with a unique city form adaptable to local climate and geographical features.

To analyze the unique city-form features of the traditional inner city, we focus on three typical development periods of the city: the Qing Dynasty, the Republic era, and today. This study analyzes the problems, patterns, and forces of the urban transformation in the three major periods. This paper also approaches the research subject from four space dimensions: urban frame, channel network, street fabric, and channel-side node. The spatial features of streets and blocks are investigated

J.-w. Huang (✉)
School of Architecture and Urban Planning, Guangdong University of Technology, Guangdong, Guangzhou, People's Republic of China 510090
e-mail: kenpost@163.com

Y. Xu
Architectural Design Research Institute of SCUT, South China University of Technology, Guangdong, Guangzhou, People's Republic of China 510640

Q. Pan, J. Cao (eds.), *Recent Developments in Chinese Urban Planning*, GeoJournal Library 114, DOI 10.1007/978-3-319-18470-8_19

Fig. 19.1 Location map of Jieyang city

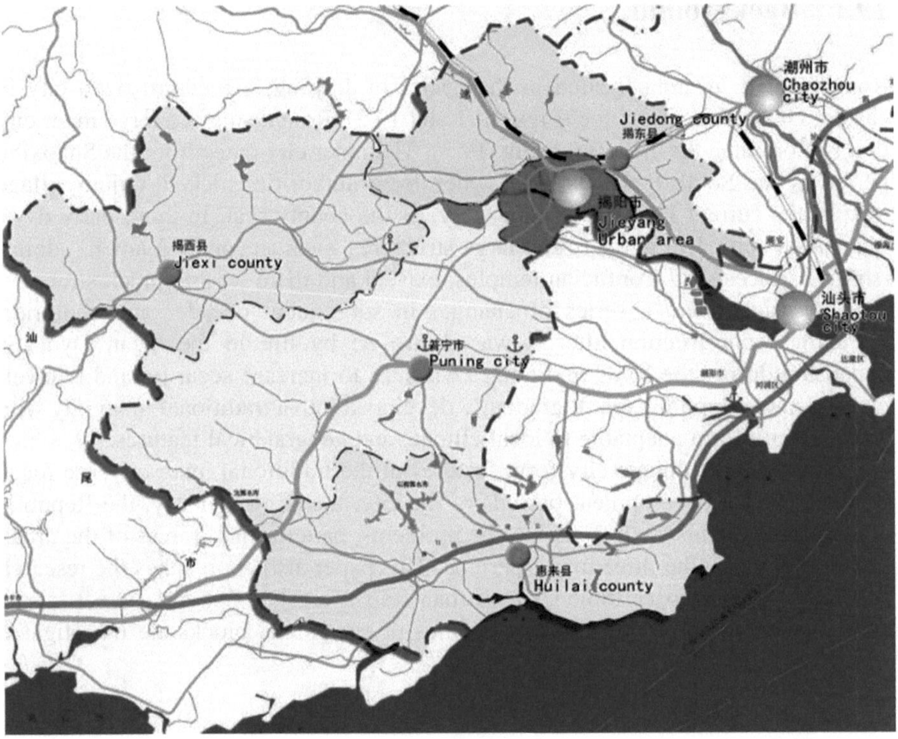

Fig. 19.2 Jieyang city administrative region and the urban area

along with the time dimension for the changes in city-form features. At the end, we elaborate on two questions: "How can the traditional city-form features be inherited?" and "What implications will it provide for city planning at the present day?"

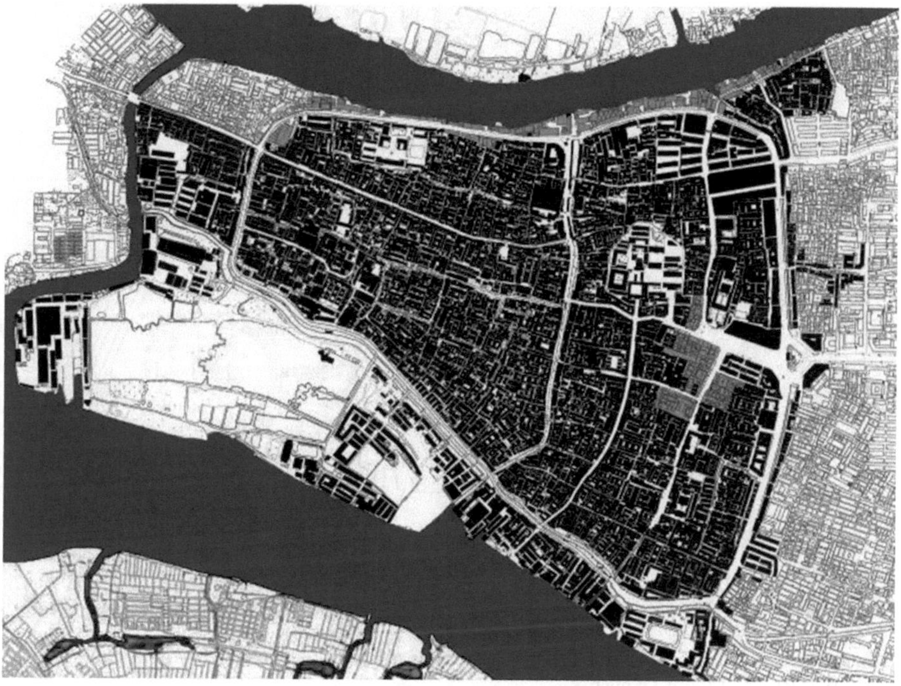

Fig. 19.3 Jieyang inner city and the building form

19.2 Analysis on the City Form Transformation of Rongcheng Area in Time Dimension and Space Dimension

19.2.1 Elements of Urban Frame

By comparing urban maps from the three study periods, (Rongcheng Township Map for the Guangxu period of Qing Dynasty; Rongcheng Area Map for the 23rd year of the Republic; Jieyang Land Use Status Map for the year 2003), we can see that the inner city of Jieyang does not change with the rise and fall of the city wall. It is built in accordance with its unique geographical features.

Rongcheng Township Map shows that the inner city of Jieyang was surrounded by water on four sides, with rivers both south and north and channels to the east and west sides (Fig. 19.4). In this way, water forged a natural defense for the city. According to records in the Qianlong county chronicle: the circumference of the city wall is 1,600 Zhang (a unit of length, $\approx$3.3 m). It is 2.3 Zhang tall and 1.6 Zhang thick. Five city gates exist on land, including North Gate, South Gate, Jinxian Gate, East Gate and West Gate. There are also four water gates: North Water Gate, South Water Gate, Mashan Water Gate, and Wuxi Water Gate (Liu Yeqin 1973). The Jinxian Gate tower faces the east and was built in 1622 (Ming

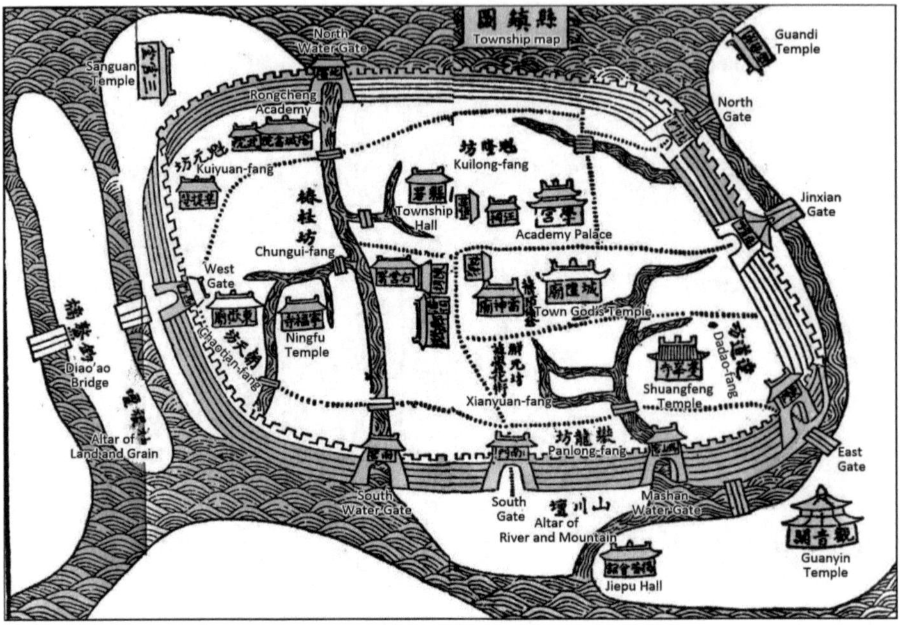

Fig. 19.4 Rongcheng Township Map (From 'Jieyang county extend chronicle county map', Guangxu period of Qing Dynasty)

Dynasty). It is so named because it leads to the Academy Palace and Jinxian means "ushering in the wise and talented" (Chinese Album Editorial Department 2008). This demonstrates the unique philosophy behind its layout. Ancient cities rarely include a fifth city gate and four water gates.

Compared with the Rongcheng Township Map, Rongcheng Area Map reveals no great changes in the urban frame from the Qing Dynasty to the Republic era (Fig. 19.5). The south and north river branches form the boundary between the west and east downtown areas. The east side of downtown has a great concentration of public buildings and infrastructure, while the west side of downtown developed more slowly. The original feature of the 'city within a city' remains unchanged, however. This inside city used to be called Golden City, but is now called Nei City or Jin City. In the Republic era, it hosted the county government and served as an important place for administration. This inside city originated in the Yuan Dynasty, functioning as a key defense fortress for the county. Later, the inside city, together with the outer city, formed two important layers of city walls. It is the only completely preserved city wall from the Yuan Dynasty that survives to the present day in Guangdong Province. The outer city wall has been less fortunate. According to official records, it was destroyed in the 1937 (the 26th year of the Republic) when the city-ring road was being built. Only the Jinxian Gate tower remains (Yang Sen 1996).

On the Jieyang Land Use Status Map, we can see that the inner city has broken the city-wall boundary and extended to both the east and the west (Fig. 19.6).

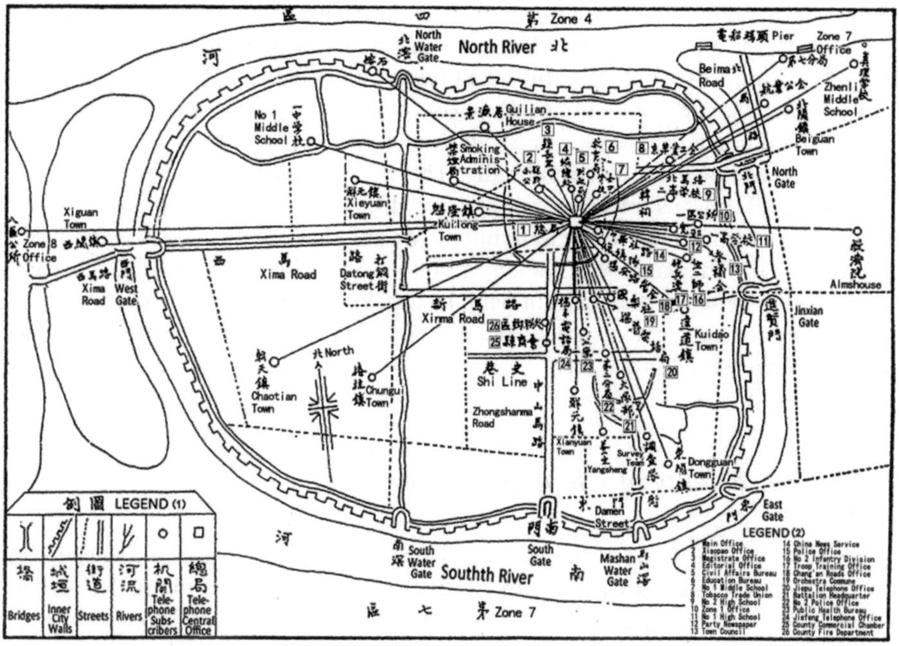

Fig. 19.5 Rongcheng Area Map (From 'Summary of Jieyang county government', the 23rd year of the Republic period)

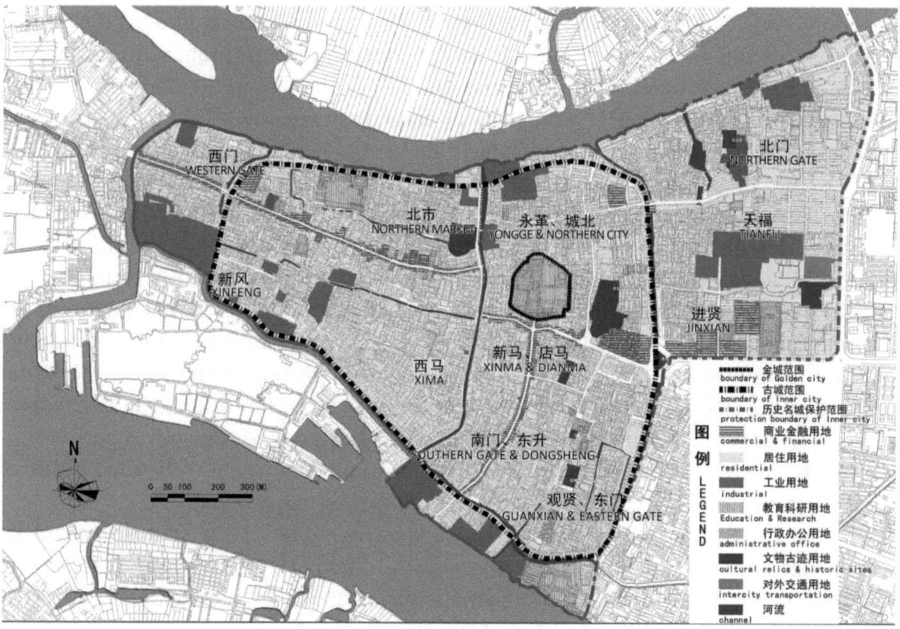

Fig. 19.6 Land Use Map of Jieyang inner city for the year 2003

However, the city form of the inner city remains stable. The core areas of the inner city consist of three adjacent nodes in the east downtown. These nodes are Golden City (for administration, education, and research), Academy Palace (for religious culture and tourism), and Jinxian Gate (for business, finance, and transportation). The surrounding areas in the east downtown are packed with historic sites, such as temples, academies and ancestral halls. The west downtown is mainly residential and is full of various ancestral temples for different families. The gourd-nozzle-shaped tip is connected to the agricultural western suburb by a suspension bridge over the river. This keeps the market fair on the West Road flourishing. Through a comparison between the current and historical layouts of the city, it can be concluded that the ancient urban framework of 'east palace and west market' has been well-preserved in the city.

19.2.2 Elements of Channel Network

Lying by the Rongjiang River, the inner city is a typical riverside township and gradually developed into a city with a well-organized channel network. Rongcheng Township Map demonstrates that the south and north branches of Rongjiang River run around the city from west to east and there are three channels linking the south and north branches of the river. In this manner, waterways and branches intertwine to form a natural water network. The total length of the network is about 130 km (Literature and History Editorial Committee of Jieyang Political Consultative Conference 1986), earning Jieyang city as the nicknames 'overwater lotus' and 'floating gourd' since ancient times (Fig. 19.7). Poets have even praised the unique character of the water city: "north branch connecting south branch, front channel connecting rear channel, follow the rise and fall of tides secretly, and surround the county both east and west"; "a majority of bamboo in the city grows by the riverside while more than half of households have their own boats" (Chen Yiyao 2013).

Due to its unique geographic situation, the inner city of Jieyang is crisscrossed with channels and rivers, which not only make the city picturesque, but also effectively regulate its water level. Such a water system has made the inner city a treasured land, safe from major flooding. Based on records from the Republic era, the wasted water was drained out through the channel network every day with the morning and evening tides. The daily circulation of water also serves to regulate the city temperature and maintain a stable amount of groundwater. Thus, the water quality of the city is guaranteed by the self-adjustment of the water system because, as the saying goes, 'Running water is never stale and a door-hinge never gets worm-eaten' (Rongcheng Literature and History Editorial Department of Jieyang Political Consultative Conference 1993).

Based on the current situation, the channel network of the inner city can be categorized into four types: river branches, channels, waterways, and drains. Among them, the north-south channels have good self-purification capacity. Between these channels are the east-west waterways that connect them. In this

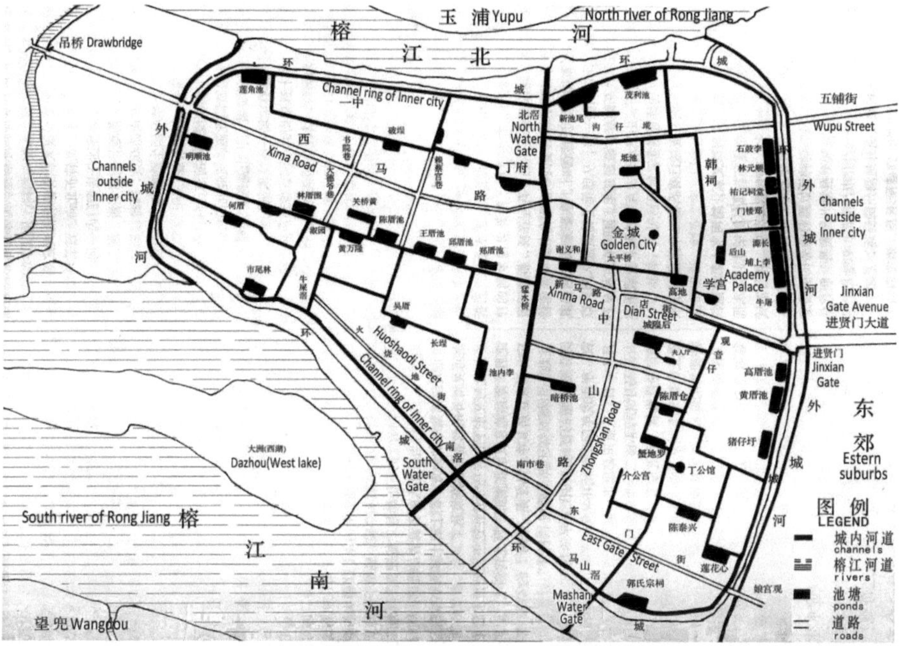

Fig. 19.7 Water network map of Rongcheng Area in the Republic era

way, a well-developed waterway transportation network is formed in the city. In the 1970s, some channels and waterways were artificially blocked, which weakened the self-purification capacity. However, major branches and channels, such as the south and north channels, Fangcuoqian River, the west segment of Mengshui River, and East City Moat remain intact (Fig. 19.8).

19.2.3 Elements of Street Fabric

The remaining architectural complex, dating from the Qing Dynasty, takes on a similar pattern: one side facing the street and the other side by the river. Thus, the streets and rivers of the inner city of Jieyang closely interweave and exist side by side to form a complete urban space. The main pedestrian streets are under the arcade buildings, whose fronts face the streets and backs face the riverside. This arrangement facilitates both land and water transportation. These streets are connected by alleys or small covered bridges so that pedestrians do not have to take umbrellas when they go out on rainy days (Fig. 19.9).

The arcade buildings on the Zhongshan Road, Xima Road, and the west section of Sixian Road date from the 1920s (early days of the Republic era). Some segments of these pedestrian streets were formerly riverbanks where vendors loaded and unloaded their cargo. They did not become pedestrian streets until the

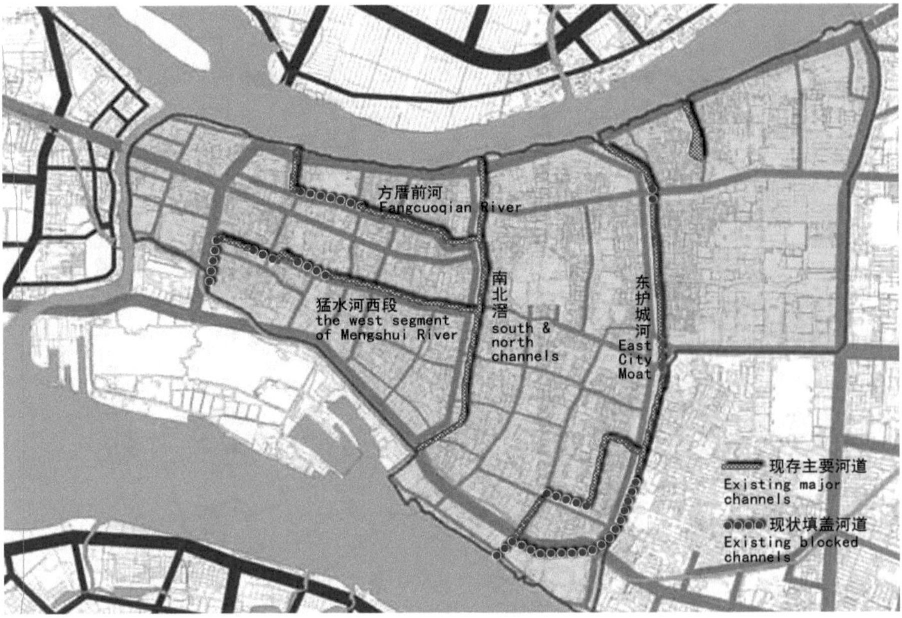

Fig. 19.8 Existing situation of the main channels in Rongcheng Area

riverbanks were blocked and filled with mud. Some other segments of the streets were originally entranceways to the Golden City, which became part of the streets after the change in functions of the Golden City. After years of redevelopment, the streets turned into what we see now. Though vendors are scattered throughout the streets, making them difficult to travel through, the major commercial and trade functions of the inner city remain unaffected.

Another feature of the street fabric is the T-shaped street layout that embodies the perfect integration of city frame and climate (Fig. 19.10). Jieyang is situated on the southeast coast of Guangdong province, often plagued by strong subtropical storms between summer and autumn. During the 268-year reign of the Qing Dynasty, there were up to 516 typhoons here. That is an average of 1.93 typhoons per year (Wu and Huang Yanhua 2011). The city road system of Jieyang is mainly comprised of T-shaped streets and ring roads. That is, T-shaped streets – including Xima Road, Datong Street, Xinma Road, Zhongshan Road, and Hanshan Road – link to the ring roads. The crossings on arcade building streets are all T-shaped crossings and no crossroads exist. When a typhoon strikes, the T-shaped crossings can reduce wind speed and mitigate winds effectively so as to ensure the safety of pedestrians.

19.2.4 Elements of Channel-Side Node

There is a special kind of channel-side node in the inner city of Jieyang, which is unique to the Chaozhou-Shantou area. As a rule, there is a rectangle square in front

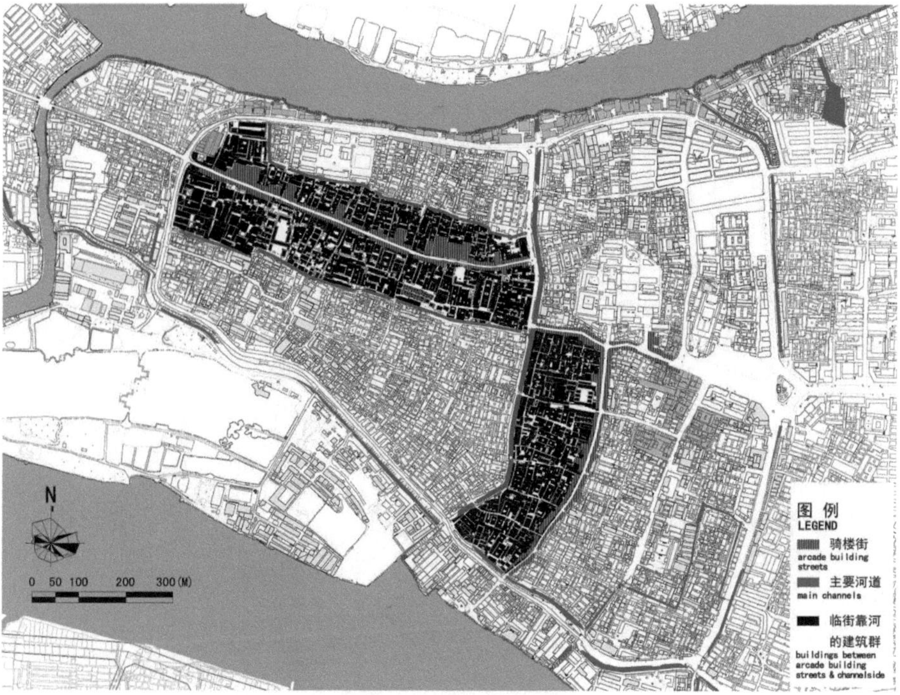

Fig. 19.9 Pedestrian alleys connecting the arcade building streets and channel-side allows for both land and water transportation

of an ancestral temple or a residential complex, usually the same length as that of the building. The square is surrounded on three sides by adjacent buildings while the fourth side borders a round or half-moon-shaped pond. The pond, square, and entrance screen wall form a separate space that the locals call 'Cheng.' The Cheng is often paved with granite and may include big trees, stone benches, and a shaft. It functions as a major site for residents to relax and engage in outdoor activities (Fig. 19.11).

Relatively large Chengs are known as 'Guang Cheng'. On festival days, fellow countrymen gather there for different activities, such as offering sacrifices to gods or ancestors or watching operas. Thus, the opera 'Guang Cheng Xi,' a style with strong local flavors, was born. The appearance of Guang Cheng Xi is closely related to the dense channel-side nodes over the network. It is an epitome of folk culture around the Chaozhou-Shantou area. It is recorded that in the Qing Dynasty, country folk would have various events to pray for favorable weather. For instance, hundreds of boats would assemble and raise their flags high, formidable sounds of gongs and drums would spread far and wide, and lanterns would be lit throughout the night. Statues of gods sit on major barges for people to worship. In addition, theatrical troupes are often invited to tour around the channel network and land on Guang Cheng to give their performances. In order to enjoy their performances,

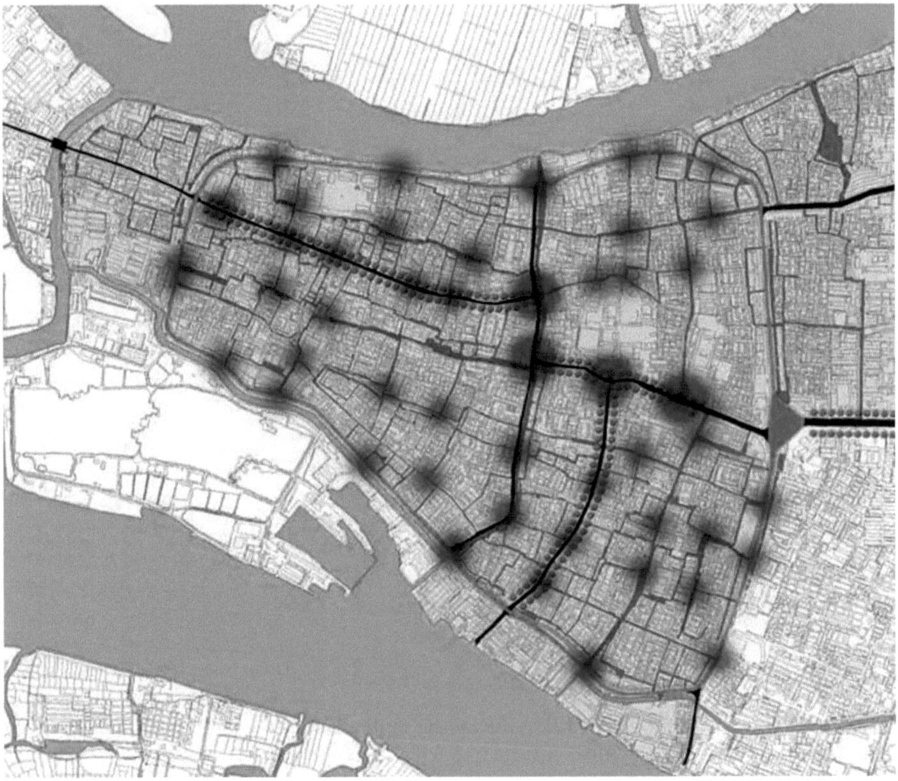

Fig. 19.10 Street fabric feature: T-shaped street layout

country folk pack into the place (Ye Chunsheng 2008). The front ponds of Guang Cheng took shape in the Qing Dynasty, thrived in the Republic era, and are scattered all over the city at present. This is evident in the contemporary Rongcheng water system diagram.

The existence of so many Chengs also helps with disaster prevention in Jieyang. Jieyang belongs to the earthquake zone where seismic intensity ranges between VII and VIII.[1] There have been several magnitude 6 or 7 earthquakes in the city's

[1] 'Seismic intensity' is a professional term quoted from The China Seismic Intensity Scale (CSIS), which classified different seismic impacts into 12 degrees of intensity. Seismic intensity VII: Slight destruction – localized destruction, crack, may continue to be used with small repairs or without repair; Collapse of river banks; frequent burst of sand and water from saturated sand layers; many cracks in soft soils; moderate destruction of most standalone chimneys. Seismic intensity VIII: Moderate destruction – structural destruction occurs, continued usage requires repair; Cracks appear in hard dry soils; severe destruction of most standalone chimneys; tree tops break; death of people and cattle caused by building destruction. (source: http://en.wikipedia.org/wiki/China_Seismic_Intensity_Scale)

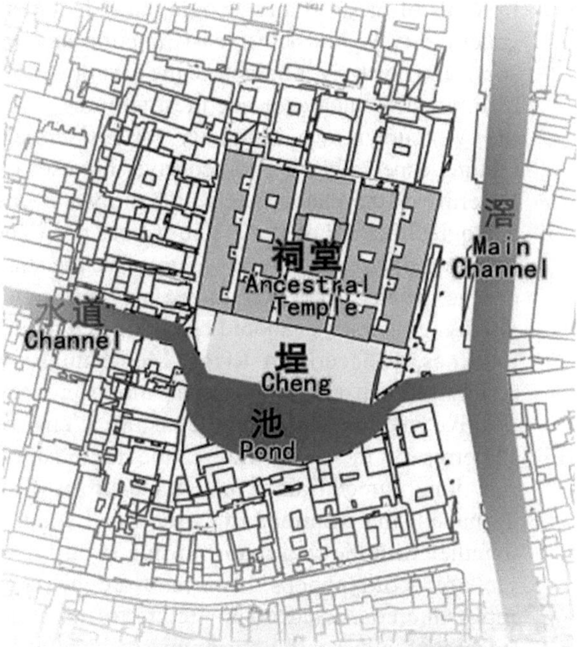

Fig. 19.11 A typical example of 'Cheng'

history. The nearby coast lies close to the earthquake cluster region of Nanao and is affected by seismic activity in the Taiwan Strait. As the eastern section of the seismic belt of the southeast coast, Jieyang is vulnerable to earthquake disasters because seismic activities here are intense and frequent (Wang Shengze 2010). As an emergency shelter in case of earthquake, Chengs make quick evacuation possible and provide ready access to water in case of fire. In conclusion, Chengs are a multi-function channel-side nodes that can be used for festival celebrations, relaxation, and disaster prevention, and are adaptable to the local environment.

19.3 Implications of the City Form Transformation for Current City Planning

The time-honored inner city of Jieyang has developed the spatial morphology of mountain blending with water and an urban framework of south and north river branches coexisting with connecting channels. This arrangement embodies many elements of city-form features that help to Jieyang adapt to the local environment and satisfy the need of disaster prevention. After a detailed analysis of several featured elements in the course of city form transformation, we come to the following implications for current city planning in Jieyang:

(i) In terms of urban frame, geographical features determine the heritage of the general city layout being 'east function and west market.' Separated by the south and north river branches, the east downtown has a great concentration of historic sites, such as academies, temples, and the Golden City. Meanwhile, the west downtown is mainly for residential use and for traditional folk culture, especially for great courtyard houses of historic significance.

(ii) In terms of the channel network, the water environment is important to the ecological beauty of 'East Yue waterland' (Yue is another name for Guangdong province). As long as we preserve the status quo and make conditions to dredge the city water system, we may restore it to the original water network. Dredging operations should be conducted mainly on the south and north river branches, Fangcuoqian River, Mengshui River, and East City Moat. The restoration can also improve the flood prevention function and may play a positive role in conserving the historical environment.

(iii) In terms of the street fabric, the regional climate has shaped the multi-functional street framework, made up of T-shaped streets and ring roads. T-shaped streets formed by Xima Road, Xinma Road, and Zhongshan Road, together with the ring roads, form a strong street network that helps mitigate the effects of typhoons. If we renovate the damaged arcade buildings, we can make interconnected commercial streets. When the commercial streets are integrated with nearby traditional residential housing, a sophisticated pedestrian district can be formed.

(iv) In terms of channel-side nodes, local geographical features accommodate the massive existence of channelside space (Guang Cheng), though the congestion problem of illegal buildings around Guang Cheng is very serious. We must demolish illegal buildings around Guang Cheng to improve traffic flow and create a hierarchical street fabric. In particular, we should exercise caution when planning new roads to avoid further man-made destruction of Guang Cheng areas. This will ensure that the multiple functions of Guang Cheng – festival celebrations, relaxation, and disaster prevention – remain intact.

Under the joint influence of geographical condition, hydrology, climate, and landscape, the elements of cityform features coexist and interact with one another. These elements fully reflect local cultures, traditions, and residential behaviors. Notably, one of the legacies of the city form is good disaster prevention capacity, including T-shaped streets for typhoon prevention, big ponds for flood prevention, and Guang Cheng for earthquake prevention. However, Jieyang city's planning and transformation experience is hard generalize to other cities, and there are therefore limitations in terms of the implications. Through this research, however, we find important practical benefits of traditional city form. Therefore, present-day city planning should emphasize the principles of protecting our excellent legacy and making positive renovations. Only when we embark on the road of innovative renewal in view of preserving its unique city-form, can we unfold the beauty of a traditional city context in a sustainable way.

Acknowledgments This research is Supported by National Natural Science Foundation of China (No. 51308130)

References

Chen Yiyao. (2013). *Ecological Jieyang – A natural beauty with mountain and water*. Beijing: Chinese Tourism News.

Chinese Album Editorial Department. (2008). *Guangdong*. Beijing: Chinese Tourism Press.

Literature and History Editorial Committee of Jieyang Political Consultative Conference. (1986). *Jieyang literature and history*. Literature and History Committee of Jieyang Political Consultative conference, Guangdong.

Liu Yeqin. (1973). *Jieyang County Records* 1–3. Taiwan: Chengwen Press.

Rongcheng Literature and History Editorial Department of Jieyang Political Consultative Conference. (1993). *Rongcheng literature and history*. Rongcheng Literature and History Editorial Department, Guangdong.

Wang Shengze. (2010). On the earthquake disaster and vulnerability analysis of Jie Shan Chao urban agglomeration of Guangdong province. *Journal of Seismological Research, 2010*(1), 82–91.

Wu Zhifeng, & Huang Yanhua. (2011). On the characteristics of and damage by Typhoons in Guangdong during the Ming and Qing Dynastie. *Journal of Wuyi University (Social Sciences Edition), 31*(1), 41–45 + 93.

Yang Sen. (1996). *A dictionary for Guangdong historic sites*. Beijing: Yanshan Press.

Ye Chunsheng. (2008). *Chinese folk customs (Guangdong folk customs)*. Lanzhou: Guansu People's Press.

Chapter 20
Urban Design Practice Towards Planning Management in China—Urban Design Guidelines in Wuhan City as Examples

Liangping Hong and Wenzhu Tao

Urban design is a type of urban planning and an important means of promoting the urban built environment in China. Urban design takes the urban public space as study object, which aims to shape the city's character, improve the quality of urban space and reflect human care. It has played an important role in China's urban planning system as a supplement of statutory plans. China's modern urban design has been borrowed from European and American countries since the 1980s. After 30 years of development abundant theoretical and practical accomplishment has been accumulated. Today, urban design has received more and more attention in the field of housing and urban and rural development. Many cities in China have conducted a large number of urban design practices. Often, however, the results of urban design formulation can't be implemented successfully—they remain only as beautiful blueprints for the future of urban construction. Why? Urban design doesn't have enough connection with planning management and development control. For one, local planning departments are often given only formal authorization on urban design work; physical and operational elements of urban design are not created in the framework of the actual planning system. For another, results of urban design formulation do not have enough connection with existing statutory planning to enter the actual planning management processes. In order to make urban design better connected with urban planning management and development control, urban design guidelines—which can transfer results of urban design into specific planning control language—have become an important component.

L. Hong
Huazhong University of Science and Technology (HUST), Wuhan, China

W. Tao (✉)
School of Architecture and Urban Planning, Huazhong University of Science and Technology, 1037 Luoyu Road, Wuhan City, Hubei Province 430074, China
e-mail: wenzhutao@gmail.com

Q. Pan, J. Cao (eds.), *Recent Developments in Chinese Urban Planning*,
GeoJournal Library 114, DOI 10.1007/978-3-319-18470-8_20

20.1 Urban Design Practice in Wuhan City

20.1.1 Overview of Wuhan City

Wuhan is the capital city of Hubei Province, China. Because of its centralized location at the confluence of north-south and east-west traffic, Wuhan has a very important position in the Chinese cities. Economic development in Wuhan city is average for China, making it a representative example. In addition, multiple rivers and lakes in Wuhan have formed a unique natural landscape, making the local planning department pay significant attention to urban design. This is another important reason we chose to study in Wuhan (Fig. 20.1).

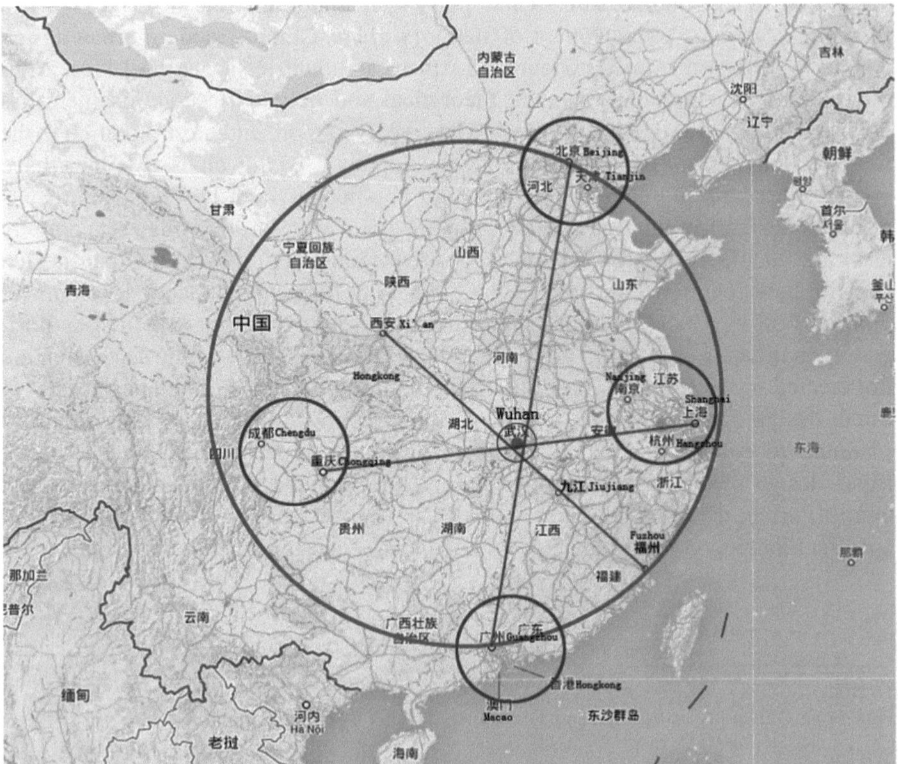

Fig. 20.1 Traffic location map of Wuhan city in China

20.1.2 Research Progress of Urban Design in Wuhan

In recent years, a series of urban design projects were formulated at all levels from the over-all city to local sites in Wuhan. Urban design system and relevant management regulations were also established based on these urban design practices. Urban design focused towards planning management was explored and practiced frequently in Wuhan, which has promoted and facilitated the optimization of urban public space image in Wuhan. Formulation of urban design guidelines is one of the most important parts of this process.

Since 2007, available linkages of urban design to statutory plans and planning management has been explored actively in Wuhan once the formulation of the master plan was completed and a new round of regulatory detailed planning formulation begun. The planning department has issued a series of regulation documents such as "Technical Regulations of Urban Design Formulation in Wuhan (*Wu han shi cheng shi she ji bian zhi ji shu gui cheng*)", "Urban Design Management Rules in Wuhan (*Wu han shi cheng shi she ji guan li ban fa*)", "Urban Design Formulation, Approval and Adjusting Rules in Wuhan (*Wu han shi cheng shi she ji bian zhi shen pi tiao zheng guan li xi ze*)", "Contents Set of Local Urban Design Guidelines in Wuhan (*Wu han shi ju bu cheng shi she ji dao ze nei rong*)", and has finished a research report "Technical Elements Database of Urban Design Formulation and Management in Wuhan (*Wu han shi cheng shi she ji bian zhi yu guan li ji shu yao su ku*)". Through the above-mentioned documents, regulation rules on formulation, approval, adjusting and implementation of urban design were clearly set—especially formulation standards of urban design guidelines for local urban design results.

The document "Technical Regulations of Urban Design Formulation in Wuhan (trail, 2008)" systematically prescribed for the first time the status, significance, and function of urban design and its formulation levels, contents, depth, and results form. It also established approval and management measures, which has been a big step towards the system construction of urban design in Wuhan. Then, the regulation documents "Contents Set of Local Urban Design Guidelines in Wuhan" and "Regulations on Formulation Results of Local Urban Design Guidelines in Wuhan" further prescribed contents of guidelines and results form in detail for the use of local urban design in planning management, which has improved the effectiveness of urban design. In addition, combination of urban design results with "the Diagram" digital planning management system and 3D-model-assistant draft approval system was actively explored.

20.2 Study on Urban Design Guidelines of Wuhan

20.2.1 Position of Urban Design in the Planning System of Wuhan

Urban design is an overall conception and arrangement of the city form, structure and space, throughout the process of urban planning. It is aimed at guiding the construction of urban stereo space form and the external space environment in accordance with urban planning. Urban design is one of the special planning components of the urban planning system in the "Technical Regulations of Urban Design Formulation in Wuhan (trail, 2008)", and that approved results of urban design is the important basis of urban planning management (Fig. 20.2).

20.2.2 Urban Design Formulation System of Wuhan

The urban design formulation system of Wuhan includes three levels: the master urban design, sub-area urban design and local urban design, all of which are in accordance with regulatory plans (i.e. the master plan, sub-area plan and regulatory detailed plan) (Fig. 20.3).

The scope of local urban design is usually defined according to regulatory detailed plan, and based on the master plan, sub-area plan, the master urban design,

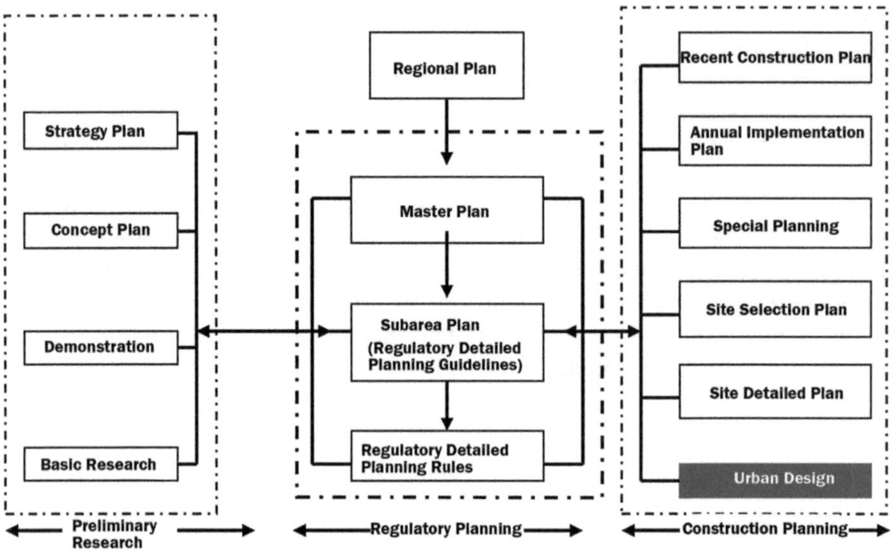

Fig. 20.2 Wuhan urban planning system framework (Source: Wuhan Urban Planning Bureau, urban and rural planning research of Wuhan)

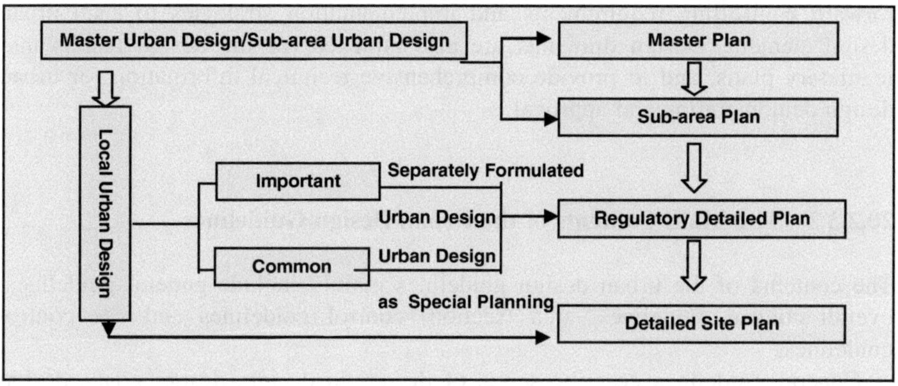

Fig. 20.3 Diagram of relation between urban design system and regulatory planning system (Source: Wuhan urban planning bureau, urban and rural planning research of Wuhan)

and sub-area urban design. Local urban design arranges and designs spatial combination, buildings, square green space, the street landscape, environment embellishments and human activity places of local area in detail, and gives specific implementation suggestions. Objects of local urban design can be divided into three types: surfaces (important area), lines (important urban streets, etc.) and points (important sites), which are basically among special-intention areas of the city. These include urban center, sub-centers, landscape areas, historic conservation areas, main city streets and intersections, ports, transportation hubs, squares, pedestrian streets, important mountains, lakes, and other areas. Results of local urban design include urban design guidelines which are combined with regulatory detailed plan in addition to design text, drawings and attachments. Local urban design guidelines briefly express the contents and key points of urban design in the form of diagrams, tables, and provisions, and transfer design schemes into specific development control briefs.

20.2.3 Overview of Wuhan Urban Design Guidelines

20.2.3.1 The Structure of the Urban Design Guideline and Basic Requirements

Urban design guidelines can be divided into four parts: general principles, overall control provisions, area control provisions and site control provisions. Overall control provisions qualitatively determine the control elements and contents, while area control provisions quantitatively carry out the control elements into each area under the guidance of overall control, and put forward regulatory and instructive requirements. Urban design guidelines concisely and precisely express the core contents of urban design in the form of normative provisions; they put forward controlling requirements and implementation strategies to each urban design element. Design drawings are also attached for the design results into regulatory plans, and to provide comprehensive technical information for urban design demonstration and approval.

20.2.3.2 The Main Contents of the Urban Design Guidelines

The contents of the urban design guidelines mainly include general guidelines, overall control guidelines, area (section) control guidelines and site control guidelines.

General guidelines describe scope of the urban design, design goals, design principles, explain the ownership department, etc. Overall control guidelines include several aspects of the design area such as functional structure, spatial structure, implementation of the master plan and related planning control requirements, the open space system and the public environmental elements. Area and section control guidelines put forward general principles of control requirements with the project characteristics based on the overall control guidelines, and detailed control requirements for those important areas and sections. Site control guidelines define graphic control requirement according to the function, spatial landscape structure, lines of sight landscape of the area (or sections), including regulatory elements and instructive elements.

20.2.3.3 Controlling Elements of Urban Design Formulation and Management

In order to formulate urban design guidelines more scientifically and normatively, the study "Technical Elements Database of Urban Design Formulation and Management in Wuhan" was started in 2010 as an important technical support for the formulation of urban design guidelines.

Referring to Beijing, Shenzhen and other cities' experiences, an urban design elements database of Wuhan was generally established for the demand of planning management and urban space characteristics in Wuhan.

This elements database of urban design formulation and management in Wuhan includes two topics: public space and building. In public spaces—such as mountain-fronts, waterfronts and historic conservation areas—special controlling elements are added. For the type of public space, there are three medium-categories elements including ten minimum-categories elements, while for the type of building, there are two medium-categories elements including four minimum-categories elements. All types of elements are classified as shown in Table 20.1.

Table 20.1 Main urban design formulation and management elements in Wuhan

Maximum-categories	Medium-categories	Minimum-categories	Elements	Types
Public pace	Road space	Block	Parcel division, block center, control indicators, mixed use, block size	Instructive
		Street wall	Street wall height	Regulatory
			Public open space, Street wall bottom setting, street wall type	Instructive
		Traffic	Motor vehicle entrance prohibited sections	Regulatory
			Road type, traffic organization, pedestrian crossing, intersection type, road barriers setting, parcel entrance, parking facilities	Instructive
	Environmental facilities	Environment	Site design, ground pavement, lighting, public art, ecological sustainable strategy	Instructive
		Facilities	Advertising signs, leadership, public service facilities, transportation facilities, municipal facilities, safety facilities and barrier-free facilities	
	Special control	Waterfront	Waterfront landscape visual corridor and open surface, building size, accessibility, public open space	Regulatory
			Waterfront skyline, waterfront building group layout, shoreline types, building form	Instructive
		Mountain front	Skyline, open degree	Regulatory
			Building form	Instructive
		Historic areas	Block texture, Building form and style	Regulatory
			Block style and features, Building diversity	Instructive
		Square	Square size, Square interface	Regulatory
			Square facility, cite control	Instructive
		Green space	Green ratio	Regulatory
			Green space type, plants configuration	Instructive
Building	Building group	Building combination	Combination of buildings, tower position, max and minimum height	Regulatory
			Building functional zoning, height zoning, color zoning	Instructive
		Building layout	Building set back and stick line, skirt building along street	Regulatory

(continued)

Table 20.1 (continued)

Maximum-categories	Medium-categories	Minimum-categories	Elements	Types
			Building entrance, combination of building to public space, Underground space use	Instructive
	Building details	Building form	Building size (height, width)	Regulatory
			Building facade, window wall ration	Instructive
		Building style	Building color, material and roof	

20.3 Case Study of Urban Design Guidelines in Wuhan

In recent years, urban design work in Wuhan was summarized carefully, which has created specific urban design formulation, approval and management by a series of standards and rules. Based on this work, several special urban designs (like "The Historical and Cultural Characteristics Study of Wuhan (*Wu han shi li shi wen hua feng mao te se yan jiu*)", "Main Open Space Planning Study of Wuhan (*Wu han shi zhu yao kai chang kong jian gui hua yan jiu*)", "Green Line Planning of Wuhan (*Wu han shi lv xian gui hua*)", etc.) at the macroscopic level have already been finished, as well as "Wuhan Urban Development Area Master Urban Design (*Wu han du shi fa zhan qu zheng ti cheng shi she ji*)" and some sub-area urban designs (like "Second Ring Road Area Urban Design of Wuhan (*Wu han shi er huan xian di qu cheng shi she ji*)", "Wuhan Two-Rivers-and-Four-Shoresides Waterfront Urban Design (*Wu han shi liang jiang si an bin shui qu cheng shi she ji*)", etc.) From there, local urban design was created to explore formulation methods of urban design guidelines, which are an important basis of urban design policies of Wuhan.

There are three types of finished local urban design projects: area, main road and site. Characteristic area urban design projects include "Wangjiadun CBD Planning and Design (*Wang jia dun zhong yang shang wu qu gui hua she ji*)", "Shahu Lake Waterfront Urban Design (*Sha hu zhou bian cheng shi she ji*)", "Yangchunhu Lake Waterfront Urban Design (*Yang chun hu di qu cheng shi she ji*)" and so on. Important road urban design projects include "Second Ring Road Area Urban Design (*Er huan xian di qu cheng shi she ji*)", "Wuluo Road- Luoyu Road Urban Design (*Wu luo lu ji luo yu lu yan xian cheng shi she ji*)", "Xiongchu Street Urban Design (*Xiong chu da jie yan xian cheng shi she ji*)", "Donghu Road Urban Design (*Dong hu lu cheng shi she ji*)", "JInghan Road Urban Design (*Jing han da dao cheng shi she ji*)" and so on (Table 20.2).

Table 20.2 Urban design projects within Second Ring Road area in Wuhan

Type	Quantity	Name of urban design
Important areas	10	① Two-Rivers-and-Four-Shoresides, ② East-west Mountains, ③ Shahu Lake Waterfront, ④ Moshuihu Lake Waterfront; ⑤ Yuehu Lake Waterfront, ⑥ Wangjiadun CBD Area, ⑦ Yiyuan Area, ⑧ Hanzheng Street Area, ⑨ Hanyang Old City Area, ⑩ Wuchang Old City Area
Important roads	9	① Second Ring Road, ② Riverside Ave, ③ River Along Ave, ④ Qinnian Road-Wuluo Road, ⑤ Hanyang Ave, ⑥ Jiangcheng Ave-Xudong Ave- Zhongbei Road, ⑦ Xiongchu Street, ⑧ Wuxian Road, ⑨Hong Kong Road- Shahu Lake Bridge
Important sites	12	① Hankou Railway Station,② XibeIhu Lake, ③ Wuhan Business square, ④Wuhan Customs, ⑤Qingtai Theater, ⑥ South Shoreside Core, ⑦ Yellow Crane Tower, ⑧ Shouyi Square, ⑨ Wuchang Railway Station, ⑩ Hongshan Square, ⑪ Moon Bay, ⑫ Provincial Museum

Resource: Second Ring Road Area Urban Design of Wuhan

20.3.1 *Design Guideline Case Study—Shahu Lake Waterfront Urban Design* (Sha hu zhou bian di qu cheng shi she ji)

The urban design of Shahu Lake waterfront, which includes characteristics areas of Wuhan, is a typical local urban design project that represents the formulation of a "characteristics area" urban design. This project was finished through the collaboration of urban planners, designers, architects, and road transportation professionals.

Shahu Lake waterfront is located on the east of the Yangtze River with a design scope of 453 ha excluding water area. In the design, Shahu Lake waterfront is defined as a "public stage and sitting room" of the Wuchang headquarter area, according to higher level plans and case studies. Comprehensive design structures and strategies were then proposed based on land use, landscape and traffic organization; in addition, two bridge sites which can best show the portal landscape of Shahu Lake were elaborately designed. After that, planners and designers proposed Shahu Lake waterfront urban design guidelines (Table 20.3, Fig. 20.4).

20.3.2 *Urban Design Controlling Elements Study*

There are 14 minimum-categories elements in the overall element control system, among which 2 minimum-categories have been chosen to study in this chapter.

Table 20.3 Shahu Lake Waterfront Urban Design results

	Urban design study report	Urban design guidelines	
Design content and depth	**Positioning and design conception**	**General principles**	
	① SWOT analysis, higher level planning analysis,	① Scope of application, design gist, ownership explanation,	
	② Design goals and design strategy.	② Design goals and design strategy.	
	Land use	**Overall control**	① Land use,
	① Proposing planning land use based on existing land use analysis and higher level planning study		② Landscape structure,
			③ Space form control,
	Landscape		④ Open space and green space control,
	① Proposing space landscape structure, put forward important sites, landscape axis and corridors,		⑤ Control of shoreline around lakes
	② Classification control of 2D and 3D space form of building group		⑥ Control of interface,
	③ Proposing building height distribution, landmark and height control requirements;		⑦ Height zoning control,
			⑧ Pedestrian and traffic control,
			⑨ Overall controlling elements guidance (Important areas, sites, signs, borders, paths, main visual corridors and public green space)
	Waterfront interface	**Area (section) control**	**Control guidelines for common areas**
	① Design of waterfront skyline, proposing relevant control requirements;		① Land use,
			② Landscape intention,
	② Proposing design requirements to waterfront view corridor and building styles and features.		③ Open space,
			④ Road interface,
			⑤ Building height,
	Public open space		⑥ Building form.
	① Distribution of important open space,		
	② Proposing shoreside patterns and design guidelines.		
	Road and traffic		**Control guidelines for important areas**
	① Proposing traffic organization, traffic facilities and parking lots locations,		① Regulatory contents: land use control and landscape control (functional structure, building set

(continued)

Table 20.3 (continued)

	Urban design study report	Urban design guidelines	
	② Width and section design of main roads,		back line, open space, public channel, entrances and building group spatial characteristic, sites, interfaces, visual corridors, etc.)
	③ Organization of pedestrian and traffic at rail stations, parking, and recreation areas		② Instructive contents: building form and environment design (building style, color, size, height, green space and advertising, etc.)
	④Organization and design requirements of walking system, activities.		
	Important sites		
	① Proposing the position of important sites and design requirements for them,		
	② Detailed arrangement of land use and building function on these sites.		
	③ concept design of important sites. Proposing construction requirements on landscape, buildings style, shape and color		
Resulting form	**Design text and drawings**		**General principles: text,**
			Others: text and drawings

Resource: Shahu Lake Waterfront Urban Design

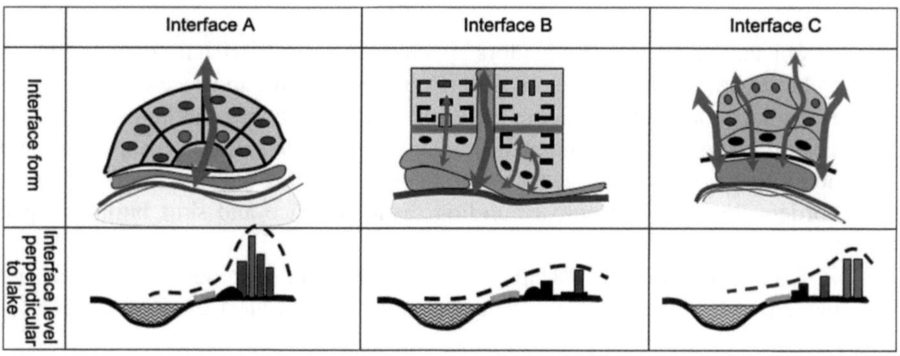

Fig. 20.4 Interface control in the Shahu Lake Waterfront Urban Design Guidelines (Resource: Shahu Lake Waterfront Urban Design)

20.3.2.1 Road Space Category Elements Example: Street Wall—Street Wall Height

Street wall height refers to the height of a wall which is closest to the road red line, from ground to the top part of a building along a road. Street wall interface is composed by the built area under 60 m of buildings along roads.

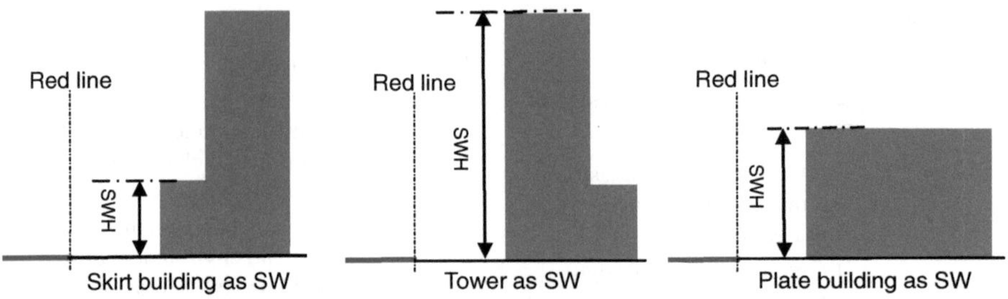

Fig. 20.5 Types of street wall

When street wall is composed of a skirt building, street wall height is usually controlled under 30 m. If it's a plate building the height is usually controlled under 60 m. If it the street wall consists of a tower the height could be controlled according to the largest range which can be actually observed by a person because tower height is not restricted (Fig. 20.5).

Neighboring buildings in the same block and same parcel ought to be set back in relation to one another and coordinate with each other in the street wall height. This can be controlled through the following four height options: 5–6 m, 15–19 m, 30 m and 60 m, in order to form a continuous wall street interface.

20.3.2.2 Special Control Category Elements Example: Waterfront—Waterfront Landscape Visual Corridor and Open Surface

Waterfront landscape visual corridors specifically refer to visual corridors which lead to landscape resources from a waterfront, or show urban space characteristics (for example, connection corridors between spaces or people and landscape). This is an important way for each independent landscape space in the urban space system to echo mutually. Open interface refers to the non-building land interface within the city borders, including principal building open interface and skirt building open interface.

Two control methods, landscape visual corridor and open degree of waterfront interface, are proposed to ensure the permeation of the waterfront landscape.

Landscape Visual Corridor Control

1. **Visual corridor on the ground**

Control of ecological green wedge: Ecological green wedges, which can improve lakes' regional climate environment and create appealing waterfront landscapes, are encouraged in waterfront areas that display urban diversity. The width of green wedge should be more than 60 m. In addition, mixed use as urban green space and leisure area is encouraged (Fig. 20.6).

Fig. 20.6 Control of ecological green wedges

Control of roads perpendicular to the river (lake): Roads perpendicular to the river (lake) include planned roads and public channels. Their design of should make sure to reflect the green landscape characters of waterfront roads, increase the building set back, enlarge the street space and form proper wind-channels (Fig. 20.7).

Control of landscape corridor: Each 180–200 m continuous waterfront area without either of the two controls above should set a landscape corridor which shall not be less than 25 m wide (Fig. 20.8).

2. **Visual corridor in the air**

Layout of waterfront buildings should be open and transparent, avoiding landscape resources being sheltered from continuous tall buildings. Each single corridor in the air should be more than 25 m wide from multiple perspectives against river or lake. Neighbor corridors should be less than 60 m in distance (Fig. 20.9).

Open Interface Control

Waterfront building is prioritized to tower form, which can reduce the disadvantage from building width and shape of a transparent waterfront environment. Buildings along roads which are 20 m wide and more should be designed transparently-open interface for skirt buildings should not be less than 30 % in principle, and for main buildings not less than 50 % of the lakes' width along the road. For other waterfront buildings, the total width of the main buildings along the water should be less than 60 % of the parcel length (Figs. 20.10 and 20.11).

Fig. 20.7 Control of roads perpendicular to river (lake)

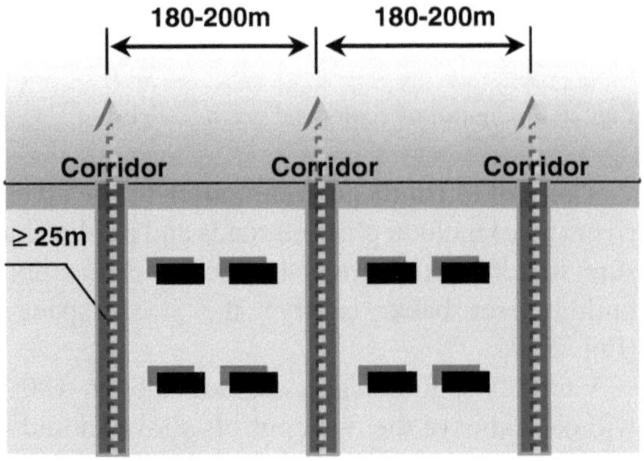

Fig. 20.8 Control of landscape corridor

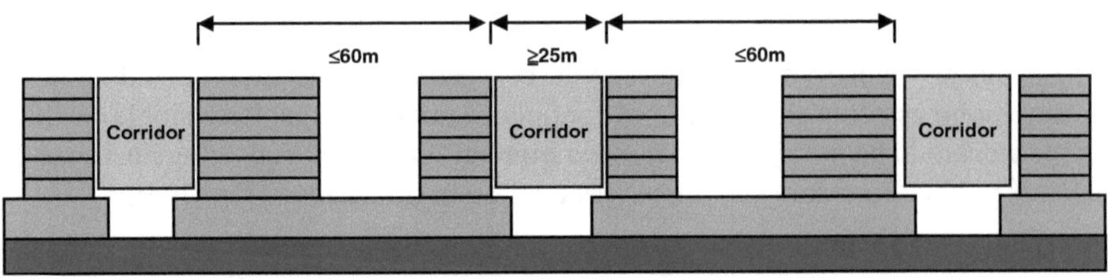

Fig. 20.9 Visual corridor in the air

20.3.3 Cases Summary

From the above cases, it seems that urban design guideline formulation has already begun to be standardized. The content of urban design guidelines are in place, and control elements are also comprehensive. Through the analysis of local urban design in Wuhan city, we can summarize the common contents and personalized contents in order to connect with planning management.

Common contents include: ① Planning positioning and design conception, ② Land use, ③ Transportation, ④ Landscape (building height, building interface, public open space, architectural form), ⑤ The overall design, ⑥ Site design.

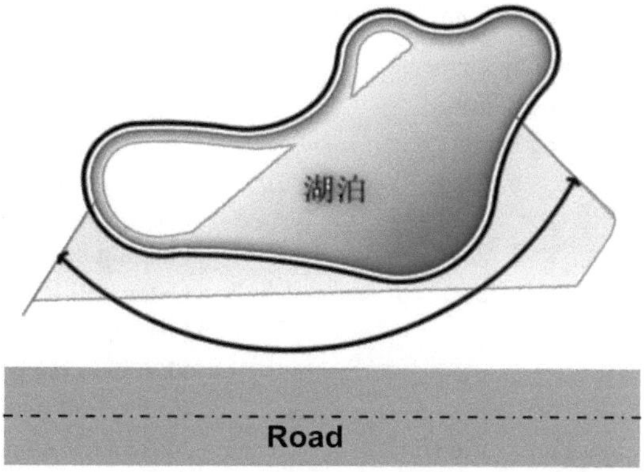

Fig. 20.10 Open interface L > 50 %

Fig. 20.11 Open interface control of waterfront buildings

Personalized contents include: Waterfront interface, sight perception landscape around lakes, overlooking landscape, night sight landscape design, public pedestrian space, road section optimization, site space types, space sequence, road grey space design, etc.

20.4 Urban Design Towards Planning Management in Wuhan

20.4.1 Link of Urban Design to Statutory Planning

Urban design of Wuhan includes the master urban design, the sub-area urban design and the local urban design which are formulated according to each level of statutory planning. Linkup between local urban design and regulatory detailed planning in the statutory planning system has been especially explored. For important areas, urban design is formulated before regulatory detailed planning and its approved result is brought into regulatory detailed planning rules for management. For ordinary areas, urban design is formulated as a special research topic at the same time of regulatory detailed planning formulation, in order to strengthen and comply with design control elements in the statutory planning (Tables 20.4 and 20.5).

Table 20.4 Some controlling elements optimizing to regulatory detailed planning

Controlling elements	Optimizing methods
Building density	Density control according to different land uses
Building height	Height zoning control of areas
	Angular control in street space
	Coordination control of width-height ratio and best building height
Building size	Outer wall line of tower building
	Plane and height ratio of tower building
	Skirt building position and height
	Roof of tall building
Floor area ratio (FAR)	Make FAR more scientific from the aspect of urban design
Green area ratio	Add contents of green space system rules and landscape guidelines
Parking space and related public facilities	
Traffic entrance	

Table 20.5 Planning management strategies of urban design control elements

Control elements of UDG		Planning management strategies
Planning control indicators	Mixed land use	Be concluded in the design briefs according to regulatory detailed plans
	Parcel division	
	Building density	
Layout control	Building set back	Be concluded in the design briefs
	Building layout	
	View corridor	
	Building distance	
	Public channel and parking	
	Traffic entrance	
	Pedestrian entrance including underground space	
Building form control	Building height	Be concluded in the design briefs
	Building width	
	Building form	
	Building characteristic and color	
Environment control	Advertising signs	Combined with special planning or guidelines
	Public identification	
	Lighting	

20.4.2 *Link of Urban Design Results to Planning Management Process*

It is stipulated in the "Technical Regulations of Urban Design Formulation in Wuhan (Trial, 2008)" that urban design is both a deepening and a complement of the statutory urban planning system, and is an important basis of urban planning management. Enforcement of the link between controlling elements in local urban design and in regulatory detailed planning (or urban design control into site planning and design guidelines (design briefs)), is the key point for cooperation between urban design and planning management. In addition, compliance with urban design control requirements should be checked before the approval of construction project schemes in important areas.

In order to transfer urban design control contents into design briefs that can be used in the actual planning management process, formulation of site design guidelines based on urban design results is needed. This process requires that regulatory and instructive design requirements are drawn out and some very blinding design control indexes are adopted such as street wall line, stick-to-line ratio of building interface, height width ratio of open space section, etc. Then, design control on public space landscape and architectural form can become available.

20.4.3 *Analysis of Urban Design Practice of Wuhan*

20.4.3.1 Merits of Wuhan's Urban Design Practice

Learn Advanced Experiences from Developed Countries and Cities

Urban design formulation and management in Wuhan is based on previous experiences from developed countries and cities. On one side, cognition of urban design and related practice is clear through summarizing domestic and foreign urban design theories. On the other side, the link of urban design to planning management is explored learning from foreign advanced experience. In England, an integrated development control and design control (with planning permission) system has had excessive discretion. In America control policies focus on design review with a basis in design guidelines, and are carried out by zoning—this process is strict relative to the Chinese rapid urbanization process. Thus, Wuhan city has explored suitable design control methods in China.

Clarify the Status of Urban Design and Establish Urban Design System and Related Institutions

In the past, urban design lacking of legal status could only be used as a technical reference document in the planning process, which greatly weakened the efficiency and effectiveness of urban design results. Clarifying the importance of urban design, creating an urban design system and establishing formulation and approval institutions by a series of documents help urban design operability. Implementing of control requirements step by step, will also improve the effectiveness of urban design.

Integrating Design Control into the Planning Management Process

Integrating design control into the planning management process has been explored on the basis of urban design formulation in criterion. In Wuhan, approved urban design guidelines are the legal basis of planning management, along with regulatory detailed planning guidelines. Control requirements of urban design are listed in design briefs of "Planning permit for construction land" *(jian she yong di gui hua xu ke zheng)*, as basis for constructional detailed planning. While issuing certifications of "Planning permit for construction project" (*jian she gong cheng gui hua xu ke zheng*), case review is used to control and guide the development of specific projects.

20.4.3.2 Existing Problems in Wuhan Urban Design Practice

The urban design practice of Wuhan in recent years is proactive and pioneering, however, some problems still exist. Firstly, urban design formulation technology needs further innovation. The resulting form of urban design guidelines is not perfect, and control elements are too broad—pertinence and practicality need to be further strengthened. Secondly, relevant institutions of urban design need to be consummated. Urban design formulation, approval, adjustment, implementation and supervision require a series of institutions, but existing institutions are relatively less involved in implementation and supervision. In addition, exploration of urban design control into planning management process needs to be further strengthened. Application methods and processes for the management of urban design guidelines need to be ensured by more related institutions. Design control should also be combined with planning review in multinational patterns. Lastly, public participation in urban design should be normalized, which has not been paid enough attention up until now.

20.5 Conclusions

The development of modern urban design in China is relatively recent, and the integration of urban design into planning management is still in the stage of exploration. Design control in western countries has been relatively mature—their advanced experiences are helpful for us to study and reference, although there are several differences in the national condition and urbanization stage. This chapter summarized urban design works of Wuhan in recent years and analyzed the formulation of urban design guidelines. These works actively explore the unification of urban design and planning management. Some of them are proactive and pioneering, which demonstrates that the effectiveness of urban design has drawn more and more attention.

After 30 years of development, many experiences and lessons learned on urban design have been accumulated in the urban construction of China Today, in the rapid urbanization age, we should make full use of urban design as public policy to regulate and control China's urban construction. We need to strengthen the research on urban design implementation strategies, which helps urban design to be implemented and therefore the quality of city will be improved.

References

Hongyan, S., & Qing, C. (2007). Urban design guidelines—Link of urban design to urban construction management. *City, 11,* 71–73.

Lingshu, X. (2008). Urban design towards planning management—study on practices of Chengdu. In *Proceedings of annual meeting of China's Urban Planning in 2008*. Dalian: Dalian Press.

Shifu, W. (2005). *The urban design towards implementation*. Beijing: China Architecture & Building Press.

Wantai, H. (2004). The relation between urban design and urban planning practice. *Journal of Chongqing Architectural University, 26*(4), 1–4.

Wuding, M. (2002). Heading towards integration of urban design into planning management. *Urban Planning, 9*, 63–67.

Wuhan Urban Planning Advisory Service Center. (2010). *Shahu Lake waterfront urban design (Draft)*.

Wuhan Urban Planning Bureau. (2009). *Technical regulations of urban design formulation in Wuhan*.

Wuhan Urban Planning Bureau. (2010a). *Urban design management rules in Wuhan*.

Wuhan Urban Planning Bureau. (2010b). *Regulations on formulation results of local urban design guidelines in Wuhan (Draft)*.

Yong, J. (2008). *Effectiveness of urban design*. Nanjing: Southeast University Press.

Hong Liangping professor of school of architecture and urban planning, Huazhong University of Science and Technology (HUST), Wuhan, China. Research field is sustainable urban design.

Tao Wenzhu PhD student of urban planning at HUST. Research field is urban design theories and methods.

MIX
Papier aus verantwortungsvollen Quellen
Paper from responsible sources
FSC® C105338

If you have any concerns about our products,
you can contact us on
ProductSafety@springernature.com

In case Publisher is established outside the EU,
the EU authorized representative is:
Springer Nature Customer Service Center GmbH
Europaplatz 3, 69115 Heidelberg, Germany

Printed by Libri Plureos GmbH
in Hamburg, Germany